One of the predicted consequences of the depletion of stratospheric ozone is an increase in the amount of ultra-violet light reaching the surface of the earth, in particular, UV-B (320–280 nm). Although the real effects are as yet unknown, this change in radiation could have profound consequences for plant growth and productivity. The need for information concerning the relationship between plants and UV-B is therefore pressing. This volume brings together authoritative contributions from leading experts in UV-B/plant studies and is unique in considering interactions at various scales, ranging from the level of the cell through to the level of the community. Information concerning ozone depletion and physical aspects of UV-B radiation complements the biological information to provide a thorough and comprehensive review of the present status of knowledge.

T0292351

SOCIETY FOR EXPERIMENTAL BIOLOGY
SEMINAR SERIES: 64

PLANTS AND UV-B
RESPONSES TO ENVIRONMENTAL CHANGE

SOCIETY FOR EXPERIMENTAL BIOLOGY SEMINAR SERIES

A series of multi-author volumes developed from seminars held by the Society for Experimental Biology. Each volume serves not only as an introductory review of a specific topic, but also introduces the reader to experimental evidence to support the theories and principles discussed, and points the way to new research.

PLANTS AND UV-B
Responses to Environmental Change

Edited by

P. J. Lumsden

Department of Applied Biology
University of Central Lancashire, Preston, UK

CAMBRIDGE
UNIVERSITY PRESS

CAMBRIDGE UNIVERSITY PRESS
Cambridge, New York, Melbourne, Madrid, Cape Town, Singapore, São Paulo, Delhi

Cambridge University Press
The Edinburgh Building, Cambridge CB2 8RU, UK

Published in the United States of America by Cambridge University Press, New York

www.cambridge.org
Information on this title: www.cambridge.org/9780521114110

First published 1997
This digitally printed version 2009

A catalogue record for this publication is available from the British Library

Library of Congress Cataloguing in Publication data

Plants and UV-B: responses to environmental change/edited by P. J.
Lumsden
    p. cm. – (Society for Experimental Biology seminar series; 64)
    Includes index.
    ISBN 0 521 57222 3 (hardback)
    1. Plants, Effect of ultraviolet radiation on.   I. Lumsden, P.J.
II. Series: Seminar series (Society for Experimental Biology (Great Britain));
64.
QK757.P53 1997
571.4´562–dc21 96 30065 CIP

ISBN 978-0-521-57222-4 hardback
ISBN 978-0-521-11411-0 paperback

Contents

Contributors

A-H. MACKERNESS, S.
Department of Molecular and Environmental Physiology, Horticulture Research International, Wellesbourne, Warwick CV35 9EF, UK

ALLEN, D.J.
Department of Biological and Chemical Sciences, University of Essex, Wivenhoe Park, Colchester, Essex CO4 3SQ, UK

BAKER. N.R.
Department of Biological and Chemical Sciences, University of Essex, Wivehoe Park, Colchester, Essex CO4 3SQ, UK

BJÖRN, L.O.
Department of Plant Physiology, University of Lund, Box 117, S-221 00, Sweden

BORNMAN, J.F.
Department of Plant Physiology, University of Lund, Box 117, S-221 00, Sweden

BRAY, C.M.
School of Biological Sciences, 3.614 Stopford Building, University of Manchester, Oxford Road, Manchester M13 9PT, UK

BRITT, A.B.
Division of Biological Sciences, Section of Plant Biology, University of California, Davis, California 95616, USA

CALDWELL, M.M.
Department of Rangeland Resources, Utah State University, Logan, Utah, UT 84322-5230, USA

CALLAGHAN, T.
Sheffield Centre for Arctic Biology, University of Sheffield, Sheffield, UK

CEN, Y-P.
APPA Division, IRRI, Manila 1099, Phillippines

CHRISTIE, J.M.
Plant Molecular Science Group, Division of Biochemistry and Molecular Biology, Institute of Biomedical and Life Sciences, Bower Building, University of Glasgow, Glasgow G12 8QQ, UK

COOP, D.J.S.
Division of Biological Sciences, University of Lancaster, Bailrigg, Lancaster LA1 4YQ, UK

CORLETT, J.E.
Annual Crops Department, Horticulture Research International, Wellesbourne, Warwick CV35 9EF, UK

FUGLEVAND, G.
Plant Molecular Science Group, Division of Biochemistry and Molecular Biology, Institute of Biomedical and Life Sciences, Bower Building, University of Glasgow, Glasgow G12 8QQ, UK

GEHRKE, C.
Department of Plant Ecology, University of Lund and Abisko Scientific Research Station, S-221 00 Lund, Sweden

GUNNARSSON, T.
Department of Chemical Ecology and Ecotoxicology, University of Lund, S-221 00 Lund, Sweden

HÄDER, D-P.
Institut für Botanik und Pharmazeutische Biologie, Friedrich-Alexander-Universität, Staudstr. 5, D91058 Erlangen, Germany

HOLMES, M.G.
Department of Plant Sciences, University of Cambridge, Downing Street, Cambridge CB2 3EA, UK

HOLMGREN, B.
Department of Meteorology, Uppsala University and Abisko Scientific Research Station, Uppsala, Sweden

JENKINS, G.I.
Plant Molecular Science Group, Division of Biochemistry and Molecular Biology, Institute of Biomedical and Life Sciences, Bower Building, University of Glasgow, Glasgow G12 8QQ, UK

JOHANSON, U.
Department of Plant Physiology, University of Lund, Box 117, S-221 00 Lund, Sweden

JONES, H.G.
Annual Crops Division, Horticulture Research International, Wellesbourne, Warwick CV35 9EF, UK

JORDAN, B.R.
Department of Crop and Food Research, Levin Research Centre, Private Bag 4005, Levin, New Zealand

McLEOD, A.R.
Institute of Terrestrial Ecology, Monks Wood, Abbots Ripton, Huntingdon, Cambridge PE17 2LS, UK

MEPSTED, R.

Bath University, Bath, Avon BA2 7AY, UK

MOODY, S.A.

Division of Biological Sciences, University of Lancaster, Bailrigg, Lancaster LA1 4YQ, UK

NEWSHAM, K.K.

Institute of Terrestrial Ecology, Monks Wood, Abbots Ripton, Huntingdon, Cambridge PE17 2LS, UK

NOGUÉS, S.

Department of Biological and Chemical Sciences, University of Essex, Wivenhoe Park, Colchester, Essex CO4 3SQ, UK. *Present address:* Departament de Biologia Vegetal, Facultat de Biologia, Universitat de Barcelona, Avgda, Diagonal 645, 0828 Barcelona, Spain

PAUL, N.D.

Division of Biological Sciences, University of Lancaster, Bailrigg, Lancaster LA1 4YQ, UK

PYLE, J.A.

Centre for Atmospheric Science, Department of Chemistry, University of Cambridge, Lensfield Road, Cambridge CB2 1EW, UK

REUBER, S.

Department of Plant Physiology, University of Lund, Box 117, S-221 00 Lund, Sweden

ROZEMA, J.

Department of Ecology and Ecotoxicology, Vrije University, De Boelelaan 1087, 1081 HV Amsterdam, The Netherlands

SNOGERUP, S.

Botanical Museum, University of Lund, S-221 00 Lund, Sweden

SONESSON, M.

Department of Plant Ecology, University of Lund, S-221 00 Lund, Sweden

STEPHEN, J.

Annual Crops Department, Horticulture Research International, Wellesbourne, Warwick CV35 9EF, UK

STERNER, O.

Department of Organic Chemistry 2, University of Lund, S-221 00 Lund, Sweden

TAYLOR, R.M.

School of Biology Sciences, 3.614 Stopford Building, University of Manchester, Oxford Road, Manchester M13 9PT, UK

THOMAS, B.

Department of Molecular and Environmental Physiology, Horticulture Research International, Wellesbourne, Warwick CV35 9EF, UK

THOMAS, B.

Department of Molecular and Environmental Physiology, Horticulture Research International, Wellesbourne, Warwick CV35 9EF, UK

TOBIN, A.K.

Plant Science Laboratory, School of Biological and Medical Sciences, Sir Harold Mitchell Building, University of St Andrews, Fife, KY16 9TH, UK

TOSSERAMS, M.

Department of Ecology and Ecotoxicology, Vrije University, De Boelelaan 1087, 1081 HV Amsterdam, The Netherlands

WEBB, A.R.

Department of Physics, UMIST, PO Box 88, Manchester M60 1QD, UK

WEISSENBÖCK, G.

Botanisches Institut de Universität zu Köln, Gyrhofstrasse 15, D-50923 Köln, Germany

WOODFIN, R.

Division of Biological Sciences, University of Lancaster, Bailrigg, Lancaster LA1 4YQ, UK

VAN DE STAAIJ, J.W.M.

Department of Ecology and Ecotoxicology, Vrije University, De Boelelaan 1987, 1081 HV Amsterdam, The Netherlands

YU, S-G.

Department of Biochemistry, University of Lund, S-221 00 Lund, Sweden

Preface

Public awareness of the dangers to health of exposure to ultraviolet (UV) light has increased in recent years. Although exposure to UV can have positive effects on humans – sunbathing generally induces a feeling of well-being, partly due to production of natural endorphins, and stimulation of synthesis of vitamin D – the dangers far outweigh the benefits. The most noticeable effect of exposure is sunburn, or erythema; more serious is the increase in risk of skin cancer, and of damage to the eyes, and these symptoms may not surface for several years. At the same time, the public has also become aware that protection from solar UV is provided by ozone in the upper atmosphere, the stratosphere, but that, over recent years, the ozone layer has suffered depletion due to the action of man-made chemicals. The predicted result of this is that more UV radiation from the sun will reach the earth's surface.

Energy from the sun covers the whole electromagnetic spectrum, from short gamma rays (10^{-5} nm) to long radio waves (10^3 m). UV light is that region of the spectrum with shorter wavelengths than blue light, between about 400 nm and 250 nm, and this is divided still further into UV-A (400–320 nm); UV-B (320–280 nm) and UV-C (280–250 nm). UV-A does not interact with ozone, since individual photons do not carry enough energy to carry out the necessary photochemical reactions; the energy in UV-B is used in breaking the bonds between oxygen atoms in molecules of ozone, which effectively results in absorption of the UV-B; UV-C is effectively absorbed by ozone/oxygen, and would still be so even under high depletions of ozone. Any reduction in the ozone layer will therefore mean that less UV-B is absorbed, and will lead to more UV-B reaching the earth's surface.

Until the early 1980s there was no evidence that the ozone in the stratosphere was actually being depleted; the first recorded evidence for a thinning in the ozone layer came in 1984 from a British team in Antarctica, led by Joe Farman. Since then, comprehensive and continuous data have come from the total ozone mapping spectrometer (TOMS) carried on the Nimbus 7 satellite; together with other instruments, this has detected thinning of ozone not just over Antarctica, but over the entire globe (see chapter by Pyle). By 1987 it was recognised that the threat to the ozone layer from various gases, particularly man-made chlorofluorocarbons (CFCs) was real, and as a result, the Montreal Protocol on Substances that Deplete the Ozone Layer was agreed by representatives of 27 of the technologically developed countries. Specified limits for a number of gases were agreed at Montreal; however, CFCs are very long-lived, so that, despite limitations being agreed, these were probably not sufficient to allow recovery of depleted ozone before well into the twenty-first century.

Although changes in stratospheric ozone would suggest that incident UV-B will increase, it is clearly vital to have accurate information as to whether increases in UV-B can actually be detected at ground level, against a background of other climate variation (see chapter by Webb). The most recent analyses show that ozone has declined globally by about 4–5% since 1979, leading to a fairly general assumption that, over the next three decades, there will be a depletion of the ozone layer of about 15%, which may result in an increase in effective UV-B of 30% over the current level. It should be noted, however, that within this range there are significant latitudinal variations in incident UV-B. For example, the amount of UV-B experienced at tropical latitudes is much greater than in temperate regions because the angle of the sun is such that there is less absorption by the atmosphere. In addition, the ozone layer itself is thinner in equatorial regions. Thus even with substantial reductions in ozone in temperate regions the effective UV-B would still not exceed that currently experienced in the tropics. However, if there were thinning of ozone in tropical regions, the levels of UV-B would exceed those previously recorded. Conversely, in regions of very high latitude, the amount of UV-B radiation is relatively low, but since it is here that thinning of the ozone layer is most marked, the *relative* increase in UV-B is potentially greatest.

Over the past few years, research has been carried out to investigate whether increases in UV-B radiation resulting from ozone depletion would have a significant impact on plants, in particular, on aspects of physiology and crop yield, and this volume of the Cambridge University Press seminar series is intended to provide a comprehensive review of our current knowledge in this area. It arose partly from a session entitled *UV-B and Plants* held at the annual meeting of the Society for Experimental Biology (SEB), Lancaster University, 1996, to which researchers from the UK, funded by the Biological Sciences and Biotechnology Research Council (BBSRC), Natural Environment Research Council (NERC), Department of the Environment (DOE) and Ministry of Agriculture Fisheries and Food (MAFF), and researchers from abroad were invited. All have contributed to the volume. Some chapters provide a fairly broad review, while others give a more detailed account of recent experimental work. Some of the current issues relating to research into UV radiation which were discussed at the meeting are summarised below, along with some detail of the contents of the chapters.

Experimental work has been carried out in laboratory growth chambers, where plants are grown under artificial white light to which is added different levels of UV-B, or outdoors, using artificial UV-B to supplement the UV-B in natural sunlight. From these studies it is now recognised that there are both *direct* and *indirect* effects of UV-B on plants. Direct effects include radiation-induced changes in photosynthesis, cell division, or other life processes of direct importance to growth and development. These effects are observed after relatively short periods of irradiation, hours or days. Damage to DNA has long been recognised as an important consequence of exposure to UV; this is discussed in the chapter by Taylor, Tobin and Bray, who describe the products formed as a result of damage to DNA, and the activity of the photolyase enzyme(s) involved in repair. Britt then describes a genetic approach of screening for mutants defective in DNA repair as a way of isolating the genes involved in the repair process. Effects of UV on photosynthesis are discussed by Baker, Nogués and Allen, the conclusion being that the primary damaging effect of UV is not on photosystem II reaction centres, but on some component of the Calvin cycle. This is demonstrated in more detail by Mackerness, Jordan and Thomas, who describe the effects of UV on the expression of photosynthetic genes and the levels of photosynthetic proteins. Clearly some of the UV-B induced down-regulation of photosynthetic genes is not simply a consequence of non-specific damage to DNA, since other defence-related genes are up-regulated, and there is developmental variation in the response to UV-B. Furthermore, protection under high light appears to be due to some component of photosynthetic activity rather than to photolyase activity. Direct responses, then, include not only *damage reactions*, but also *adaptive responses*, including the switching on of a range of defence mechanisms to afford protection against UV radiation. The best studied is the production of pigments, especially flavanoids, which will act to screen out the UV-B. Bornman *et al.* describe how the production of polyphenols is stimulated, with differential production of certain types, and that the largest concentration of pigments is located in the epidermis, effectively reducing the penetration of UV-B deeper into the mesophyll cells of the leaf. This type of response involves the stimulation of expression of particular genes, implying specific UV-B light detection systems and signal transduction processes which lead to the regulation of transcription. The question of perception of UV, and the relationship between perception of UV-A, UV-B and blue light is discussed by Jenkins, Fuglevand and Christie.

For these sorts of studies, essentially at the molecular and cellular level, the absolute amount of UV radiation does not need to equate closely to ambient levels, since the aim is to investigate the individual mechanisms associated with response to UV. It appears that responses to UV-B represent a new area of photomorphogenesis that has, to date, received little attention, and where new questions are consequently arising. For example, is there a 'simple' UV-B receptor, or are there several types of receptor, as appears to be the case with blue light responses? Is it possible to determine whether a response is a consequence of a 'damage' reaction, or is an adaptive response following specific perception? (Britt suggests that this discrimination might be achieved by treating a mutant which is defective in recovery; if there is still a response, then the effect is one of damage, and is not a function of a 'receptor'.) Understanding the molecular basis of perception and signal transduction is likely to be of direct relevance to a scenario of global climate change.

Effects seen at the whole-plant level, especially under natural conditions, are likely to involve both direct effects and *indirect* effects. Indirect effects are those mediated by radiation-induced changes in the plant environment, or changes in the plant which are of importance mainly in relation to other organisms. Examples include effects on other plants which compete with the plant under consideration (see chapter by Caldwell), effects on nutrient mobilisation, effects on herbivores and micro-organisms of importance to the plant. To understand these effects of UV requires the study of several components of natural ecosystems, and only recently have such studies been started. Unlike studies of direct effects, where responses occur quite rapidly, studies of indirect effects need to be carried out over a long period of time, since effects are likely to be of a slow, cumulative nature. To date, more effort has been given to understanding the direct effects, and indeed these are probably the most important ones in crops, growing as they do in a partly human-controlled environment, and mostly in monoculture. Indirect effects, however, may be more important for wild plants in a natural environment.

Outdoor studies require the construction of suitable irradiation facilities; the UK now has four sites with such facilities, at Lancaster University, Sheffield University, Monks Wood (Institute of Terrestrial Ecology) and Wellesbourne (Horticulture Research International). At these sites, supplementary UV irradiation is provided in a modulated form; most researchers carrying out experiments in the field recognise that modulated systems, where lamp outputs are adjusted to give a supplement which is a constant value above ambient (changing with cloud cover, for example) are preferable to square wave additions (where lamps are simply on or off). These outdoor facilities in the UK currently surpass those elsewhere in the world.

For outdoor studies, supplementary radiation should be related to realistic ozone depletion scenarios. To achieve this, it is crucial to have accurate information about the *action spectrum* of the response being studied. An action spectrum indicates the relative effectiveness of different wavelengths of radiation in bringing about a particular response; the relative effectiveness, derived from the action spectrum, can then be multiplied by the irradiance at each wavelength, and summed over the appropriate wavelength range to give the important function, *biologically effective radiation*. This term is most important in calculating predicted doses of UV resulting from different ozone depletion scenarios, but is also important in determining the dose provided by supplementary lamps, the spectrum of which is quite different from that of sunlight (see chapter by Paul). Throughout this volume the action spectrum used to calculate the biologically effective dose of supplementary UV will be quoted. There are several that can be used, the most common being derived from the Caldwell generalised plant action spectrum, normalised at 300 nm (PAS300). However, this spectrum is a composite from those available in the early 1970s, and it is quite likely that particular responses will have action spectra which are slightly or substantially different from this. The implications of this are discussed in the chapter by Holmes, where evidence is also presented that, when UV is given over an extended period of time, the action spectrum for overall growth has a significant 'tail' in the UV-A region. The reason for this is that, whereas short-term exposures will give spectra characteristic of absorption by a particular molecule (for example DNA), over an extended period, a number of processes may be involved, each with a slightly different action spectrum, added to which the plants will produce screening pigments which may alter the action spectrum.

The use of appropriate controls for outdoor arrays is a further issue also now receiving some attention. Tubes used to provide supplementary UV-B emit some energy at wavelengths below 280 nm, in the UV-C region. To exclude this, tubes are wrapped with cellulose acetate, which does not transmit below 290 nm. Control treatments have usually been unenergised tubes, providing the same shade as operational tubes. However, the UV-B tubes do emit a certain amount of UV-A, so that treatments with supplementary UV-B will also have a certain supplementation of UV-A above ambient. This can be overcome by an additional control treatment in which tubes are covered with polyester foil, which excludes wavelengths < 315 nm, thus providing a treatment which allows the effects of UV-B and UV-A to be separated. Since there is some evidence that UV-A has biological activity (see chapter by McLeod & Newsham), such an additional treatment is desirable. Since there is good evidence that responses to UV-B continue into the UV-A part of the spectrum, meters used should cover the UV-A as well. Detailed measurements, calibration of meters against standardised spectroradiometers and characterisation of the radiometric information, allowing comparisons to be made between different experimental systems, are also crucial.

Effects on growth and interactions between ecosystem components are reviewed in this volume for a range of species and habitats. Aquatic ecosystems, which are responsible for half of the biomass production of the planet, are reviewed by Häder, who discusses the penetration of solar UV radiation into the water column, the biological sensitivity of aquatic organisms, and the effectiveness of mitigating and protective measures in aquatic organisms. It is suggested that aquatic systems are already under UV stress, and that further depletion may result in changes in species composition, with significant consequences for productivity.

Effects on crops are reviewed in the chapter by Corlett *et al.*; although glasshouse and controlled environment experiments suggest that UV causes significant damage, evidence from field studies suggests that, even with a large ozone depletion, the impact on commercial monocultures may be limted, changes in morphology and pigment concentrations being more likely than large changes in productivity. Rozema, van de Staaij & Tosserams draw similar conclusions with respect to natural ecosystems, namely that effects on canopy architecture, competitive relationships and litter decomposition are as important as direct effects on productivity. A detailed description of how changes in species balance may come about, owing to slight changes in morphology, is given in the chapter by Caldwell.

Less research has been carried out into effects on more remote systems, such as upland moors and Arctic ecosystems. In Arctic regions, radiation levels are less than at temperate latitudes, but changes in the ozone layer are greater, giving the possibility of greater relative increases in UV. Moorlands and Arctic ecosystems are characterised by vegetation that is slow growing and perennial in nature, allowing effects to accumulate. Particularly in the Arctic, the low temperature may also impair repair processes. The chapters by Björn *et al.* and by Moody,

Coop and Paul review effects on plant growth and on other components of these ecosystems, such as microbial activity, decomposition and herbivory. Moody *et al.* show that, in UK upland moorlands, the effects of UV-B on plant growth are generally small in comparison with the effects of CO_2, and that their opposing effects tend to cancel each other. As a result, biomass under high CO_2 plus high UV-B treatments is almost the same as that for ambient UV-B plus ambient CO_2. Effects of UV-B on microbial–plant interactions are complex; changes in plant chemical composition can reduce microbial activity, while a direct effect of UV-B during decomposition is to decrease colonisation by fungal decomposers. Increased UV-B may therefore lead to a slowing down of nutrient recycling.

Impacts on forest ecosystems are thoroughly reviewed by McLeod and Newsham, who consider the distribution of UV through forest canopies, direct effects on growth and physiology, where there is more evidence for stimulation of pigment accumulation than reductions in growth, and also impacts on leaf decomposition, micro-organisms, insects and nutrient recycling. Interestingly, it appears that increases in leaf thickness are due to UV-B, while an increase in lammas shoot length and some aspects of herbivory may be specifically due to elevated UV-A.

Finally, some specific examples of pathogen/plant interactions are reviewed in the chapter by Paul. Again, there is significant variation between species: *Septoria tritici* appears to be sensitive to UV-B, in that spore germination is inhibited by elevated UV-B, although strains from lower latitudes (where ambient UV-B is higher than in the UK) are more resistant. In other cases, the interaction is more complex; changes in plant chemical composition may alter the activity of fungal pathogens. Paul also considers in detail the consequences of using different action spectra in determining the UV-B doses supplied from supplementation with artificial lights, highlighting the very large differences in biologically effective dose that can be derived.

Despite the significant progress made in recent years, or perhaps more accurately, because of the progress made, a number of questions arise. At the climate level, is the thinning of the ozone layer having a real effect on terrestrial levels of UV-B? At the biological level, is it possible to separate damage and adaptive responses, and further, is it possible to link responses at the molecular level with those occurring at the eco-system level? Plant responses to UV-B is an exciting area; new insights into fundamental mechanisms are being gained, while the subject has very real significance in the light of changes in climate. However, many of the studies need to be carried out over longer time period to provide definitive answers to questions such as cumulative effects of UV-B, effects of UV-B at the ecosystem level, and interactions of elevated UV-B with other stress factors. Further information on UV-B, and other aspects of plant photobiology can be found across the Worldwide Web. A good starting point is the Plant Photobiology page, the address for which is http://cc.joensuu.fi/photobio/photobio. html

Peter J. Lumsden
HRI Wellesbourne
Warwick
1996

PART I

The ozone layer and UV-B radiation

J.A. PYLE

Global ozone depletion: observations and theory

Introduction

It is now well known that ozone concentrations in the stratosphere have declined during the last 25 years as a consequence of the emission into the atmosphere of a variety of chlorine- and bromine-containing compounds. This decline in the thickness of the ozone layer could have important consequences. Firstly, ozone is an important climate gas. Absorption of solar radiation by ozone controls the temperature of the stratosphere and leads directly to the stable stratification of the stratosphere. Changes in ozone can be expected to lead to changes in the climate of the stratosphere; these changes may also be important for the circulation of the troposphere and hence for weather. In addition, ozone is a greenhouse gas, absorbing radiation in the infra-red region of the spectrum. A decrease of ozone in the lower stratosphere can act to cool the climate system, counteracting the impact of the growth of carbon dioxide and other greenhouse gases. Secondly, because ozone absorbs solar ultra-violet radiation, any reduction in the stratospheric ozone abundance will lead to enhanced intensity of ultra-violet radiation at the surface. Biological systems are particularly sensitive to short wavelength radiation at less than about 300 nm. Increases in the intensity at these wavelengths may have an influence on the health of humans, of plants (see later chapters in this volume) and of other animals.

Ozone change will be a controlling influence on changes in surface ultra-violet radiation. In this chapter the evidence for ozone change will first be presented and the causes of the ozone decline will be discussed. Since the ozone decline is related to the accumulation in the stratosphere of man-made chlorine and bromine compounds, the reduction in the emission of these compounds should lead to an eventual recovery of the ozone layer. Factors controlling the rate of recovery will be presented. Finally, some outstanding issues will be raised.

Ozone changes in the stratosphere

The possibility that chlorine compounds could significantly perturb the stratospheric ozone layer was first proposed by Molina and Rowland (1974). They showed that a build-up of chlorine compounds in the stratosphere would be the necessary consequence of the substantial growth during the 1960s and early 1970s in the production and emission of a number of chlorinated compounds, which were being used as aerosol propellants and in foam blowing and refrigeration. As they move upwards through the stratosphere, these compounds, which are unreactive in the troposphere, would be subject to increasingly intense energetic solar radiation which could dissociate them. This would lead to the liberation of free chlorine atoms which could then take part in very efficient catalytic chemical cycles to destroy ozone. In their theory, the largest ozone depletions would occur in the upper stratosphere at around 40 km.

The first clear evidence for ozone depletion came with the discovery of the Antarctic 'ozone hole' (Farman, Gardiner & Shanklin, 1985). Data collected at the British Antarctic Survey base at Halley Bay since 1957 showed that a substantial decline in springtime ozone values had occurred during the 1980s. The result was surprising in that the details of the loss were different from the predictions of Molina and Rowland (1974). Nevertheless, it was soon confirmed that the growth in concentrations of chlorine compounds was the cause of the ozone loss (see next section).

The discovery of the 'ozone hole' prompted a massive research effort. This has involved detailed studies of the chemical and physical mechanisms responsible for the depletion. In addition, data sets of the ozone concentration from both ground-based and satellite instruments have been carefully analysed to determine trends in ozone not just in Antarctica but at all latitudes. Figure 1 shows the trends in ozone calculated during the 1980s from the TOMS satellite instrument (Stolarski *et al.*, 1991). Analyses with different data sets show the same basic features. The most recent analyses were reviewed by WMO (1995) and SORG (1996). Here, only the salient features will be indicated. While the largest depletion is evident in Antarctica in the spring, there is also evidence of a substantial ozone decline in high latitudes of the Northern Hemisphere in winter and spring. In both hemispheres the data show that there has been a loss of ozone since the 1970s. Although the Figure indicates that there has also been a decline in tropical latitudes, the changes there are small and are not statistically significant.

The most recent analyses (see SORG, 1996) show that ozone has

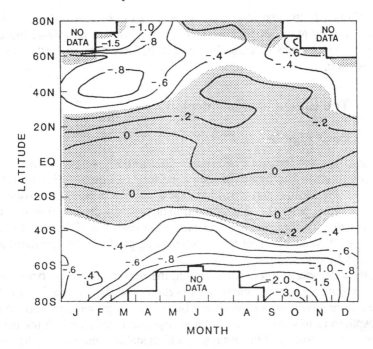

Fig. 1. Total ozone trend (% per annum) determined from TOMS satellite data as a function of latitude and season. The lightly shaded areas indicate regions where the trends are not statistically significant. Poleward of the heavy solid lines are the Arctic and Antarctic polar nights where TOMS was unable to make measurements. (Figure from Stolarski *et al.*, 1991).

declined globally by about 4–5% since 1979. In middle latitudes, the annually averaged loss in the same period is about 7%. At northern mid-latitudes, including the UK, in winter and spring the accumulated loss since 1979 is now about 11%.

In Antarctica, the 'ozone hole' occurs each spring. Values for the ozone column (a measure of the sum of all the ozone molecules between the surface and the top of the atmosphere) in October are now about one-third of those observed in the 1960s. Each spring, nearly all the ozone at altitudes between 15 and 20 km is removed in a period of about six weeks. While there are interannual variations in the depth of the ozone hole, it is clear that the effect is close to saturation (once all the ozone is removed, no further depletion can occur).

Evidence for the ozone decline in northern high latitudes has strengthened in the last few years with, for example, record low values being reported at various stations during the winters of 1994/1995 and 1995/96. The causes of these losses will be investigated further in the next section.

The causes of ozone depletion

Polar regions

The discovery of the Antarctic ozone hole initially surprised the scientific community. The nature of the ozone loss was quite different to that suggested by Molina and Rowland (1974). It was occurring in southern polar latitudes, with the loss predominantly below 20 km altitude. The loss was mainly confined to the polar spring season and developed very rapidly each year. None of these details was predicted in Rowland and Molina's theory.

After a decade of intensive research, the reasons for polar ozone loss have become reasonably clear. It is certain now that ozone is depleted by reactions involving chlorine and bromine compounds, present in the atmosphere following a variety of anthropogenic activities. In the troposphere the chlorine and bromine are held in stable compounds, the chlorofluorocarbons and halons. In the stratosphere these compounds can be dissociated. The majority of the chlorine and bromine is then transformed into other compounds ('reservoirs') which themselves do not destroy ozone. However, it is now known that reactions on surfaces can turn these reservoirs (for example, HCl and $ClONO_2$) into active forms (such as Cl and ClO) which, in the presence of sunlight, do destroy ozone. The polar lower stratosphere provides the conditions for these processes to occur. During winter, a strong westerly circumpolar flow (the polar vortex) is established. Within the polar vortex temperatures become very low and polar stratospheric clouds (PSCs) can form. These PSCs provide the surfaces on which the reservoirs are turned into active forms. When sunlight returns to the polar regions in springtime, rapid ozone depletion can occur. Thus the sequence leading to ozone depletion requires that, first, temperatures are low, so that PSCs can form and release active chlorine, and, secondly, sunlight is present to drive the photochemical destruction cycles. (For a more complete description of the relevant processes see, for example, WMO, 1995.)

The conditions in the Antarctic lower stratosphere are particularly favourable for ozone depletion. Each winter and spring, temperatures drop sufficiently for PSCs to form. The low temperatures persist into the spring when sunlight returns. In contrast, during winter the lower stratosphere over the Arctic is warmer than its southern hemisphere counterpart and temperatures for PSC formation are less widespread. The lowest temperatures may occur during the darkness of polar night. For these reasons, depletion in the north was not expected to be as severe as in the south. Nevertheless, in the late 1980s it became clear that conditions conducive to ozone loss do occur in the north and a number of detailed research programmes, including the European Arctic Stratospheric Ozone Experiment (EASOE) in the winter of 1991/92 (see Pyle *et al.*, 1994) and the Second European Stratospheric Arctic and Middle latitude Experiment (SESAME) in the winters of 1993/94 and 1994/95 (see, for example, Pyle *et al.*, 1995), have investigated northern polar loss.

It is now evident that chemical destruction of ozone has also occurred in the Arctic lower stratosphere, where a wide variety of measurements have shown that active chlorine is present following low temperatures. Studies using, for example, ozone sonde measurements have confirmed that local losses in the lower stratosphere approaching 2% per day occurred during 1994/95 (Rex *et al.*, 1996), with an accumulated loss throughout the winter amounting to about 50% at 20 km inside the polar vortex. Numerical modelling studies have shown that large interannual variability can be expected in the magnitude of the ozone depletion. The model studies confirm that quite significant losses can be expected in cold stratospheric winters like 1994/95 and 1995/96 (Chipperfield, Lee & Pyle, 1996).

As discussed above, there are important differences between the Arctic and Antarctic polar vortices. In the north, the vortex is more variable and more mobile. In the south, the vortex is usually centred close to the pole but in the north the location of the vortex is quite variable and can be influenced strongly by weather systems in the troposphere. In consequence, the vortex is often located over northern Europe. On these occasions, the air in the lower stratosphere above populated regions of Europe can be characteristic of the chemically perturbed vortex. Thus, in March 1996, record low values of ozone were reported at the UK measurement sites in Cambourne and Lerwick when the polar vortex passed overhead. In the context of the middle latitude decline, discussed below, it is important to realise that these low ozone events are relatively short-lived and that the ozone-depleted air will subsequently move away from the UK as the vortex sweeps back towards more northerly latitudes.

Middle latitudes

The mechanisms leading to losses of ozone in polar latitudes are now quite well understood. Much more controversial is the cause of the decline in ozone in middle latitudes of the northern hemisphere, amounting to about 7% (annual average) since 1979. As discussed, polar vortex air can sometimes be seen over middle latitudes where low ozone can then be measured. However, it is clear that this cannot be the main explanation of the annually averaged trends in middle latitudes, and other theories have therefore, been advanced. Most theories implicate the halogen compounds, but the precise details of the processes involved remain in dispute. One possibility is that the ozone depletion occurs in polar latitudes and this air is then mixed irreversibly into middle latitudes causing a general 'dilution' of ozone levels there. A second possibility is that air is primed for ozone depletion by the reactions on PSCs in polar regions, but is then transported southward, and possibly mixed, before the depletion occurs. If this process operates continuously (like a 'flowing processor', see McIntyre, 1995), then large ozone loss might occur in mid-latitudes. A further possibility is that the chlorine is activated *in situ* in middle latitudes, possibly on sulphate aerosol, followed by local ozone depletion.

Studies in the last few years suggest that all three mechanisms are involved. It is clear that polar air can be mixed into middle latitudes after filaments of air are stripped from the edge of the polar vortex (see, for example, Pyle *et al.*, 1995). However, this may not be the major factor, and increasing evidence is accumulating for the importance of the *in situ* reactions on sulphate aerosol which can lead to the initiation of ozone destruction cycles, similar to those seen in polar latitudes but operating more slowly in middle latitudes (see SORG, 1996).

What is now clear is that both chlorine and bromine compounds are involved in the middle latitude decline. However, we only have a qualitative, and not a quantitative, understanding of the loss. The lack of quantitative understanding means that we cannot predict accurately the course of a recovery of ozone amounts in middle latitudes.

Ozone recovery – some outstanding questions

International regulations to control the emissions of certain ozone-destroying substances have been established since the discovery of the 'ozone hole' and the development of a clear understanding of the mechanisms of the ozone loss. Production of the major carriers of chlorine to the stratosphere, the CFCs, are now phased out with a few exceptions for developing countries and essential uses. Regulations also control the emissions of the chlorine-containing transitional compounds which have replaced the CFCs. Similarly, production of the Halons, which can carry bromine to the stratosphere, has now ceased and some of the uses of methyl bromide are to be controlled. The international action began with the Montreal Protocol in 1987, followed by amendments to the Protocol in London (1990), Copenhagen (1992) and Vienna (1995). The consequence is that accumulated chlorine loading of the troposphere has already reached its maximum; maximum chlorine levels in the stratosphere are expected within the next few years. Peak concentrations of bromine in the stratosphere are expected to occur slightly later, perhaps by about 2010.

Stratospheric ozone levels should start to recover once the concentrations of chlorine and bromine begin to fall. Thus, the next decade or so represents the period of greatest risk when, for example, large ozone loss could continue to occur in the Arctic vortex. The lifetime of the CFCs is long, so that their removal from the stratosphere will be slow, and concentrations of chlorine compounds are expected to remain high enough in the stratosphere to produce the springtime Antarctic 'ozone hole' at least until the middle of the next century. Prediction about the rate of recovery in middle latitudes is made difficult by our lack of a quantitative understanding of the observed depletion.

Thus, if chlorine and bromine concentrations fall, and all else remains the same, the ozone layer is expected to recover, albeit slowly. However, all else may not remain the same. For example, a downward trend in the temperatures in the lower stratosphere has been reported (see, for example, Oort & Liu, 1993; SORG, 1996). This could lead to more widespread conditions for the formation of polar stratospheric clouds and, hence, to the conditions for chemical destruction of ozone. The conditions inside the Arctic vortex during the last two winters are particularly intriguing in this context. Both 1994/95 and 1995/96 saw the establishment of record low temperatures in the lower polar stratosphere. We do not know whether these records were the result of purely natural variability or whether they represent some kind of trend (for example, caused by increasing concentrations of greenhouse gases or reduced concentrations of ozone). The next few winters will provide the answer. What is quite clear is that if the Arctic lower stratosphere continues to cool, the conditions will become more like those found in Antarctica and will favour ozone loss. Any cooling would inevitably delay the recovery in ozone expected to follow the decreases in chlorine and bromine loading.

Conclusions

Ozone depletion is widespread across both polar and middle latitudes and will affect both the climate system and the penetration of ultraviolet radiation to the earth's surface. The causes of polar loss are well understood in terms of a chemical depletion involving chlorine and bromine compounds. In recent Arctic winters, unequivocal evidence has emerged for a chemical loss similar to that seen in the Antarctic. For example, between January and March 1996, about 50% of the ozone in the Arctic lower stratosphere was destroyed chemically. A quantitative understanding of the observed decline in ozone amounts in middle latitude is still not available, although it is now clear that, again, chlorine and bromine compounds play a role.

International regulations, beginning with the Montreal Protocol in 1987, have begun to have an impact on the levels of ozone-destroying compounds in the atmosphere. However, recovery will be slow. Thus, large ozone losses in the Arctic, comparable with those seen in the last two winters, can be expected during the next decade during cold stratospheric winters. Meanwhile the 'ozone hole' should remain a feature of the Antarctic springtime stratosphere until the middle of the next century. Recovery may be made even slower if other conditions in the stratosphere change. Thus, if the very low temperatures in the last two Arctic winters represent part of a trend, instead of being purely the result of natural variability, then continuing large ozone loss can be expected and the pace of recovery will be slowed.

References

Chipperfield, M.P., Lee, A.M. & Pyle, J.A. (1996). Model calculations of ozone depletion in the Arctic polar vortex for 1991–92 to 1994–95. *Geophysical Research Letters*, **23**, 559–62.

Farman, J.C., Gardiner, B.G. & Shanklin, J.D. (1985). Large losses of total ozone in Antarctica reveal seasonal ClO_x–NO_x interaction. *Nature*, **35**, 207–10.

McIntyre, M.E. (1995). The stratospheric polar vortex and sub-vortex: fluid dynamics and middle latitude ozone loss. *Philosophical Transactions of the Royal Society, London A*, **352**, 227–40.

Molina, M. & Rowland, F.S. (1974). Stratospheric sink for chlorofluoromethanes: chlorine atom catalyzed destruction of ozone. *Nature*, **249**, 810–12.

Oort, A.H. & Liu, H. (1993). Upper-air temperature trends over the globe, 1958–1989. *Journal of Climate*, **6**, 292–307.

Pyle, J.A., Chipperfield, M.P., Kilbane-Dawe, I., Lee, A.M., Stimpfle, R.M., Kohn, D., Renger, W. & Waters, J.W. (1995). Early modelling results from the SESAME and ASHOE campaigns. *Faraday Discussions*, **100**, 371–87.

Pyle, J.A., Harris, N.R.P., Farman, J.C., Arnold, F., Braathen, G., Cox, R.A., Faucon, P., Jones, R.L., Megie, G., O'Neill, A., Platt,

V., Pommereau, J.-P., Schmidt, V. & Stordal, F. (1994). An overview of the EASOE campaign. *Geophysical Research Letters*, **21**, 1191–5.

Rex, M.P., von der Gathen (& 35 co-authors) (1996). Chemical ozone loss in the Arctic winters 1991/92 and 1994/95 (Match), Proceedings of the Third European Symposium on Polar Ozone Research, Schlierse, 18–22 September, 1995, *European Commission Air Pollution Research Report*, **56**, 586–9.

SORG. (1996). *Stratospheric Ozone 1996*, United Kingdom Stratospheric Ozone Review Group. Fifth report. HMSO, London.

Stolarski, R.S., Bloomfield, P., McPeters, R.D. & Herman, J.R. (1991). Total ozone trends deduced from Nimbus 7 TOMS data. *Geophysical Research Letters*, **18**, 1015–18.

WMO. (1995). Scientific Assessment of Ozone Depletion: 1994, WMO Global ozone research and monitoring project – report no. 37, Geneva.

A.R. WEBB

Monitoring changes in UV-B radiation

Introduction

The biological consequences of ozone depletion, mediated through an increase in ultraviolet-B (UV-B) radiation, have been cause for concern, prediction and speculation for many years. Estimating the potential effects of ozone depletion involves several steps:

1. Estimating the ozone depletion that might realistically be expected over a given region of the world (this requires assumptions about, for example, compliance with the Montreal Protocol, or not). Alternatively, observed ozone depletions to date can be used to assess the changes already experienced.
2. Calculating changes in UV-B due to changes in ozone. The assumption is that all else remains unchanged and these calculations are usually made for clear-sky conditions. If changes in UV-B irradiances have been observed, then they may be used instead, but they cannot necessarily be attributed solely to changes in ozone.
3. The exposure of biological systems to the available UV-B must be assessed. For plants growing at a single location, this can be assumed to remain unchanged, but for mobile systems (animals, fish, and especially people) adaptive behaviour is possible.
4. The biological (or chemical) results of exposure to the (changed) UV-B must be predicted, based on experiment and observation. Response may depend upon accumulated dose, upon reaching some threshold dose, and upon possible protective mechanisms, for example, the build-up of UV-absorbing pigments (melanin in humans, flavonoids in plants).

It is clear that step number 2, assessment of changes in UV radiation, straddles the gap between atmospheric cause and biological effect, and in doing so becomes the focal point for two different questions:

(a) *Atmospheric.* What is the expected change in UV-B due to ozone depletion (and can it be observed, and if not, why not?), or the corollory of this, can any observed change in UV-B be attributed solely to ozone depletion?

(b) *Biological.* What is the real change in available UV-B in the biosphere, regardless of atmospheric processes?

In search of an answer to both questions, the same data and the same predictive techniques are used, and as there is not enough information available in either respect there is still doubt as to whether UV is changing over much of the earth's surface. This does not contradict the basic premiss that less ozone in the atmosphere means more UV-B at the ground, rather it indicates how many other factors have to be considered in trying to apply this otherwise simple relationship.

Atmospheric factors affecting UV radiation reaching the biosphere

The solar zenith angle, determined by season and latitude, controls the underlying diurnal and annual changes in radiation of all wavelengths of solar radiation reaching the earth's surface. However, zenith angle and other astronomical factors such as earth–sun distance are considered constants in considering possible changes in UV-B radiation on a time-scale of decades and will not be discussed any further.

Ozone is the chief selective absorber of UV-B radiation in the atmosphere, but it is by no means the only variable attenuator. Clouds and aerosols can have a vast effect on UV at the surface on short timescales (minutes to years) and result in a very noisy long-term (decades) pattern of surface irradiances from which to try and extract an ozone-related change which, at its worst, is often small compared to the noise. The uncertainty is compounded by the fact that ozone itself is highly variable on a day-to-day basis, and dynamics (at least in mid-latitudes) often conspires to provide low ozone and clear skies together (high pressure systems) or high ozone and cloud together (low pressure systems). A sunny 'anticyclonic' summer may thus provide a far higher dose of UV then a preceding dull summer of passing depressions, while having nothing at all to do with ozone depletion. Over a sufficiently long time period (30 years +) this natural variability in the weather can be treated statistically to produce an expected range of conditions, a climatology. Thus we have an expected range of ozone column thicknesses for a given location, and excursions outside this normal range may be single extreme events (the ozone equivalent of a 100-year storm), or if persistent may represent a changing climatology (attributed in this case to chemical ozone depletion). Such ozone trends (see chapter by Pyle) might be expected to lead to similar observed changes in UV-B, if every other aspect of the climate and the environment remains the same. Should cloud climatology change, one hypothesis associated with global warming (IPCC, 1990), its effect could be either to enhance or offset ozone-related changes in UV radiation, depending on how the cloud increased or decreased. Aerosols and pollution, including low-level ozone, decrease the UV reaching the surface. Thus, increasingly poor air quality in urban and industrial areas over several decades may again disguise the effects of small ozone losses in the stratosphere. As methods to reduce air pollution become more advanced, declines in pollution and stratospheric ozone could act in the same direction to produce an increase in UV radiation.

One further aspect of change which could be significant in some regions is the land surface or surface use. Surface reflectivity (albedo) is low for UV wavelengths and most vegetative surfaces. It is somewhat higher for rocks and concrete, and can be extremely high, over 90%, for clean snow (Madronich, 1993). From an atmospheric point of view, changing albedo alters the amount of radiation which is reflected back to the atmosphere from the ground and is then available to undergo further back-scattering and return to ground level. Increasing albedo generally increases the overall surface irradiance. Biologically, the effect of increasing albedo is likely to be even greater. Not only is an organism exposed to greater irradiance from above, but it will also receive directly reflected radiation from below, and at angles which may enable exposure of otherwise protected areas. The greatest contrast in albedo is between snow and vegetation from season to season (normal) or climatologically – another hypothesis of global warming.

To try to predict the outcome of the combination of ozone depletion, global warming and a cleaner air policy on UV radiation at the surface would be highly speculative, yet all may have influenced the UV measurements made to date, while ozone depletion is often the only factor which can be confidently considered in calculating ozone depletion effects. Our knowledge of changes in cloud, aerosol, albedo, and their detailed effects on UV radiative transfer, is so scant that predictions are almost always restricted to clear-sky calculations where the only variable allowed to change is ozone. This gives a clear-cut answer which is not often observed. However, the lack of evidence of an undisputed trend in UV-B radiation over much of the globe is not due solely to the confusions wrought by cloud and aerosol, but also to the UV instrumentation and its very sparse deployment for any significant length of time.

UV monitoring equipment

The two basic categories of UV monitoring equipment are the broadband radiometer and the spectroradiometer. A third category, the multi-filter instrument lies part-way between the other two categories, providing data for a discrete number of narrow spectral bands. There is only one report in the literature from a multi-filter instrument in use for a significant period of time (Correll *et al.*, 1992), and therefore this type of instrument will not be discussed in detail.

The broadband instrument is the one that has been in operational use for the greatest time. It is simpler, cheaper and more robust to operate than the spectroradiometer, and is therefore also the most widely deployed type of instrument. There are several different versions of the broadband instrument, available from different manufacturers, but the majority of them purport to measure the same quantity – the erythemally effective UV radiation incident on a horizontal surface. The erythemally effective radiation (EER) is defined as the solar spectrum (I_λ) weighted with the erythemal action spectrum (A_λ, the effectiveness of each wavelength in producing erythema or sunburn) and integrated over the entire solar UV waveband (280–400 nm), thus:

$$EER = \int I_\lambda \, A_\lambda \, d\lambda$$

In reality, each instrument measures the integral of the solar spectrum and the individual instrument's response spectrum (R_λ). The response spectra vary somewhat from manufacturer to manufacturer and even from instrument to instrument of the same make, and none of them exactly matches the erythemal action spectrum (of which there are several, the McKinlay–Diffey (1987) or CIE response being the one most commonly used). The response spectrum can also change with age as the optical elements of the instrument degrade, so the stability must be checked if long-term monitoring is the intended use of the equipment. Not all broadband instruments measure EER; some of them have been designed to match a different action spectrum, or to measure unweighted UV across a designated waveband, but long-term records are not generally reported from such instruments in the context of ozone depletion and will not be discussed.

Spectroradiometers measure the UV irradiance at a series of discrete wavelengths to provide a solar spectrum showing the change of intensity with wavelength. As with broadband instruments, there are many makes of spectroradiometer capable of measuring at different wavelength intervals (resolution), and with varying degrees of compliance to the ideal of discrete wavelength measurements. All spectrometers have a finite slit width through which radiation of a designated wavelength is directed on its way to the detector. A degree of radiation from neighbouring wavelengths (near straylight) inevitably passes through the slit at the same time and also reaches the detector, to be falsely attributed to the nominal wavelength of measurement. The intensity of the ground-level solar UV radiation drops by several orders of magnitude across the narrow UV-B part of the spectrum, so reducing this near straylight becomes vitally important, the more so towards shorter wavelengths. Random photons bouncing around in the instrument from any of the incoming radiation (far straylight) must also be suppressed and prevented from reaching the detector. To achieve suitable straylight rejection, a double monochromator is generally employed.

It is clear that there is no standard instrument for solar UV measurements, and that there is variety even between the types of instrument, which, even assuming perfect calibrations and other common features (e.g. cosine response), will give differences in the measurements made with individual instruments. The importance of these differences depends upon the way in which the data is to be used.

In looking for a change in UV radiation with time the most important property of any instrument is its proven stability over a long time period. This can be achieved by regular calibration of the equipment, the required frequency depending on experience and the inherent stability of the components of the instrument. Once internal site-system stability limits have been determined, the data record from that site can be used to identify any changes in solar UV which may be greater than the uncertainty in stability. This holds regardless of the relationship of the site instrument to any other instruments, or the calibration reference to other calibration references. So long as there is internal consistency maintained for the duration of the data record, the possibility to detect change at a single site exists.

Site stability can only be assessed and maintained by careful and sufficiently frequent calibrations. The method of calibration, and the frequency with which it is required, will depend on the type of instrument in use. Broadband instruments are usually calibrated either against other broadband instruments or against spectroradiometers. Spectroradiometers are calibrated against standard lamps, and all absolute irradiance calibrations should eventually revert back to a National Standards Laboratory reference. The greater the number of transfer calibrations between the Standards Laboratory and the site calibration, the greater the uncertainty in the absolute irradiance measurements on site. For a single site interested only in change (stability) this uncertainty is not an initial problem (Fig. 1): the absolute irradiance may only be known to ±10%, but a lamp should have much greater stability at its unknown position within this range. However, when some link in the calibration chain has to be changed (as inevitably happens when lamps age and degrade), then great care must be taken that a shift in the reference irradiance level is not introduced in all following calibration transfers.

If the data at one site are to be compared with data collected at other sites, then more attention must be paid to the agreement between the different monitoring instruments, to ensure that any differences identified in UV radiation are correctly attributed to the site location and are not really a consequence of unmatched instruments or their calibrations. At this point, position on the absolute irradiance scale does become important (Fig. 1), in addition to stability which should also be maintained with the same vigour as for an isolated site. Achieving absolute agreement between different instruments depends upon both the instruments' characteristics and the reference standards to which their calibrations are referred. Calibrated lamps from National Standards Laboratories have uncertainties of 2–3% at the short UV-B wavelengths,

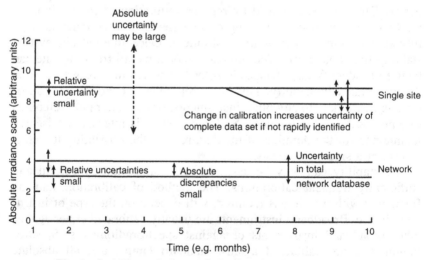

Fig. 1. Stability, relative and absolute irradiance standards for single and multiple monitoring sites.

reducing to approximately 1% in the UV-A. Each transfer of the absolute standard to another lamp or instrument introduces additional uncertainty. Thus, in the unfortunate situation when uncertainties might be additive, two well-calibrated instruments may differ in their absolute UV-B measurements of the same source by 4–5%. Further differences in broadband response functions, spectroradiometer slit functions, straylight, and cosine responses (both types), can add further to the discrepancies. This point has been amply illustrated by a number of international intercomparisons of both spectroradiometers and broadband radiometers (Gardiner & Kirsch, 1993, 1994, 1995; Koskela, 1994; Seckmeyer *et al.*, 1995; Leszczynski *et al.*, 1996). The best agreement that has been achieved to date is at the 5% level during intercomparisons. It has to be hoped that stability with location and with time enables this level of agreement to be maintained when instruments return to their home sites.

The easiest comparisons of data from site to site come from assessing the outputs of instruments of the same type. Most spectroradiometers take a finite time (minutes) to scan across the UV spectrum, and during monitoring will usually take a scan at regular designated times or zenith angles. Broadband meters, by contrast, record the instantaneous EER, which is usually sampled frequently and integrated over a chosen period of time to give the total EER dose received, for example, in half-hourly periods. The two types of data record have different uses and are often difficult to compare directly unless the conditions of measurement are very stable (clear skies or uniform, stable overcast). There are also many ways in which data records can be analysed, depending on the type of UV instrument used for the monitoring, and the availability of supporting data from other equipment at the same site, for example, solarimeters, devices for measuring other solar spectral bands, turbidity data, ozone, etc.

UV records

There is no coherent, global UV data record from which to assess whether, and where, there might have been changes in UV radiation over the past two decades. Instead, data are available from individual sites and a small number of regional networks, which are unevenly distributed around the globe (Fig. 2). Many of the sites shown in Fig. 2 did not become operational until after the announcement of the Antarctic ozone hole (Farman, Gardiner & Shanklin, 1985), and monitoring efforts in place in the decade before this announcement had often ceased or been scaled down by 1985. Thus there are very few unbroken UV

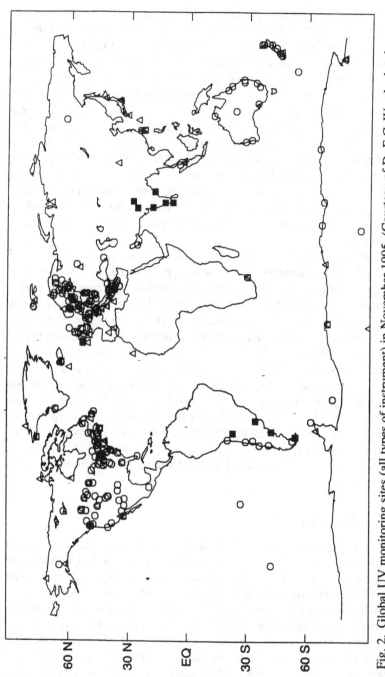

Fig. 2. Global UV monitoring sites (all types of instrument) in November 1995. (Courtesy of Dr E.C. Weatherhead. Information can be obtained from betsy@srrb.noaa.gov)

records extending from the pre-ozone depletion era to the present day. This situation automatically precludes any statements about changes in UV climatology or long-term trends because the data records are simply not long enough to make such assessments. However, the data records that are available do provide useful information. It must be realised that the data is generally site (or region) specific and results cannot be applied directly to other locations because ozone depletion is not globally uniform, and neither are the other confounding factors discussed above. None the less the evidence from different sites can be used to improve our understanding of UV irradiances at the surface and their relation to ozone depletion at different locations.

The clearest link between reduced ozone and increased UV-B radiation comes from Antarctica where the springtime appearance of the ozone hole is so dramatic that the corresponding increases in UV radiation are easily identified. Since 1988 a network of three spectral UV monitoring instruments have been deployed on the Antarctic continent by the American National Science Foundation (Booth *et al.*, 1994). The UV-B at each of the three sites varies tremendously from year to year depending on the position of the polar vortex and whether the lowest ozone values coincide with clear skies and the higher solar elevations which come as the season progresses. Despite these complications there is no doubt that lowering ozone by such extreme amounts gives correspondingly huge increases in UV-B. The collection of spectral data enables the changes in irradiance at UV-B wavelengths to be compared with changes at UV-A or visible wavelengths (unaffected by ozone) under the same conditions. If only the UV-B wavelengths change, and the change increases as wavelength decreases, mirroring the ozone absorption spectrum, then change in ozone can be identified as the cause of change in irradiance. If all wavelengths in the UV/visible region increase or decrease, then cloud, aerosol, the instrument, or a combination of factors is a more probable explanation for the observations. This same technique has been used at other locations to show ozone-related changes in UV-B radiation on a short timescale. The NSF network expanded after 1988, first to Ushuaia in southern Argentina, and then to Barrow, Alaska and San Diego, California. Southern Argentina can intermittently come under the influence of the springtime Antarctic ozone hole as the position of the polar vortex wobbles around the southern high latitudes, and when under the vortex UV-B radiation received in Ushuaia can exceed that expected for the latitude by 45% (Frederick *et al.*, 1993). The year-to-year variability in absolute UV doses (integrated by month) in Ushuaia is as much a function of the variability in cloud as changing ozone (WMO, 1995), but if cloud effects are removed then there is an upward trend in UV-B matching the ozone depletion (WMO, 1995).

Away from the Antarctic region, ozone depletion has been less severe. In mid-latitudes there has been a general decline in ozone, of the order of a few per cent per decade, with a seasonality to the change: it is greatest in winter/spring and least in the summer/autumn. Extracting evidence of trends in UV-B under these circumstances, with a short data record and against a background of noise from natural variability, is impossible. However, short-term UV-B changes can be clearly associated with periods of low ozone, as, for example, in northern mid-high latitudes during 1992 and 1993. These years represent a special situation when ozone was lower than previously observed over this region, and for a prolonged period. Much of this anomaly was due to a combination of aerosol in the stratosphere following the eruption of Mount Pinatubo in 1991 (the aerosol acting to increase ozone destruction), and unusually persistent springtime high pressure regions over parts of the North American continent and Northern Europe (tending to give low ozone for dynamic reasons rather than because of ozone depletion). A four-year record of spectral UV data from Toronto, Canada showed a significant increase in UV-B radiation in these years (1989–1993), with the spectral signature associating the change with ozone (Kerr & McElroy, 1993). The NSF instrument at Barrow, Alaska provided supporting evidence of this short-term change, reporting elevated springtime UV-B levels in 1993 compared to measurements in the previous two years (Booth et al., 1993). From 1992 to 1994 low ozone (as much as 40% depletion) over Finland in the early spring was also picked up by broadband UV measurements, but it was only in 1993, when the low ozone extended into April and May, that the increased absolute irradiances became sufficiently large to appreciably influence the seasonal and annual doses of UV measured in Finland (Jokela et al., 1995). Spectral analysis of data from Garmisch-Partenkirchen in Germany also showed increases at short UV wavelengths during 1993 compared to the preceding year (Seckmeyer et al., 1994). However, in nearby Innsbruck, broadband measurements of UV dose showed no increase in the total seasonal doses of UV for 1993 compared to the average dose for the 1981–1988 period (Blumthaler et al., 1994). This seeming contradiction is also seen at other locations. For example, in the UK, spectral measurements made at Reading clearly show that, when ozone decreases, there is a corresponding increase in UV-B (Fig. 3), yet there is no identifiable change in seasonal or annual UV doses (EER) measured by a network of broadband instruments spread across the UK since 1988 (UMIRG, 1996). The explanation is straightforward:

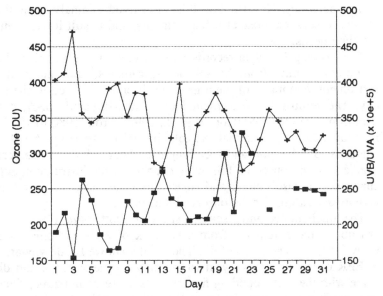

Fig. 3. The ratio of erythemally effective UV-B to UV-A radiation (■) measured at noon at Reading (51.5° N), and the corresponding ozone (+), for the month of March 1995. Using the ratio of the two wavebands removes some of the influence of cloud and aerosol (affects both wavebands) but not the ozone effect (UV-B only). Inspection of the spectral scans on the 16th and 20th March revealed that sky conditions had changed significantly during the scan, negating the analysis, a fact clearly seen in the Figure.

over the UK (and other mid-latitude regions) ozone depletion has not been severe enough to enable a change in the total UV dose (an average over all conditions including a large proportion of cloudy days) to be identified from a short data record which includes great natural variability. The EER is also less affected by ozone than the shorter UV-B wavelengths because it includes a UV-A component that does not respond to changes in ozone. Spectral data, on the other hand, allow the shortest and most vulnerable UV-B wavelengths to be inspected, and also enable comparison of different wavelength bands (UV-A and UV-B) to help remove some of the confusion caused by cloud and other atmospheric constituents (Fig. 3). If ozone decline is severe enough, then it can also be picked up by broadband meters on a short timescale: in the south west of England and Ireland in March 1996, exceptionally low ozone coincided with clear skies for a few days, giving unprecedented UV doses for the time of year (C. Driscoll, personal communication). Nevertheless this short-lived anomaly is unlikely to affect the seasonal dose at these locations, and is still lower than summertime doses.

The existing UV data records which exceed those mentioned above in duration come from broadband instruments, and one multi-filter instrument, and many of them have not been maintained to the present day. The multi-filter instrument located in Maryland, USA, provides data from 1975 to 1990 (Correll et al., 1992). From 1980 to 1987, there is evidence of a substantial increase in UV-B especially for the filter channels at the shorter wavelengths. After 1987, a decrease in irradiances was observed following changes to the instrument, although changes in aerosol, cloud and solar activity are also suggested by the authors as reasons for the observed decreases.

The much quoted and discussed study of Scotto et al. (1988) reported broadband measurements from a network of eight radiometers in the USA for the years 1974–1985. The results showed a downward trend in EER of 0.5 to 1% per year (depending on the station), and did not agree with the corresponding figures for a decrease in ozone. There has since been much investigation and reanalysis of the data from this network, and while the evidence available indicates that the instruments were stable during the decade in question, there may have been a step change in the calibration in 1980 which would account for the observed downward trend (WMO). Other investigations of individual sites in the network show that, for a subset of the data (clear skies in summer), the UV-B trend was consistent with the ozone data measured at the same sites, but the rest of the data were contradictory (Frederick & Weatherhead, 1992). The final interpretation of this US data set is still unclear. Further analysis of broadband data from Atlanta, Georgia also revives a previous explanation for the Scotto data, that changes in pollution at the monitoring sites could be responsible for the downward trend in UV-B. Justus and Murphy (1994) indicate that changes in aerosol at some broadband meter sites can mask expected increases in UV-B because of the response of the instruments to UV-A as well as UV-B.

Elsewhere in the world, long-term broadband UV records are available from Russia, Switzerland and New Zealand. The Russian data (Garadzha & Nezval, 1987) comes from city sites and shows the same sort of downward trend as the US data for a similar time period, and is open to similar questions. In Switzerland, measurements have been made at the Jungfraujoch High Altitude Research Station since 1981, the clean air site avoiding any question of aerosol influences on the data record. This is also the only broadband data set to be corrected for the temperature coefficient of the instrument used. The data have not been gathered on a continuous basis but for discrete (and changing) periods each year, giving coverage of every season over the monitoring period. The UV data, expressed as a ratio of the total solar radiation, showed an increase of 0.7±0.3% per decade from 1981–1989 (Blumthaler & Ambach, 1990) calculated for the year as a whole. This trend continued through 1991, but data for 1992 matched the ratios observed in the early 1980s (WMO, 1995).

Another incomplete broadband UV record is reported form Invercargill, New Zealand by Zheng & Basher (1993). In this case there is a gap in the record, but extrapolating across the dataless period gives the expected pattern of UV changes, mirroring the ozone changes at the same site.

UV records of any duration are sparse in the middle and high latitudes, while in the tropical regions they are non-existent. However, there has been no statistically significant ozone depletion in the tropics (WMO, 1995), so no related changes in UV-B would be expected.

Finally, the number of UV monitoring sites has mushroomed tremendously over the past five years, although there is still a significant lack of sites in tropical regions, Russia, Asia and the Middle East (Fig. 2). Both broadband and UV spectral monitoring is increasing, as are efforts to improve the instrumentation and calibration both at and between individual sites. Some of these efforts are at a national network level, but regional and international collaborations are also playing a valuable role in improving UV measurements (Gardiner & Kirsch, 1995; Koskela, 1994; Seckmeyer *et al.*, 1995). The need to centralise standards and provide a comprehensive UV database has been recognised by WMO who have established a Scientific Steering Committee to address this task, so that in the future there will be more, and more readily available, UV data.

Sorting out causes and effects

It is clear from the discussion above that, while there is proof that low ozone means high UV-B, the degree of ozone depletion to date, at all but high latitudes, is hard to detect in the UV data records available. It is possible to speculate that a cumulative dose of UV-B over several years at a mid-latitude site where there has been an average 5% per decade decline in ozone will be higher than the same cumulative dose would have been without ozone depletion (regardless of other changes in the atmosphere). Instantaneous measurements show the ozone–UV link holds, giving reasonable support to this argument. However, the variability of the whole atmosphere system prevents any such proof of increasing cumulative dose being obvious in the UV records, and from the biological standpoint this is the important factor: a UV effect will only be observed if UV exposure changes, regardless of the cause. Trying to separate the causes of UV attenuation in the atmosphere helps to explain the UV observations made to date, and how UV exposures might have changed subtly, if not in a total dose manner. It also provides for better predictive models in the future, models being the only way to estimate what might be happening in the large areas where there are no data.

Individual causes can be investigated at a single site where the other variables of interest are also measured, and with sufficient data different effects may be separated from each other. For example, Bais *et al.* (1993) showed that high concentrations of SO_2 can significantly influence UV irradiances. An alternative method of studying different factors affecting UV irradiances at the surface, and/or the associated changes in real dose, is to make measurements with matched instruments at different sites at the same or equivalent times, knowing specific differences between the sites. An example of simultaneous measurements is given by Blumthaler *et al.* (1994) who used two pairs of spectroradiometers to measure UV at sites in the German Alps separated by 1 km vertically but only 5 km horizontally. In cloud-free conditions the difference in UV irradiances between the sites was then due only to the aerosol and ozone in the 1 km depth of atmosphere, and this was sufficient to produce a change in UV of 24% at 300 nm, 11% at 320 nm and 9% at 370 nm (unaffected by ozone). For EER calculated from the spectral measurements, the 1 km change in altitude produced a 14% change in UV. These altitude effects were independent of zenith angle (in the 30–70° range) and were in agreement with the range of model values calculated using the measured 4 DU (Dobson units) ozone in the 1 km layer, and aerosol profiles which were reasonable estimates for the region. The total altitude effect was made up of a combination of Rayleigh scattering (known), ozone (known) and aerosol (unknown) attenuation throughout the 1 km depth. The influence of the aerosol could then be assessed from the residual of the total minus the known (calculated) effects. For a typical continental aerosol the aerosol attenuation was constant for UV-B wavelengths at 1.8% per km, reducing slightly to 1.5% at 360 nm. For an urban-type aerosol there was rather more dependency on wavelength, from 4.1% to 3.4%. The actual aerosol was probably somewhere between the two types.

Equivalent time measurements are reported by Seckmeyer *et al.* (1995) from 12 stations in the Northern and Southern Hemispheres using spectral (and a few broadband) measurements related through calibration to four instruments which had been successfully inter-compared in Garmisch-Partenkirchen in summer 1993. Six of the sites were the NSF sites, four were in Australia and run by the Australian Radiation Laboratory, one was at Lauder in New Zealand, and one at Garmisch-Partenkirchen in Germany. The three Northern Hemisphere sites covered latitudes from 32 to 71° N, and the nine Southern Hemisphere sites ranged from 12 to 90° S. The mean daily erythemal dose for each of the four summer months (May – August, NH, and November – February, SH) was calculated for each site and then compared as a function of latitude and hemisphere. The geographical differences in UV doses were larger than the uncertainties in the measurements and showed that there was a distinct difference between the two hemispheres. In mid-latitudes this study reinforces the observations of Seckmeyer & McKenzie (1992) who compared data from Lauder and Germany and showed that for clear skies the differences were bigger than expected from the effects of different sun–earth distances and different ozone amounts. In the early study, the heavier aerosol loading in the Northern Hemisphere was cited to explain the extra hemispheric effect. This effect is now shown to hold in cloudy as well as clear conditions, and at other latitudes. In the Northern Hemisphere the expected decline in UV dose with latitude is observed, but in the Southern Hemisphere the latitudinal gradients are much weaker, especially in the spring, reflecting the influence of the more dramatic ozone depletions at this time in the Southern Hemisphere. In the summer months the daily doses at high southern latitudes exceed those at mid-northern latitudes due to the combined effects of low ozone, daylight for 24 hours, high albedo, low cloud and aerosol extinction, and at the south pole high altitude. In the winter months when the poles are in darkness, latitudinal gradients must become more pronounced, thus annual integrated doses will also show more latitudinal dependence than just the summer doses.

The latter study illustrates the importance of knowing what sort of change in UV irradiances is important for a given biological effect. Is it the annual dose, a particular seasonal dose, or a threshold maximum dose that may be reached for a given period of time? It is also important not to extrapolate results of either absolute irradiances, or change, from one site to another.

Is there a change at the plant level?

This question cannot be addressed until it is stated in more exact terms, and even then the answer is often not clear-cut. First, the location where a change is sought must be specified, and then the type of change: annual, seasonal, dose or threshold.

In the tropics there has been no significant change in stratospheric ozone, therefore one would not expect any change in UV-B radiation, although there are no measurements to prove this. In Antarctica and neighbouring regions, which may periodically come under the influence of the ozone hole, there are instances of extreme UV-B irradiances when there is severely depleted ozone overhead. These are highly elevated threshold doses, and if time beneath the ozone hole is prolonged, and/or thick cloud is not present, they contribute to elevated seasonal doses and even annual doses. Similar but less dramatic increases in short-lived threshold doses and seasonal doses are observed at southern mid-latitudes as well. In the Northern Hemisphere the evidence is less clear. There have certainly been instances of high UV doses on a short-term basis, days or a single season, but these are not always apparent in all types of data record and the short-lived changes are not sufficient to influence longer-term doses. The year 1993 produced several reports of unusually high UV irradiances in mid-high northern latitudes, extending through into the early summer, but this pattern has not been repeated since and must be considered an anomaly rather than an impending trend. In general, the complications of cloud and aerosol, the less dramatic ozone changes, and the insufficient data records allow only the speculative comments made at the beginning of the previous section.

References

Bais, A.F., Zerefos, C.S., Meleti, C., Ziomass, I.C., & Tourpali, K. (1993). Spectral measurements of solar UV-B radiation and its relations to total ozone, SO_2 and clouds. *Journal of Geophysical Research*, **98**, D3, 5199–204.

Blumthaler, M. & Ambach, W. (1990). Indication of increasing ultra-violet-B radiation in alpine regions. *Science,* **248**, 206–8.

Blumthaler, M., Webb, A.R., Seckmeyer, G., Bais, A.F., Huber, M. & Mayer, B. (1994). Simultaneous spectroradiometry: a study of solar UV irradiance at two altitudes. *Geophysical Research Letters*, **21**, 2805–8.

Booth, C.R., Lucas, T., Morrow, J., Weiler, S. & Penhale, P. (1994). The United States National Science Foundation/Antarctic Program's Network for Monitoring Ultraviolet Radiation. In *Ultraviolet Radiation and Biological Research in Antarctica, AGU Antarctic Research Series.* (Weiler, S. & Penhale, P., eds). AGU, Washington DC.

Booth, C.R., Mestechkina, T., Lucas, T., Tusson, J. & Morrow, J. (1993). Contrasts in Southern and Northern hemisphere high latitude ultraviolet irradiances. In *Proceedings International Symposium on High Latitude Optics*, European Optical Society, Tromso, July 1993.

Correll, D.L., Clark, C.O., Goldberg, B., Goodrich, V.R., Hayes, D.R., Klein, W.H. & Schere, W.D. (1992). Spectral ultraviolet-B radiation fluxes at the Earth's surface: long-term variations at 39° N, 77° W. *Journal of Geophysical Research*, 97, 7579–91.

Farman, J.C., Gardiner, B.G. & Shanklin, J.D. (1985). Large losses of total ozone in Antarctica reveal seasonal ClO_x–NO_x interaction. *Nature*, 315, 207–10.

Frederick, J.E., Soulen, P.F., Diaz, S.B., Smolskaia, I., Booth, C.R., Lucas, T. & Neuschuler, D. (1993). Solar ultraviolet irradiance observed from Southern Argentina: September 1990 to March 1991. *Journal of Geophysical Research*, 98, D5, 8891–7.

Frederick, J.E. & Weatherhead, E.B. (1992). Temporal changes in surface ultraviolet radiation: a study of the Robertson–Berger meter and Dobson data records. *Photochemistry and Photobiology*, 56, 123–31.

Garadzha, M.P. & Nezval, N.I. (1987). Ultraviolet radiation in large cities and possible ecological consequences of its changing flux due to anthropogenic impact. *Climate and Human Health World Climate Applications WCAP No. 2. Proc. Symp.* Leningrad Sept. 22–26, 1987. World Meteorological Organisation/World Health Organisation/United Nations Environment Panel. pp. 64–8.

Gardiner, B.G. & Kirsch, P.J. (1993). European Intercomparison of Ultraviolet Spectroradiometers. *CEC Air Pollution Research Report 38.*

Gardiner, B.G. & Kirsch, P.J. (1994). Second European Intercomparison of Ultraviolet Spectroradiometers. *CEC Air Pollution Research Report 49.*

Gardiner, B.G. & Kirsch, P.J. (1995). Setting Standards for European Ultraviolet Spectroradiometers. *CEC Air Pollution Report 53.*

Intergovernmental Panel on Climate Change (IPCC) (1990). *Climate Change*. Cambridge University Press, New York.

Jokela, K., Leszczynski, F., Visuri, R. & Ylianttila, L. (1995). Increased UV exposure in Finland in 1993. *Photochemistry and Photobiology*, 62, 101–7.

Justus, C.G. & Murphy, B.B. (1994). Temporal trends in surface irradiance at ultraviolet wavelengths. *Journal of Geophysical Research*, 99, D1, 1389–94.

Kerr, J.B. & McElroy, C.T. (1993). Evidence for large upward trends of ultraviolet-B radiation linked to ozone depletion. *Science*, 262, 1032–4.

Koskela, T. (ed.) (1994). The Nordic Intercomparison of Ultraviolet and Total Ozone Instruments in Izana from 24 October to 5 November 1993. Final report, Finnish Meteorological Institute, *Meteorological Publications 27*.

Leszczynski, K., Jokela, K., Visuri, R. & Ylianttila, L. (1996). Calibration of the broad band radiometers of the Finnish solar ultraviolet monitoring network. *Meteorologia*, **32**, 701–4.

Madronich, S. (1993). The atmosphere and UV-B radiation. In *Environmental UV Photobiology*. (Young, A.R., Bjorn, L.O., Moan, J. & Nultsch, W., eds). Plenum Press, New York, London.

McKinlay, A.F. & Diffey, B.L. (1987). A reference action spectrum for ultraviolet induced erythema in human skin. *Commission Internationale d'Eclairage Journal*, **6**, 17–22.

Scotto, J., Cotton, G., Urbach, F., Berger, D. & Fears, T. (1988). Biologically effective ultraviolet radiation: measurements in the United States 1974–1985. *Science*, **239**, 762–4.

Seckmeyer, G. & McKenzie, R.L. (1992). Increased ultraviolet radiation in New Zealand (45° S) relative to Germany (48° N). *Nature*, **359**, 135–7.

Seckmeyer, G., Mayer, B., Erb, R. & Bernhard, G. (1994). UV-B higher in Germany in 1993 than in 1992. *Geophysical Research Letters*, **21**, 577–80.

Seckmeyer, G., Mayer, B., Bernhard, G., McKenzie, R.L., Johnston, P.V., Kotkamp, M., Booth, C.R., Lucas, T., Mestechkina, T., Roy, C.R., Gies, H.P. & Tomlinson, D. (1995). Geographical differences in UV measured by intercompared spectroradiometers. *Geophysical Research Letters*, **22**, 1889–92.

UMIRG (1996). Ultraviolet Measurements and Implications Review Group, UK Department of Environment. HMSO, London (in press).

WMO/UNEP Scientific Assessment of Ozone Depletion (1995). WMO Report No. 1994.

Zheng, X. & Basher, R.E. (1993). Homogenisation and trend detection analysis of broken series of solar UV-B data. *Theoretical and Applied Climatology*, **47**(4), 189–203.

M.G. HOLMES

Action spectra for UV-B effects on plants: monochromatic and polychromatic approaches for analysing plant responses

Introduction

Small increases in solar ultraviolet-B (UV-B; 280–320 nm) radiation can have substantial effects on the growth and development of many plant species. As a result of a reduction in stratospheric ozone, UV-B radiation has been increasing over Europe for at least a decade, and current evidence points to a gradual increase in incident solar UV-B over Europe at a rate of about 1% per annum (Blumthaler & Ambach, 1990; Ambach & Blumthaler, 1991; WMO Ozone Report Summary, 1994). Over the course of the year, ozone depletion is variable, with the main decrease occurring in late winter and early spring; although total UV-B is much less at this time than during summer, the proportional increase is greatest during spring and is therefore mainly of threat to crops and other plants growing at this time.

Understanding the impacts of this increase on plants requires appropriate action spectra. In approximate terms, an action spectrum indicates the relative effectiveness of different wavelengths of radiation in bringing about a particular response. The relative effectiveness, derived from the action spectrum, can then be multiplied by the irradiance at each wavelength, and summed over the appropriate wavelength range to give the important function, *biologically effective radiation*. This term will be found throughout this volume; usually it will have been derived from the Caldwell generalised plant action spectrum and be normalised at 300 nm (PAS300). It should be noted that other action spectra are sometimes used.

A convenient term for the non-photobiologist is the *radiation amplification factor* (RAF), which describes the relative increase in biological response for a given amount of ozone depletion (Rundel, 1983). However, it must be borne in mind that RAFs are of dubious scientific value unless their many limitations are understood. The reader is referred to Coohill (1994) for a concise summary of different RAFs and for the restricted conditions under which they apply.

The accuracy of action spectra is crucial for any studies which attempt to assess the consequences for vegetation of current increases in solar UV-B and for determining the biological effectiveness of artificial lamp systems used to simulate increased levels of solar UV-B radiation. Any error in the action spectrum will not only be reflected in the accuracy of subsequent empirical studies but also in the conclusions which are drawn. In this chapter, emphasis is placed on the correct methodology for action spectroscopy and the limitations resulting from factors such as the suitability of the plant assay, the treatment duration, the irradiation system, screening, reciprocity, and the analysis of the dose–response curves. The suitability of currently available action spectra is discussed and new, long-term polychromatic action spectra are compared with currently available monochromatic action spectra.

Action spectra vs response spectra

Action spectra and response spectra differ substantially. Action spectroscopy can be described as dosimetry with light in which the effect of a specific response is tested by varying the fluence rate of known wavelengths, often over several orders of magnitude. In other words, the variables are the number of photons and the energy of the photons. The classical action spectrum is then obtained by plotting the reciprocal of the photon fluence required to produce the specific response. Carried out correctly, action spectra can provide identification of photoreceptors and accurate weighting functions. This requires detailed dose–response curves, knowledge about reciprocity, and in some instances, screening effects. These aspects are described in greater detail below.

A response spectrum is a general description of the wavelength sensitivity of a certain response. It may compare the effects of equal fluence rates at several wavelengths, or the total fluence may be varied over a limited range by altering either the duration of irradiation, or the fluence rate. It differs most from the action spectrum in that the essential knowledge about the shape of the response curves and the limits of reciprocity are unavailable. Measuring a true action spectrum is therefore a much more demanding exercise than measuring a response spectrum. Nevertheless, response spectra are useful if they can be related to a true action spectrum.

Response spectra

Response spectra provide a useful and fast approach for screening the spectral sensitivity of different plants if comparable action spectra have already been measured. In contrast to the action spectrum, where the

fluence rate of a single wavelength is varied, response spectra tradition-
ally involve only one irradiance for each wavelength. However, this
approach is usually not appropriate. Two irradiances must be used for
the following reasons. First, the irradiance sensitivity is not necessarily
known with enough accuracy, and two points which are on the approximately linear portion of the dose–response curve must be chosen in
order to cover a value of irradiance which can be compared at all wavelengths. Secondly, and this is related to the first reason, the range of
sensitivity varies enormously between wavelengths, especially below
about 320 nm.

From these data, it is possible to interpolate sensitivity at comparable
irradiances. However, it is important to note that this can only be done
over a very limited wavelength range. The wavelength range can be
chosen to match the lower wavelengths currently considered to threaten
plant growth and the irradiance range be chosen on the basis of background information gained from action spectra.

Action spectra

Dose–response curves

Action spectra produced under conditions using brief periods (minutes,
hours) of radiation tend to be straightforward to interpret. They typically
produce non-parallel response curves with a common origin and they
obey the Weber–Fechner reciprocity law ($R = k\log(Nt)$, where R =
response, k = proportionality factor, N = number of photons, t = time).
(See below for a deeper discussion of reciprocity.) However, a wide
variety of response curves has been observed, and some of these are
shown schematically in Fig. 1.

Action spectra measured over long periods (days, weeks, months)
and with prolonged irradiation are more complex to interpret because
the response curves can differ in shape, and reciprocity tends to fail.
Under these conditions, changes in the response of the organism arise
as a result of various metabolic processes. In particular, long periods
allow for DNA and protein turnover (synthesis and destruction) and
repair of DNA, the rate constants of which will depend on temperature.
In addition, the balance between damage and repair is affected by time.

Dose–response curves can provide a large amount of both direct and
circumstantial information. At very low levels of added radiation, no
effect is observed because repair mechanisms operate faster than the
destructive effects of UV radiation. The lowest fluence rate which
causes statistically significant inhibition of growth provides information

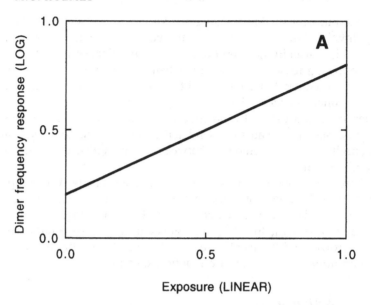

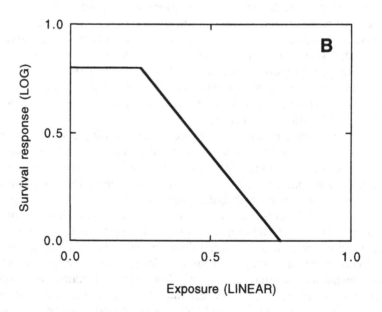

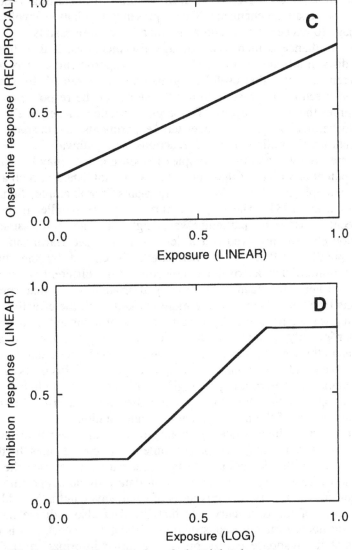

Fig. 1. Schematic examples of ultraviolet dose–response curves. A, pyrimidine dimer frequency in alfalfa seedlings (Sutherland, Quaite & Sutherland, 1994); B, survival kinetics of *Escherichia coli* (Tuveson, Peak & Peak, 1983); C, reciprocal mean tumour onset time in mice (Forbes *et al.*, 1982); D, inhibition of hypocotyl elongation in *Sinapis alba* (Holmes & Schäfer, 1981).

on the relative effectiveness of each wavelength in starting to compete with the repair mechanisms. This represents a radiation level causing damage to exceed repair, and a significant response results.

As irradiance is increased, damage also increases. It is at this stage that distinct differences in the shape of the response curves may become apparent and also that useful information can be gained. In many UV action spectra it is possible to use the slopes of the response curves to construct the action spectra. This typically reflects a relatively simple photochemical process. In longer-term experiments, and in studies using polychromatic radiation, more reactions and pathways are likely to become involved and more complex response curves may be expected. The potential variety of these curves is well established in both animals (for example, Coohill, 1984) and in plants (for example, Schäfer & Fukshansky, 1984). The response curves may be parallel if the same molecules absorb that particular wavelength range and if the same trans- duction chain is involved (for a detailed review see Hartmann, 1977). The parallel shift between wavelengths is caused by the different absorption efficiencies of those molecules at the different wavelengths.

A further consequence of prolonged irradiation is that screening effects (see below) can become more marked, with the effectiveness of shorter wavelengths usually becoming less pronounced as a result of proportionately greater production of screening pigments at shorter wavelengths (see, for example, Wellmann, 1975). The extent of this effect can only be quantified empirically. Apart from its intrinsic importance in determining biologically effective UV radiation in the various species, this should be a major reason for measuring the spectral characteristics of the plants, including pigmentation.

At a certain fluence rate (generally lower with lower wavelengths), the response curve may flatten. This inflexion point indicates the fluence rate at which the damaging effects of the radiation saturate the repair reactions. Taken together, the inflexion data provide supporting infor- mation for the relative effectiveness of each wavelength. An additional advantage of response curves is that the data also indicate naturally realistic fluence rates for construction of the action spectra. It is essen- tial that the response curves provide accurate information at fluence rates expected to be received by the plants in natural conditions.

Reciprocity

Reciprocity (that is, the product of the number of photons multiplied by the duration of the irradiation) is an important consideration in action spectroscopy. Diagnosis of the range of fluence rates and durations over which reciprocity holds and fails helps not only an understanding of the response but also determines the applicability of the data to natural radiation levels.

The reciprocity test is important because it evaluates the validity of using a particular fluence rate range and duration to construct an action spectrum. It can also give insights into possible mechanisms. Reciprocity can only be measured after the irradiance dependency of the response has been determined and the appropriate dosage:response values are known. Reciprocity for UV effects has been demonstrated on several occasions (for example, Sutherland, Quaite & Sutherland, 1994) and, if the experimental procedure is designed so that only photochemical reactions are involved, reciprocity might be expected to hold.

There are three main reasons why failure of reciprocity may be expected in UV studies. The first reason is that the shorter wavelengths induce damage and relatively low doses are required to cause this. Longer wavelengths also cause damage, but a much higher dose is needed to cause the same response; typical action spectra measured to date indicate that dosages for the same response at 290 nm and 360 nm can differ by several orders of magnitude. If reciprocity is tested by varying time, the shortest wavelengths will require seconds of irradiation and the longest will require hours of irradiation. The result is that, unless this is allowed for in the experimental protocol, repair mechanisms have time to operate and the negative effectiveness of the longer wavelengths will typically be underestimated. Related to this point is the fact that the longer wavelengths also induce photoreactivation (see chapter by Taylor *et al.*, this volume) with the result that the short and the long wavelength effects are not truly comparable and the negative effectiveness of the longer wavelengths will again be underestimated. Coohill (1994) gives a more detailed analysis.

The third reason why reciprocity is likely to fail is that, unless a specific response which is closely related to the direct effect of the radiation is being studied (for example, cyclobutyl pyrimidine dimers), the several steps in the transduction chain leading to the measured response are liable to lead to non-photochemical responses which may be rate limiting. This is highly likely in a long-term assay such as the analysis of dry weight.

Screening

Extrapolation from micro-organisms (for example, DNA strand breaks in *Bacillus subtilis*; Peak & Peak, 1982) to higher plants, or even from *in vitro* assays on higher plants (for example, chloroplast Hill reaction activity; Jones & Kok, 1966) is very unreliable because this fails to take into account the potential screening effects of waxes, hairs and UV-B absorbing compounds surrounding much of the sensitive zones of higher plants. Wavelength-neutral screening affects the overall effectiveness of radiation arriving at the photosensitive area.

Potentially more serious for the understanding of action spectra is wavelength-selective screening, which alters not only the slope of the UV-B effectiveness curve, but can also affect its shape. As an example, we can consider the striking effect of leaf cuticular waxes on the radiation arriving at the target area. Certain *Eucalyptus* leaves reflect almost 30% of the incident radiation at 290 nm. If the epicuticular wax is removed, reflectance of this radiation is reduced to about 5%; however, the proportion of 360 nm radiation or PAR (photosynthetically active radiation) is only slightly affected by wax removal. It is also important to bear in mind that the strong differential attenuation which occurs as radiation starts to penetrate leaves may have a large effect on any measured spectral response.

In vitro studies can have limitations for extrapolation to the performance of the whole plant. There can be many causes for this, but the problem can be illustrated by noting that the Jones & Kok (1966) *in vitro* spectrum for inhibition of the Hill reaction in isolated spinach chloroplasts does not produce an accurate estimate of photosynthesis *in vivo* in *Rumex* (see Rundel, 1983). In addition, the spectral characteristics of whole plants differ enormously from those of extracts (Vogelmann, 1994), especially in the UV-B waveband, so *in vitro* studies are highly unlikely to be applicable to whole plants.

Analysis of the spectral characteristics of the plant material is also essential if the effects of dynamic screening are to be accounted for. The spectral characteristics of tissues tend to change quite dramatically over a relatively narrow waveband in the UV-B region when they are irradiated with UV radiation (Wellmann, 1975; Cen & Bornman, 1993). This can have substantial effects on the plant response because it is the amount of radiation reaching the photoreceptive site, not the amount of radiation reaching the exterior of the plant, which is important. A concise review of action spectroscopy under prolonged irradiation conditions is given by Schäfer and Fukshansky (1984). Figure 2 shows four different UV action spectra. These have been normalised to a value of 1 at 300 nm; note the differences in slope at the short wavelengths, and the significant 'tail' of some spectra into the UV-A region (see chapter by Paul, this volume, for some practical consequences of these differences).

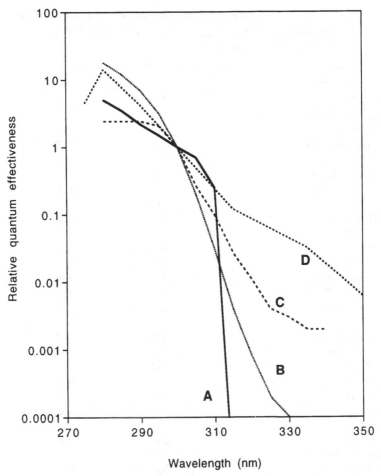

Fig. 2. Typical UV 'action spectra'. A, generalised plant damage (Caldwell, 1971); B, DNA damage (Setlow, 1974); C, erythema (McKinlay–Diffey, 1987); D, dimer production in alfalfa (Quaite, Sutherland & Sutherland, 1992).

Differential action spectra

Differential polychromatic action spectroscopy centres on a background of broad-band white light from artificial sources or natural daylight to which are added various wavelengths between about 280 and 360 nm. Possible lamp and filter combinations are shown in Table 1 and some examples of resultant spectra are shown in Fig. 3. This polychromatic

Table 1. *Examples of lamp and filter combinations for
providing different* λ_{max}, $\lambda_{central}$ *and HPBW for large-scale
spectral irradiations*

Lamp + filter	λ_{max}	$\lambda_{central}$	HPBW
TL/12	313	312	42
TL/12+WG280	313	312	41
TL/12+A	313	312	42
TL/12+WG295	313	313	40
313	313	315	43
313+WG280	313	315	43
313+WG295	313	315	42
313+A	313	315	42
TL/12+C	313	318	37
313+B	313	320	37
313+C	320	322	37
313+D	322	322	38
3×313+340+D	328	331	50
2×313+340+D	330	334	52
3×313+340+E	331	335	46
313+340+B	335	337	57
313+340+D	335	339	55
313+340+E	335	341	51
340+A	341	347	53
340+A	341	347	53
340+A	341	347	53
340+B	341	348	51
340+D	345	348	51
340+E	348	350	50
313+340+F	345	352	44
351+D	351	353	42
351+F	355	356	37
351+H	356	358	34
365+F	365	366	42
TL/05+H	367	377	59

Note: in practice, specifications vary slightly according to lamp age and glass melt batch. See Fig. 3 for example spectra.
Key: A = OX/02; B = 3 mm Sanalux; C = 4 mm Sanalux; D = 3 mm Optiwhite; E = 6 mm Optiwhite; F = 4 mm window glass; G = 6 mm window glass; H = 10 mm window glass; HPBW = half-power bandwidth; $\lambda_{central}$ = central wavelength; λ_{max} = maximal wavelength. TL/12 and TL/05 lamps are from Philips and 313, 340, 351 and 365 are from Q-Panel.

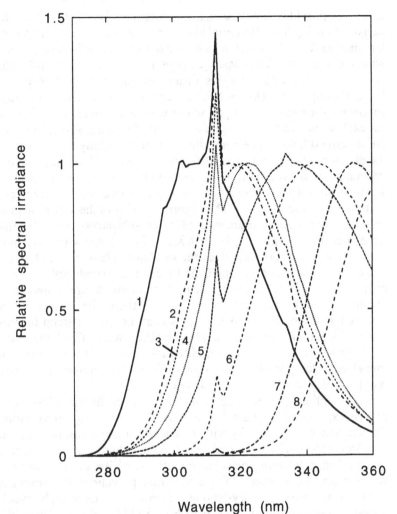

Fig. 3. Examples of the measured spectral quality of the radiation produced by some of the lamp and filter combinations described in the Table. From left to right the combinations are: 1. Q-Panel UVB-313 (Q-Panel Co., Cleveland, OH, USA) + OX/02; 2. TL/12 + 4 mm SAN; 3. Q-Panel 313 + 4 mm SAN; 4. Q-Panel 313 + 3 mm OPT; 5. Q-Panel 313 + 340 + 3 mm OPT; 6. Q-Panel 340 + 3 mm OPT; 7. Q-Panel 351 + 4 mm window glass; 8. Philips TL/05 + 10 mm window glass.

approach provides the true action spectrum. The action spectra are derived by using the differential wavelength, $\lambda_1 - \lambda_2$ etc., where λ is wavelength, and $R_1 - R_2$, where R is the response. This effectively describes bandpass radiation. The basic technique is described in detail by Rundel (1983). However, our analysis differs in one important respect from Rundel's approach. This is because Rundel was considering only one irradiance for each wavelength, whereas we are using several irradiances for each wavelength; the essential reasons for using several irradiances are described below. The consequence is that, for any two wavelengths, the differential irradiance can differ in wavelength (both λ_{max} and $\lambda_{central}$), depending on the irradiances subtracted. This situation is depicted in Fig. 4 which shows the differential bandpass of two lamp + filter combinations; these are real data. The Figure also shows the simplest situation in which the highest irradiance of λ_2 is subtracted from the highest irradiance of λ_1, continuing down to $\lambda_{1min} - \lambda_{2min}$. Even with this approach, it can be seen that not only does the irradiance change, but also that the central wavelength can change. At longer UV wavelengths ($>$ ca. 340 nm), these small changes in central wavelength are unlikely to be significant. At shorter wavelengths ($<$ ca. 320 nm) the changes in central wavelength become increasingly important because a steep increase in sensitivity with decreasing wavelength takes place. The lamp and filter combination is therefore very carefully chosen in order to take account of this effect. One possibility is to use different thicknesses, or a different number of layers of cellulose diacetate.

An essential point to note is that the change in central wavelength can be exploited to advantage because, by comparing all variations in the irradiance of λ_1 and λ_2 with R_1 and R_2, a large combination of $\Delta\lambda$ and ΔR can be derived for analysis, thereby reducing the statistical error involved in delineating between the effects of irradiance and those of wavelength. An example of a differential polychromatic action spectrum for inhibition of dry-weight accumulation over eight weeks in oilseed rape seedlings is shown in Fig. 5 (M.G. Holmes, unpublished observations).

Miscellaneous considerations

The duration of the irradiation period used for action spectroscopy can have a marked effect on the observed response. Short irradiation periods (hours rather than days, or longer) lead to errors in using the data for interpretation in longer-term studies. In the case of the action spectrum for net photosynthesis in *Rumex* leaves (for example, Rundel, 1983, using data of Caldwell), irradiation periods of between 1 and 16 hours

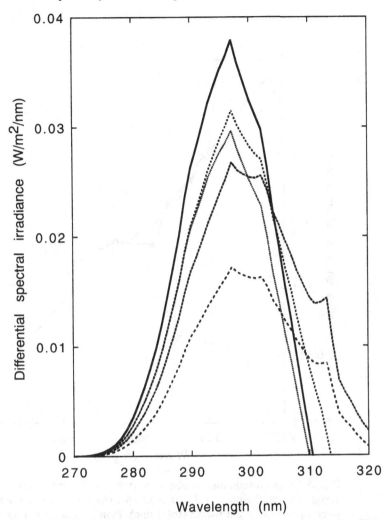

Fig. 4. The differential irradiance resulting from two lamp plus filter combinations. Two lamp frames with different spectral compositions provided five irradiances. Note that both the total differential irradiance and the central wavelength can shift according to the actual irradiances chosen. For the solid line, λ_{max} is 297 nm and the half-power bandwidth is 17 nm.

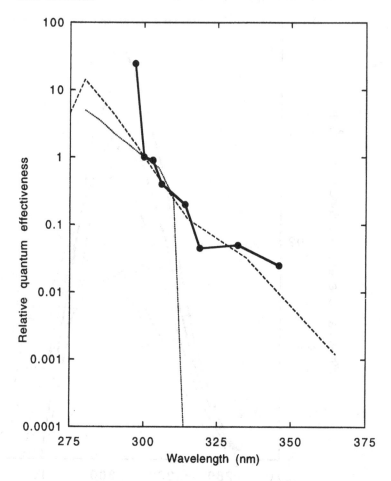

Fig. 5. A polychromatic action spectrum for the inhibition of dry weight accumulation in oilseed rape (*Brassica napus* L.) seedlings over a period of eight weeks (solid line). Plants were grown outdoors and received daylight plus five irradiances from artificial sources such as those depicted in Fig. 3 (M.G. Holmes, preliminary unpublished data). The monochromatic spectra for generalised plant damage (Caldwell, 1971; dotted curve) and for dimer production (Quaite, Sutherland & Sutherland, 1992; dashed line) are shown for comparison.

were used. This duration is too short for longer-term damage, repair and adaptive effects to develop. Also, specific responses can change with time; for example, short-term UV-B (hours) tends to cause stomatal closure, whereas long-term irradiation (days) has been reported to maintain stomata open; also, screening pigments can develop during the course of long-term irradiation, thus altering the response and misrepresenting the plant's true response to the radiation.

Both the plant material chosen, and the assay which is used, may influence the spectral response obtained. Studies of yield using realistic levels of background white light have shown that there can be large differences in sensitivity to UV-B between species. It is important to note that other factors affect interpretation of data from this type of study. In particular, relatively high UV-B enhancements produce strikingly large differences between treatments (for example, Biggs & Kossuth, 1978; Biggs *et al.*, 1984), whereas modest increases in UV-B produce little or no difference in yield between species (for example, Becwar, Moore & Burke, 1982; Dumpert & Knacker, 1985). In solar UV-B exclusion studies, the extent of the response is species dependent (for example, Bartholic, Halsey & Garrard, 1975; Becwar *et al.*, 1982). These, and other studies, indicate the importance of considering several species because one species cannot be expected to be representative of all plants. It must also be borne in mind that plant sensitivity to UV-B radiation can change with age and can be modified by the amount of UV-B previously received.

The angle of incidence of radiation used in an action spectrum is important for two reasons. First, both the potentially damaging UV-B and the longer-wavelength radiation should arrive at the plant in a carefully controlled manner which allows approximation to natural conditions. Secondly, plant orientation relative to the angle of incidence of the radiation is important. A plant with a primarily horizontal orientation is much easier to irradiate artificially in a controlled manner than a plant with primarily vertical orientation.

It is important to ascertain the relevance of the response being studied. Whereas the selection of a single parameter (for example, dimer formation (damage), pigment formation (protection), Rubisco activity (metabolism)) can give information on specific aspects of plant responses to UV-B, none of these provide information on the cumulative effects. An assay such as dry-weight accumulation is considered to be a useful measure of the integrated effect because it ensures that any rate-limiting effects of UV-B will be incorporated in the response.

It has been mentioned earlier that knowledge of the amount of longer wavelength radiation is an essential part of a polychromatic action spectrum for two reasons. The first is because it has an obvious direct effect on net photosynthesis. The second reason is that high levels of longer wavelengths provide increased protection against UV-B radiation (for example, Mirecki & Teramura, 1984; Cen & Bornman, 1990; Kumagai & Sato, 1992; Adamse & Britz, 1992; Deckmyn, Marten & Impens, 1994). Although photosynthesis is probably directly involved (Adamse & Britz, 1992), a definitive explanation is not yet available. PAR is not spectrally equivalent to the photorepair action spectra, but it provides a cheap, convenient and reliable estimate of photoreactivating wavelengths.

Very little work has been carried out on the interaction between temperature and UV-B effects, but on the basis of the limited information available, it needs to be noted that increasing temperature from 28 ° to 32 °C can ameliorate the negative effects of UV-B radiation on parameters such as height, leaf area, weight and photosynthesis (Teramura *et al.*, 1991; Mark & Tevini, 1996). Recent evidence with *B. napus* L. indicates that low temperature markedly enhances the deleterious effects of UV-B radiation.

In these considerations of the target material and of the response which is to be measured, no attention has been paid to the importance of the measurement of the treatment, i.e. the radiation. There are many pitfalls in this area. Many instruments are inaccurate to the extent which makes the values virtually meaningless. The success of any study of UV-B effects on plants depends ultimately on the accuracy and precision of the radiation measurements and measurement of UV radiation requires much attention to detail (see chapter by Webb for more details of measurement of UV radiation). Errors involved in the measurement of UV radiation tend to fall into four categories. These are instrument error (including the use of an inappropriate instrument), indication error, calibration error, and operator error. These errors overlap, and all contribute to the total inaccuracy of the measurement. Instrument error derives mainly from receiver and detector error. Calibration error usually results from lack of calibration rather than poor calibration. The total error involved in making a radiation measurement can be defined by the square root of the sum of squares for each individual error. The most important error (operator error) is best defined by a comprehensive description of the instrumentation and the techniques used for the radiation measurement.

Conclusions

Although largely resulting from technical and financial reasons, it is noteworthy that no reliable action spectra are currently available for designing studies of long-term (weeks or longer) effects of UV-B on plants. It is essential that such action spectra are measured if:

(a) the biological effectiveness of both natural and artificial UV-B radiation is to be correctly assessed;
(b) current UV-B treatments used in research are to be correctly designed;
(c) meaningful studies comparing plant species and varieties are to be made;
(d) accurate risk models are to be developed.

The accuracy of both theoretical and empirical studies on the consequences of increased solar UV-B radiation on vegetation depend critically on the accuracy of the action spectra which are employed. Without accurate information on the spectral biological effectiveness, UV-B studies relating to enhanced solar UV-B can be difficult to interpret at best, and meaningless at worst. It is not surprising that the various 'action spectra' which have been measured differ so greatly (see chapter by Paul for a detailed practical example). Their only common feature is that the extent of the damage by UV-B radiation increases with decreasing wavelength. The more important questions of 'What is the magnitude of the damage over realistically long periods?' and 'Does this apply to natural conditions?' remain unanswered.

References

Adamse, P. & Britz, S.J. (1992). Amelioration of UV-B damage under high irradiance. I: Role of photosynthesis. *Photochemistry and Photobiology*, **56**, 645–50.

Ambach, W. & Blumthaler, M. (1991). Further increase in ultraviolet B. *Lancet*, **338**, 393.

Bartholic, J.F., Halsey, L.H. & Garrard, L.A. (1975). Field trials with filters to test for effects of UV radiation on agricultural productivity. In *Climatic Impact Assessment Program (CIAP), Monograph 5.* (Nachtwey, D.D., Caldwell, M.M. & Biggs, R.H. eds), pp. 61–71. US Department of Transportation, Report Number DOT-TST-76–55. National Technical Information Service, Springland, VA.

Becwar, M.R., Moore, F.D. & Burke, M.J. (1982). Effects of depletion and enhancement of ultraviolet-B (280–315 nm) radiation on plants grown at 3,000 m elevation. *Journal of the American Society of Horticultural Science*, **107**, 771–4.

Biggs, R.H. & Kossuth, S.V. (1978). Effects of ultraviolet-B enhancement under field conditions on potatoes, tomatoes, corn, rice, southern peas, peanuts, squash, mustard and radish. In *UV-B Biological and Climatic Effects Research (BACER). Final Report*, US EPA, Washington, DC.

Biggs, R.H., Webb, P.G., Garrard, L.A., Sinclair, T.R. & West, S.H. (1984). The effects of enhanced ultraviolet-B radiation on rice, wheat, corn, soybean, citrus and duckweed. *Year 3 Interim Report, Environmental Protection Agency Report 808075-03*, EPA, Washington, DC.

Blumthaler, M. & Ambach, W. (1990). Indication of increasing solar ultraviolet-B radiation flux in alpine regions. *Science*, **248**, 206–8.

Caldwell, M.M. (1971). Solar ultraviolet radiation and the growth and development of higher plants. In *Photophysiology, Vol. 6* (A.C. Giese, ed.), pp. 131–77. Academic Press, New York.

Cen, Y.-P. & Bornman, J.F. (1990). The response of bean plants to UV-B radiation under different irradiances of background visible light. *Journal of Experimental Botany*, **41**, 1489–95.

Cen, Y.-P. & Bornman, J.F. (1993). The effect of exposure to enhanced UV-B radiation on the penetration of monochromatic and polychromatic UV-B radiation in leaves of *Brassica napus*. *Physiologia Plantarum*, **87**, 249–55.

Coohill, T.P. (1984). Action spectra for mammalian cells *in vitro*. In *Topics in Photomedicine*. (K.C. Smith, ed.), pp. 1–37. Plenum, New York.

Coohill, T.P. (1994). Exposure response curves, action spectra and amplification factors. In *Stratospheric Ozone Depletion/UV-B Radiation in the Biosphere. NATO ASI Series Volume I 18*. (Biggs, R.H. & Joyner, M.E.B., eds), pp. 57–62. Springer-Verlag, Berlin.

Deckmyn, G., Martens, C. & Impens, I. (1994). The importance of the ratio UV-B/photosynthetic active radiation (PAR) during leaf development as determining factor of plant sensitivity to increased UV-B irradiance: effects on growth, gas exchange and pigmentation of bean plants (*Phaseolus vulgaris* cv. Label). *Plant Cell and Environment*, **17**, 295–301.

Dumpert, K. & Knacker, T. (1985). A comparison of the effects of enhanced UV-B radiation on some crop plants exposed to greenhouse and field conditions. *Biochemica Physiologia Pflanzen*, **180**, 599–612.

Forbes, P.D., Davies, R.E., Urbach, F., Berger, D. & Cole, C. (1982). Simulated stratospheric ozone depletion and increased ultraviolet radiation: effects on photocarcinogenesis in hairless mice. *Cancer Research*, **42**, 2796–805.

Hartmann, K.M. (1977). In *Biophysik* (Hoppe, W., Lohmann, W., Mark, H. & Ziegler, H. eds.), pp. 197–222. Springer-Verlag, Berlin, New York.

Holmes, M.G. (1996). Interception of light and light penetration in plant tissues. In *Light as an Energy Source and Information Carrier in Plant Photophysiology*. (Jennings, R., Zuccheli, G., Ghetti, F. & Colombetti, G., eds). Plenum Press, New York.

Holmes, M.G. & Schäfer, E. (1981). Action spectra for changes in the 'high irradiance reaction' in hypocotyls of *Sinapis alba* L. *Planta*, **153**, 267–72.

Jones, L.W. & Kok, B. (1966). Photoinhibition of chloroplast reactions. I. Kinetics and action spectrum. *Plant Physiology*, **41**, 1037–43.

Kumagai, T. & Sato, T. (1992). Inhibitory effects of increase of near-UV radiation on the growth of Japanese rice cultivars (*Oryza sativa* L.) in a phytotron and recovery by exposure to visible radiation. *Japan Journal of Breeding*, **42**, 545–52.

McKinley, A.F. & Diffey, B.L. (1987). A reference action spectrum for ultraviolet induced erythema in human skin. In *Human Exposure to Ultraviolet Radiation: Risk and Regulations*. (Pschier, W.F. & Bosnajokovic, B.F.M., eds), pp. 83–7. Elsevier, Amsterdam.

Mark, U. & Tevini, M. (1996). Combination effects of UV-B radiation and temperature on suflower (*Helianthus annuus* L., cv. Polstar) and maize (*Zea mays* L., cv. Zenit 2000) seedlings. *Journal of Plant Physiology*, **148**, 49–56.

Mirecki, R.M. & Teramura, A.H. (1984). Effects of ultraviolet-B on soybean. V. The dependence of plant sensitivity on the photosynthetic photon flux density during and after leaf expansion. *Plant Physiology*, **74**, 475–80.

Peak, M.J. & Peak, J.G. (1982). Single strand breaks induced in *Bacillus subtilis* DNA by ultraviolet light: action spectrum and properties. *Photochemistry and Photobiology*, **35**, 675–80.

Quaite, F.E., Sutherland, B.M. & Sutherland, J.C. (1992). Action spectrum for DNA damage in alfalfa lowers predicted impact of ozone depletion. *Nature*, **358**, 576–8.

Rundel, R.D. (1983). Action spectra and estimation of biologically effective UV radiation. *Physiologia Plantarum*, **58**, 360–6.

Schäfer, E. & Fukshansky, L. (1984). Action Spectroscopy. In *Techniques in Photomorphogenesis*. (Holmes, M.G. & Smith, H., eds), pp. 109–29. Academic Press, New York.

Setlow, R.B. (1974). The wavelengths in sunlight effective in producing skin cancer. A theoretical analysis. *Proceedings of the National Academy of Sciences, USA*, **71**, 3363–6.

Sutherland, B.M, Quaite, F.E. & Sutherland, J.C. (1994). DNA damage action spectroscopy and DNA repair in intact organisms: Alfalfa seedlings. In *Stratospheric Ozone Depletion/UV-B Radiation in the Biosphere. NATO ASI Series Volume I 18*. (Biggs, R.H. & Joyner, M.E.B., eds), pp. 97–106. Springer-Verlag, Berlin.

Teramura, A.H., Tevini, M., Bornman, J.F., Caldwell, M.M., Kulandaivelu, G. & Björn, L.-O. (1991). Terrestrial Plants. In *Environmental Effects of Ozone Depletion: 1991 Update*, pp. 25–32. UNEP Environmental Effects Panel Report, Nairobi, Kenya.

Tuveson, R.W., Peak, J.G. & Peak, M.J. (1983). Single-strand DNA breaks induced by 365 nm radiation in *Escherichia coli* strains differing in sensitivity to near and far UV. *Photochemistry and Photobiology*, **37**, 109–12.

Vogelmann, T.C. (1994). Light within the plant. In *Photomorphogenesis in Plants* (Kendrick, R.E. & Kronenberg, G.H.M., eds), pp. 491–535. Kluwer, Netherlands.

Wellmann, E. (1975). UV dose-dependent induction of enzymes related to flavonoid biosynthesis in cell suspension cultures of parsley. *FEBS Letters*, **51**, 105–7.

WMO Ozone Report Summary (1994). *Executive Summary, UNEP, WMO, NASA, NOAA*, 19 August 1994.

PART II Effects of UV-B on plants at the cellular level

R.M. TAYLOR, A.K. TOBIN and C.M. BRAY

DNA damage and repair in plants

Introduction

DNA is a highly reactive molecule that is sensitive to damage from a wide range of both physical and chemical agents. Lesions alter the structure of DNA and consequently interfere with critical aspects of DNA metabolism such as transcription, replication and recombination. The maintenance of genetic integrity is essential for cellular survival, but the eukaryotic genome constantly incurs modifications which are potentially cytotoxic or mutagenic. Such changes can result from cellular exposure to genotoxic compounds present in the environment and from normal products of cellular metabolism. An additional source of genome modifications, particularly relevant to organisms exposed to chronically high levels of natural sunlight, is the ultra-violet radiation contained within this polychromatic light source. UV-B (280–320 nm) is considered to be the most biologically effective radiation within sunlight, although longer wavelengths (UV-A, 320–390 nm, and visible light radiation) do play their part in lesion formation but to a lesser degree (Tyrrell, 1993). (More damaging UV-C radiation is screened by the atmosphere and does not penetrate to the earth's surface.)

DNA is considered to be the primary absorbing chromophore in the cell in the UV-B region of the spectrum. The peak absorption of DNA is dictated by its component nucleotides and occurs at around 260 nm, and there is a sharp drop in photons absorbed by DNA through the UV-B range and any absorbance at wavelengths greater than 320 nm is minimal (Davies, 1995).

UV-B induced photoproducts

The DNA photoproducts formed by UV-B radiation can be split into two classes, the dimeric photoproducts which make up the bulk of the lesions formed after UV irradiation, and monomeric photoproducts. The major dimeric photoproducts are cyclobutane pyrimidine dimers (CPDs) and pyrimidine (6–4) pyrimidone photoproducts or (6–4) photoproducts (Fig. 1); both are toxic and mutagenic lesions. CPDs form between adjacent pyrimidines on the same strand of DNA which become linked covalently by the formation of a four-membered ring structure. These have a stable conformation which is resistant to extremes of pH and temperature (Friedberg, 1985).

(6–4) photoproducts are also formed and produce alkali-labile lesions at the positions of cytosine (and much less frequently thymine) 3′ to pyrimidine nucleosides. The 6–4 photoproducts are themselves readily photoisomerised into the Dewar photoproduct by UV-A radiation (Matsunaga *et al.*, 1993). Together, the CPD and (6–4) photoproduct lesions form the majority of lesions induced by UV-B radiation and can be detected in leaf tissue as a result of exposure to UV-B radiation (Fig. 2; Taylor *et al.*, 1996*a*).

Monomeric photoproducts make up the minor fractions of other DNA lesions formed by UV damage and include thymine glycols, pyrimidine hydrates and 8-hydroxyguanine. Although these lesions may be minor

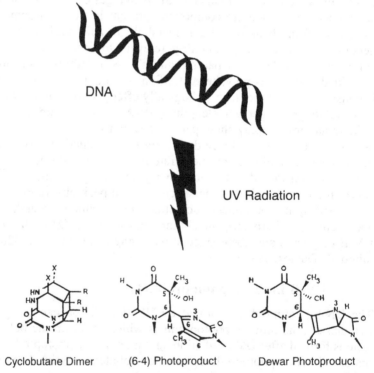

Fig. 1. The major dimeric photoproducts formed in DNA as a result of UV radiation (Mitchell & Karentz, 1993).

in number, their relative cytotoxicity has yet to be determined, and clearly the ability of cells to repair these lesions determines their biological importance to a large degree.

Other lesions can be formed in DNA molecules by UV radiation and include strand breaks (single and double), DNA–protein crosslinks and also larger-scale genetic alterations such as chromosome breakage, sister chromatid exchanges and chromatid aberrations (McLennan, 1987a). The work discussed here will concentrate on the pathways responsible for the removal of dimeric photoproducts.

Protective mechanisms

Plants are constantly exposed to sunlight due to their static lifestyle and as a result have acquired tolerance mechanisms during evolution and development to limit the amount of DNA damage resulting from solar UV-B irradiation. Several mechanisms are employed which essentially involve two basic strategies, either shielding or repair.

Shielding

Mechanisms to shield plants from UV-B damage include the production of UV-absorbing compounds and epicuticular waxes (Robberecht & Caldwell, 1978). In most plants that have been studied, reflectance from the leaf surface is relatively low (<10%) and attenuation by UV-B absorbing pigments appears to be the primary means of filtering out this harmful radiation (Caldwell, Robberecht & Flint, 1983).

The vast majority of higher plants exhibit the characteristic accumulation of UV-absorbing pigments, notably flavonoids, in the upper epidermal layers of the leaves following irradiation with UV-B. These phenolic compounds attenuate the damaging solar UV-B radiation but transmit photosynthetically active radiation (PAR) through the epidermal layers of the leaves to the underlying palisade and mesophyll tissue where the majority of photosynthesis takes place. These pigments are water-soluble, colourless flavonoids including flavones, flavonols and isoflavonoids and absorb light between 230 nm and 380 nm (see chapter by Bornman et al., this volume). Anthocyanins also readily accumulate in response to UV-B irradiation and, although they absorb maximally at around 530 nm, when they are esterified to cinnamic acid, they can provide UV-B protection (Markham, 1982; Strid & Porra, 1992). There is evidence to support the theory that the presence of these pigments is crucial for normal growth and the protection of cellular macromolecules. *Arabidopsis thaliana* mutants deficient in chalcone synthase (tt4 mutant) or chalcone isomerase (tt5 mutant), enzymes which catalyse the first steps of the flavonoid biosynthetic pathway, are much more

a

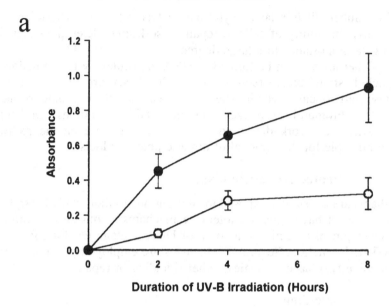

b

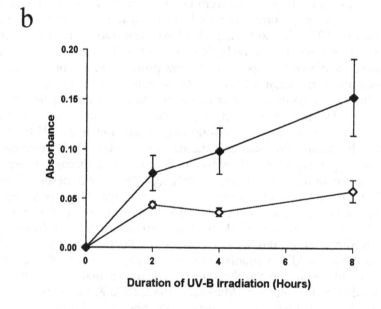

susceptible to UV damage (Li *et al.*, 1993) than the wild type. Stapleton and Walbot (1994) have also found that in *Zea mays* (maize) UV-B induced DNA damage was reduced in plants that contained anthocyanins compared with isogenic lines which lacked the regulatory genes for anthocyanin biosynthesis and so were deficient in these pigments.

DNA repair pathways

The fact that plants are unavoidably exposed to the UV-B wavelengths within sunlight suggests that they must have an effective system of DNA repair capable of coping with the effects of UV-B radiation, the levels of which may increase in the future due to the predicted depletion of the stratospheric ozone layer (see chapter by Pyle, this volume). All cellular life forms possess DNA repair enzymes which recognise chemically modified bases including those formed by UV radiation, and the most detailed studies have been carried out on *Escherichia coli*, yeast and mammalian cells (Sancar & Tang, 1993). Based on these studies, cells appear to have evolved a variety of biochemical mechanisms to restore the integrity of the genetic material after DNA damage and retain its stability. These processes, defined as DNA repair, are organised into a number of pathways that are functionally distinguished by the type of chemical modification which they remove.

The systems of repair responsible for the removal of damage induced by UV-B radiation (for example, CPDs) from cellular DNA include photoreactivation (PHR), excision repair, including nucleotide excision repair (NER) and base excision repair (BER), recombinational repair and post-replication repair (Sancar & Sancar, 1988).

Photoreactivation (PHR)
A large number of species contain a direct-acting light-dependent enzyme, DNA photolyase, which utilises photons of light in the UV-A or blue light region of the spectrum (300–500 nm) as an energy source to

---

Fig. 2. The production of DNA lesions in the third leaf pair of light-grown 12 day-old pea seedlings exposed to UV-B radiation in the absence (closed symbols) and the presence of white light (open symbols). The formation of thymine dimers (a) and (6–4) photoproducts (b) were measured using the appropriate monoclonal antibodies (described in Mizuno *et al.*, 1991, Mori *et al.*, 1991) in an ELISA assay. (Absorbance values equate to antibody binding).

convert pyrimidine dimers directly into their intact monomer bases by a mechanism involving photoinduced electron transfer (Sancar, 1994). In effect, therefore, the formation of lesions by UV-B is subsequently repaired by processes which are dependent upon longer wavelengths of light (UV-A and visible wavelengths). The precise wavelengths utilised by the photolyase enzyme in reversing DNA damage can vary with the source of the enzyme and reflect the presence of the different chromophores associated with the photolyase. The plant photolyases characterised to date require wavelengths of light between 370 and 450 nm, and studies on *A. thaliana* photolyase activation showed a broad wavelength dependency of 375–400 nm (Pang & Hays, 1991).

There have been contradicting reports of attempts to isolate a plant homologue of the photolyase gene. A number of groups have attempted to isolate cDNA clones representing plant photolyases and two candidates, the HY4 (*A. thaliana*) gene product and SA-phr1 (*Sinapis alba*) gene product both show high homology to microbial photolyases in their chromophore-binding regions (Ahmad & Cashmore, 1993; Batschauer, 1993). However, when these proteins were over-expressed in *E. coli* as fusion proteins, neither possessed any photoreactivating activity (Malhotra *et al.*, 1995). This observation, along with evidence using mutants defective in the gene of interest and the composition of chromophores incorporated within the proteins, suggests that both genes may, in fact, encode blue light photoreceptors rather than photolyases. Recently we have isolated a putative CPD-photolyase from *A. thaliana* which has been characterised by nucleotide sequence analysis (Taylor *et al.*, 1996*b*). However, more functional information is needed to confirm this finding.

Nucleotide excision repair (NER)

Nucleotide excision repair (NER) is considered to be a second major pathway for the removal of UV photoproducts from cellular DNA. This pathway is also the most versatile strategy for the repair of DNA damage, responding to a diverse range of both chemical and photochemical lesions, including the UV-induced CPD and (6–4) photoproduct lesions.

NER involves a five-step pathway that can eliminate a variety of structurally unrelated helix distorting, 'bulky' lesions by the replacement, rather than the direct reversal of damage as seen with photolyase enzymes. The sequence of events involves recognition of the damage, local unwinding of the DNA by a helicase action, endonucleolytic incision and removal of an oligonucleotide containing the damage. This is followed by re-synthesis of a stretch of DNA using the complementary DNA strand as template and ligation of the newly synthesised DNA to the existing DNA chain (Friedberg, 1985). This type of repair has been demonstrated to be widespread amongst species from bacteria through to mammals.

In mammals, the complete process is known to involve at least 20 different polypeptides and in humans, defects in these proteins are associated with various forms of inherited disorders characterised by hypersensitivity of the skin to sunlight. As the understanding of the NER system in mammals has increased, it has become clear that there is a particularly strong link between excision repair and transcription, especially the protein complex TFIIH. Components of this complex are common to repair, transcription and cell cycle control, illustrating how, in recent years, the components of the NER proteins in the mammalian systems have been shown to be shared by a wide range of key cellular functions (for review see Lehmann, 1995).

Base excision repair (BER)

An alternative excision repair process that may be less important with regard to UV damage repair is base excision repair (BER), a process first proposed by Lindahl (Lindahl, 1976). BER is made up of a collection of damage-specific DNA glycosylases which remove modified bases from DNA and generate an abasic site which must be subsequently processed by an apurinic/apyrimidinic (AP) endonuclease and other BER components including DNA polymerase and ligase (Seeberg, Eide & Bjoras, 1995).

The key reaction in BER is the cleavage by hydrolysis of the N-glycosylic bond joining the base to the sugar. This reaction mechanism gives BER its unique features: high specificity and low energy cost to the cell. However, this also gives BER a very limited substrate range because of the requirement for specificity of the glycosylases. The BER pathway seemingly evolved to protect the cell against the effects of endogenous DNA damage, but the pathway also appears to be essential for the resistance to DNA damage from exogenous damaging agents such as UV-B, and BER is known to be involved in the repair of minor UV-induced photoproducts such as thymine glycols (a direct but minor product of UV-B irradiation). Some organisms (*Micrococcus luteus* and bacteriophage T4) also contain an additional BER N-glycosylase-AP-lyase, also known as 'UV-endonuclease', which is able to recognise and cleave DNA at sites of CPD lesions. No evidence has yet been presented that eukaryotes contain an equivalent or similar enzyme (Lloyd & Linn, 1993).

Other excision repair pathways also exist but are operative within a more limited range of organisms. For example, a separate type of excision repair has recently been shown to be operative in *Schizosacch-*

aromyces pombe. Yeasts have an NER pathway but *S. pombe* also appears to contain an alternative pathway (*S. pombe* DNA endonuclease (SPDE)) which may be capable of initiating an alternative DNA excision repair pathway to remove UV-induced dimeric damage (Bowman *et al.*, 1994). This pathway also appears to be found in *N.crassa* in which there is no evidence for the existence of an NER pathway (Yajima *et al.*, 1995).

The strategies employed for the removal of DNA lesions show a high degree of evolutionary conservation between micro-organisms and humans, and CPD photoproduct repair in most organisms is due in large part to different combinations of the repair pathways described above. For example, UV photoproduct repair in *E. coli* is thought to be due to NER and PHR whereas, in UV photoproduct repair in humans, only the NER pathway has been identified as capable of removing CPDs.

Until recently, few investigations have focused upon DNA repair mechanisms in plants where it has been assumed that similar mechanisms to those in other organisms operate. Now, however, there is good evidence that both PHR and excision repair are used in plant systems.

Excision repair in plants and the role of DNA ligase

Although the two categories of excision repair have been assumed to be present in higher plants the pathways are still poorly characterised compared to other organisms. Excision repair processes are generally referred to as 'dark repair' processes to distinguish them from other light-dependent DNA repair processes. In some cases, such as in *Ginko* (Trosko & Mansour, 1969) and *Nicotiana* (Trosko & Mansour, 1968), no evidence for the excision of CPDs could be detected. In carrot protoplasts only low levels of CPD excision could be measured (Eastwood & McLennan, 1985), and in *A. thaliana* the relatively slow repair of CPDs in the absence of light (compared to the rate in the light) was assumed to be due to excision repair (Pang & Hays, 1991). In wheat, the rate of removal of both thymine dimers and (6–4) photoproducts is significant in the dark but takes place at a much slower rate than in the presence of photoreactivating light (Fig. 3). Quaite *et al.* (1994) suggested that the relative contributions of the pathways involved in repair depends upon the initial levels of damage incurred to the DNA. At higher levels of damage to alfalfa seedlings, both types of repair made significant contributions to the removal of CPDs, but at lower damage levels only photoreactivation could be detected. Thus it seems that plants do have the capability to excise UV-photoproducts but, in the removal of CPDs at least, photorepair may be the favoured method for repair. However,

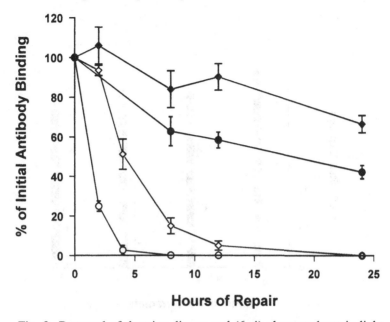

Fig. 3. Removal of thymine dimers and (6–4) photoproducts in light grown wheat leaf tissue. Light-grown seedlings (6 days) were subjected to UV-B damage by a 3-hour irradiation (incident dose rate 1.58 W m^{-2}) and then transferred to the light or dark, and the level of lesions was followed over a subsequent 24-hour period. The level of PAR for the light repair period was measured at 241 mmol m^{-2} s^{-1}. The formation of thymine dimers (circles) and (6–4) photoproducts (diamonds) was measured using monoclonal antibodies in an ELISA assay. Error bars are 1 SD and values are expressed as a percentage of the original amount of damage detected. Open symbols represent the level of lesion in tissue allowed to repair in the light, closed symbols represent the level of lesion in tissue allowed to repair in complete darkness.

it is conceivable that some of the lesions produced by UV light may not be substrates for a photolyase-type repair mechanism, and in these cases excision repair may be the only method of removal of these photolesions.

Several of the enzyme activities thought to be involved in BER have been partially purified from plant tissue, for example, a uracil DNA glycosylase has been partially purified from carrot cells (Talpaert-Borle & Liuzzi, 1982) and from wheat germ (Blaisdell & Warner, 1983). More recently a cDNA clone was isolated from *A. thaliana*

a

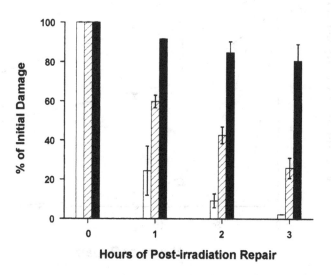

b

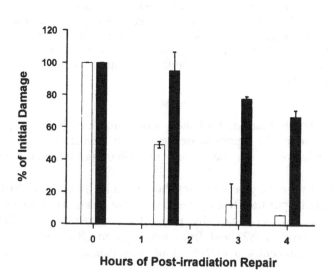

which restored resistance to alkylating agents when introduced into an *E. coli* mutant deficient in 3-methyl adenine-DNA glycosylase (Santerre & Britt, 1994).

Few components of the NER pathway, which is possibly more significant than BER for the removal of UV lesions, have been identified or isolated from plant sources. In our laboratory we have cloned an *A. thaliana* homologue of the mammalian DNA ligase I gene, the protein product of which is thought to play a role in several key cellular processes including the final step of NER and DNA replication. Experiments measuring the steady-state level of the ligase transcript in *A. thaliana* seedlings exposed to enhanced levels of UV-B showed that there was no significant increase in the levels of ligase transcript in the UV-treated seedlings compared to the control seedlings. Thus, the high levels of UV-B used and the accompanying induction of DNA damage do not appear to induce increased levels of the DNA ligase transcript which is in contrast to the situation in UV-C treated mammalian cells (Montecucco *et al.*, 1995). This finding is confirmed by the demonstration that there is no increase in the level of the apparent plant homologue of the DNA ligase I protein in the UV-B treated plants. However, we have demonstrated that, when protein extracts of sections of the wheat leaf are probed with a polyclonal antibody recognising a conserved region of the human DNA ligase I protein, higher levels of the ligase I protein are associated with actively dividing cells in the meristematic region of the wheat leaf (R.M. Taylor, A.K. Tobin & C.M. Bray, unpublished results).

Light-dependent repair of UV-induced dimeric photolesions in plants

The repair of CPDs by light-dependent photoreactivation is the best characterised of the repair systems in plants (McLennan, 1987*a*) and

---

Fig. 4. The removal of DNA damage (thymine dimers, CPDs) induced by a 3-hour UV-B irradiation in the leaves of light-grown (a) 6-day-old wheat seedlings and (b) 12-day-old pea seedlings. The levels were followed over a subsequent 4-hour period under different lighting regimes. PAR levels used during the repair period were 672 μmol m^{-2}s^{-1} (open bar), 2 μmol m^{-2} s^{-1} (hatched bar) and under conditions of complete darkness (solid bar). Points represent an average of at least two independent experimental values, error bars are 1 SD. Values are expressed as a percentage of the original amount of damage detected.

has been demonstrated to be operative in several species, including monocotyledonous and dicotyledonous species (Figs. 4*a, b*). The effect of white light on the elimination of thymine dimers (a type of CPD) in both light-grown pea (a dicotyledonous species) and wheat (a monocotyledonous species) seedlings shows that within 4 hours of post-irradiation repair in the light there is almost complete removal of thymine dimer antibody binding sites in the DNA extracted from pea and wheat leaves. However, in the absence of light during the same repair period the DNA from irradiated leaf tissue still contains more than one-half of the initial thymine dimer antibody binding sites.

Excision repair is in theory independent of light intensity and thus equates to the measured rate of repair in the absence of light (i.e. 'dark repair'). This is much lower than the combined activities, assumed to be photorepair and dark repair processes, observed in the light. The rapid rate of removal of thymine dimers in the light compared to the drastically reduced rate of dimer removal in the absence of a polychromatic light source suggests that the majority of this lesion (in these tissues at least) is removed by a light-requiring photolyase mechanism. Whilst the removal of thymine dimers in the primary wheat leaf is very rapid in the light (for example at light levels of 670 μmol m^{-2} s^{-1} PAR) it is also clear that the wheat leaf photolyase system is a highly efficient repair system, since even at low light intensities (2 μmol m^{-2} s^{-1} PAR, Fig. 4) the removal of thymine dimers is still much more efficient than in the dark.

In addition to the cases reported above, photoenzymatic repair (PER) activity to reverse CPD formation has been shown to be present in extracts from *A. thaliana* (Pang & Hays, 1991) and *Phaseolus vulgaris* seedlings (Langer & Wellmann, 1990) and the enzyme photolyase has been partially purified from maize pollen and several types of bean (McLennan, 1987*b*).

In recent years an associated light-dependent activity for the removal of (6–4) photoproducts has been identified in *Drosophila melanogaster* cell extracts (Todo *et al.*, 1993; Kim *et al.*, 1994). Plants also appear to contain this second photolyase activity. In light-grown wheat and pea tissue, the antibody binding sites for (6–4) photoproducts are removed faster in the presence of light than in the dark suggesting that this type of lesion can be repaired by a photolyase-like activity (Figs. 3, 5). The (6–4) photoproduct is also able to undergo photoisomerisation to the Dewar form when exposed to UV-A light. This conversion of the (6–4) photoproduct to its Dewar isomer may give the false impression that the original lesion is undergoing photorepair; however, experiments designed to measure the levels of the Dewar form of the lesion did not

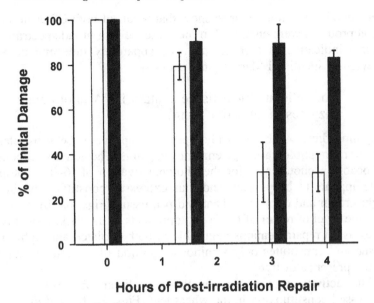

Fig. 5. The light-dependent loss of (6–4) photoproducts in light-grown pea leaf tissue. Light-grown pea seedlings (12 days) were subjected to UV-B damage by a 90-minute irradiation (incident dose rate 1.58 W m^{-2}) and then transferred to the light or dark and the level of lesions was followed over a subsequent 4-hour period. The level of PAR for the light repair period was measured at 450 μmol m^{-2} s^{-1}. Data points represent an average of at least three independent experimental values, and in the case of the zero timepoint an average of at least five independent experimental values. Error bars are 1 SD and values are expressed as a percentage of the original amount of damage detected. Open bars represent the level of lesion in tissue allowed to repair in the light, solid bars represent the level of lesion in tissue allowed to repair in complete darkness.

demonstrate any increase over the repair period, so that this is unlikely to be the main reason for the decrease in the lesion and photorepair of the (6–4) photoproduct appears the most likely explanation.

The light-enhanced rate of the removal of both thymine dimers and (6–4) photoproducts suggests that wheat and pea leaf tissue contains a light-dependent pathway for the removal of both CPDs and (6–4) photoproducts in addition to the light-independent (dark repair) pathway(s). There is also evidence in *A. thaliana* that a similar light-dependent activity for the elimination of (6–4) photoproducts exists (Chen, Mitchell & Britt, 1994). This study used an *A. thaliana* mutant defective in excision repair to show that repair of both CPDs and (6–4) photoproducts were enhanced in the presence of a broad spectrum light source indicative of a repair pathway apparently independent of the previously identified 'dark repair' pathway.

Does etiolated leaf tissue contain DNA photorepair enzymes in an active form?

Thymine dimers and (6–4) photoproducts appear to be substrates for one or more photolyase type enzymes in plants. The activity responsible (probably a photolyase) for the photoreactivation of (6–4) photoproducts appears to be present, and thus expressed constitutively, in both light-grown and etiolated wheat and pea tissue (Figs. 6a, b). It is clear that the level of repair of (6–4) photoproducts in the dark, probably due to excision repair, remains negligible in both etiolated and light-grown tissue and constitutes only a minor proportion of the total repair seen in the presence of light.

The activity responsible for thymine dimer photorepair is also expressed constitutively in the wheat leaf (Figs. 7a, b), and the activity in etiolated tissue does not differ significantly from that seen in light-grown tissue. However in the dicot pea, in contrast to the removal of (6–4) photoproducts, etiolated tissue appears to require a prior induction treatment of white light to induce the thymine dimer repair activity and an initial lag of between 1 and 2 hours is observed when no significant decrease in antibody binding sites for the thymine dimers can be measured. The result with pea leaf tissue agrees with previous observations in other dicotyledonous species such as *S. alba* (Buchholz, Ehmann & Wellmann, 1995), *Phaseolus vulgaris* (Langer & Wellmann, 1990) and

---

Fig. 6. The removal of (6–4) photoproducts induced by a pulse of UV-B radiation in the etiolated and light-grown (a) pea leaf and (b) wheat leaf. Seedlings were etiolated (circles) or grown under light conditions at a PAR of 450 μmol m^{-2} s^{-1}(squares) and then subjected to UV-B damage by a 90-minute irradiation (incident dose rate 1.58 W m^{-2}) (light-grown) or a 60-minute irradiation (etiolated) and placed under light conditions (PAR 450 μmol m^{-2} s^{-1}) (open symbols) or in the dark (solid symbols) for the indicated period of repair. Points represent an average of at least three independent experimental values, error bars are 1 SD. Values are expressed as a percentage of the original amount of damage detected.

a

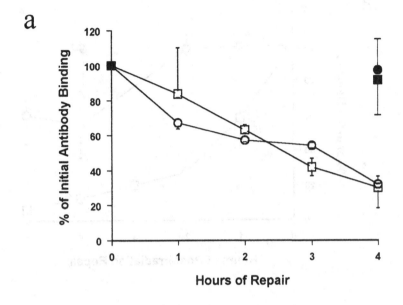

b

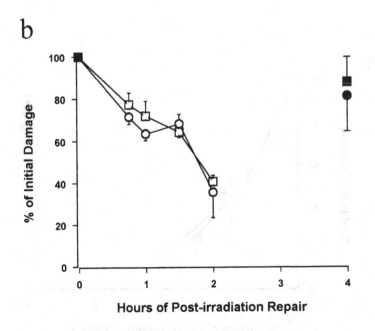

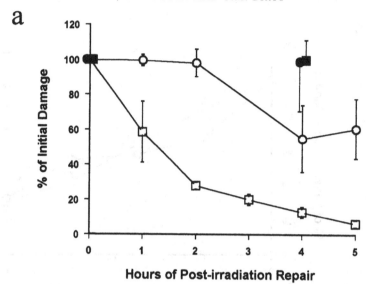

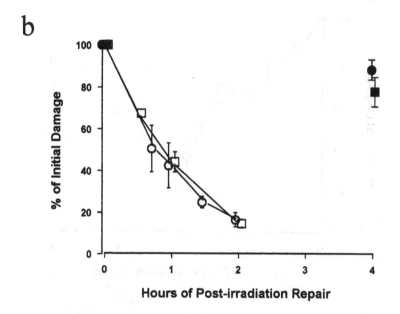

A. thaliana (Chen *et al.*, 1994) where a prior light treatment was required for the induction of the ability to remove CPDs from the DNA in the leaves of these species. However, the monocot wheat does not require this induction and is one of the first demonstrations of the constitutive photolyase-catalysed repair of thymine dimers in higher plants.

Chen *et al.* (1994) observed that (6–4) photoproducts were photo-repaired constitutively in *A. thaliana* whereas the activity responsible for CPD photoreactivation needed both a prior pulse of visible light to induce the activity (photoinduction) along with exposure to white light during the repair process. This observation suggests, that in dicotyl-edonous species, the two lesions are repaired by two distinct photo-lyases. This may not be the case with wheat where a single photolyase enzyme could be responsible for the reversal of both thymine dimer and (6–4) photoproduct damage due to the fact that both lesions are photorepaired by a constitutively expressed activity (summarised in Table 1).

Developmental regulation of the capacity for light-dependent repair of UV-induced dimeric photolesions

The 7-day-old wheat leaf is an ideal experimental system in which to investigate whether the ability to photorepair thymine dimers and (6–4) photoproducts is under developmental control since the capacity for cellular DNA repair can be studied in conjunction with the developmental state of the cell. The wheat leaf has a natural developmental gradient of maturing cells with immature dividing cells being located at the base

---

Fig. 7. The removal of thymine dimers induced by a pulse of UV-B radiation in the etiolated and light-grown (a) pea leaf and (b) wheat leaf. Seedlings were etiolated (circles) or grown under light conditions at a PAR of 450 μmol m^{-2} s^{-1} (squares) and then subjected to UV-B damage by a 90-minute irradiation (incident dose rate 1.58 W m^{-2}) (light grown) or a 60-minute irradiation (etiolated) and placed under light conditions (PAR of 450 μmol m^{-2} s^{-1}) (open symbols) or in the dark (solid symbols) for the indicated period of repair. Points represent an average of at least three independent experimental values, error bars are 1 SD. Values are expressed as a percentage of the original amount of damage detected.

Table 1. *The capacity to photorepair DNA lesions in leaf tissue from wheat and pea seedlings*

	Growth conditions	(6–4) photoproduct repair	Thymine dimer repair
Wheat (monocot.)	Etiolated	Yes	Yes
	Light-grown	Yes	Yes
Pea (dicot.)	Etiolated	Yes	No
	Light-grown	Yes	Yes

of the leaf and fully mature, non-dividing terminally differentiated cells at the tip (Leech, 1984). The cell age gradient has been explored in microscopic and biochemical studies of cell differentiation in some detail, and so the leaf provides a well-characterised system to analyse the relationship between the developmental status of the cells and the potential to repair DNA damage after UV-B irradiation.

Etiolated leaf tissue was used in the experiments discussed here so that developmental effects could be studied independently of any light-induced effects which might be present with light-grown tissue. In the 7 d etiolated wheat primary leaf the intercalary meristem which marks the base of the leaf is positioned at 2 mm from the base of the seedling. The zone of elongation extends 30 mm up the leaf towards the tip (Fig. 8).

Developmental regulation of the capacity for DNA repair was investigated by taking 25 mm sections from the base of the seedling through to the leaf tip and measuring the capacity of the different sections to remove DNA lesions after UV-B irradiation of the leaf. There is a reduced capacity for the repair of thymine dimers in the basal sections (containing both dividing meristematic cells and elongating epidermal cells) in the etiolated wheat leaf compared to the mature cells found towards the leaf tip, although all cells do possess some capacity to photorepair thymine dimers (Fig. 9a). This observation suggests that thymine dimer lesions appearing in fully differentiated cells of the etiolated wheat leaf are removed more rapidly than the same lesions in the immature cells at the base of the leaf and indicates that the appearance of specific thymine dimer photorepair activity may be under developmental control. The level of dark repair of thymine dimers is minimal

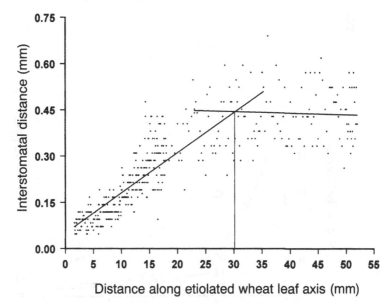

Fig. 8. Estimation of the zone of cell elongation in the etiolated primary wheat leaf. Interstomatal distances were measured along the 7-day-old etiolated wheat leaf and analysis of the data performed as described by Paolillo, Sorrells & Keyes (1991). The data were divided at 35.2 mm and two linear regressions were fitted to the subsets of data and the intercept of the lines was taken as the distal limit of the leaf extension zone.

compared to the level of light-dependent repair in all the leaf sections studied (Fig. 9a) eliminating the possibility that differences in the capacity for dark repair are responsible for the differences in the ability to repair thymine dimers in the light. Cloning of the gene encoding the enzyme responsible for this repair would allow us to ascertain at what level this control is exerted. Pang and Hays (1991) reported a similar observation in *Arabidopsis*, where the level of photolyase activity responsible for CPD removal in seedling extracts increased from being absent in seeds to a maximum level in 10 d seedlings, after which time no further increase was observed.

In contrast to the repair of thymine dimers, no significant differences in the rate of removal of (6–4) photoproducts in the different sections of wheat leaf could be detected, suggesting that the capacity for photo-repair of this lesion may not vary in wheat leaf cells at different developmental stages (Fig. 9b).

Taken as a whole, the evidence discussed here suggests that the

a

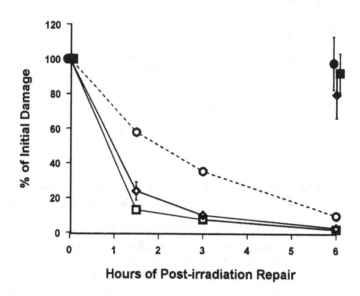

b

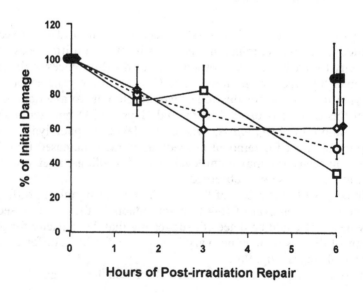

activity responsible for the photorepair of the (6–4) photoproduct is expressed constitutively in all tissues in both monocots and dicots and that the presence of this activity is independent of both light and development. However, there appears to be a major difference between monocots and dicots with respect to the activity responsible for the photorepair of CPDs (if the primary wheat leaf is taken as a typical monocot). The activity specific for CPD photorepair in monocot leaves does not appear to be under the control of white light induction, as has been reported to be the case in dicots, but is possibly under developmental control so that its activity increases as the leaf cell matures.

References

Ahmad, M. & Cashmore, A.R. (1993). HY4 gene of *A. thaliana* encodes a protein with characteristics of a blue-light photoreceptor. *Nature*, **366**, 162–6.

Batschauer, A. (1993). A plant gene for photolyase; an enzyme catalyzing the repair of UV-light-induced DNA-damage. *Plant Journal*, **4**, 705–9.

Blaisdell, P. & Warner, H. (1983). Partial purification and characterisation of a uracil-DNA glycosylase from wheat germ. *Journal of Biological Chemistry*, **258**, 1603–9.

Bowman, K.K., Sidik, K., Smith, C.A., Taylor, J.-S., Doetsch, P.W. & Freyer, G.A. (1994). A new ATP-dependent DNA endonuclease from *S. pombe* that recognises cyclobutane pyrimidine dimers and (6–4) photoproducts. *Nucleic Acids Research*, **22**, 3026–32.

Buchholz, G., Ehmann, B. & Wellmann, E. (1995). Ultraviolet light inhibition of phytochrome-induced flavonoid biosynthesis and DNA

Fig. 9. The removal of thymine dimers (a) and (6–4) photoproducts (b) induced by a UV-B pulse in 7-day-old etiolated wheat leaf sections. Seedlings were grown in complete darkness. After 7 days the coleoptile was removed from the seedlings and the etiolated primary wheat leaf tissue was subjected to UV-B damage by a 30-minute side irradiation (incident dose rate 1.0 W m^{-2}) and then placed in the light (PAR 450 μmol m^{-2} s^{-1}) (open symbols) or in the dark (closed symbols) for the indicated period of repair. Leaf sections were harvested thus; the basal 25 mm section (section 1, circles, dashed line), 25–50 mm (section 2, diamonds), and 50–75 mm (section 3, squares). Points represent an average of at least three independent experimental values, error bars are 1 SD. Values are expressed as a percentage of the original amount of damage detected.

photolyase formation in Mustard cotyledons (*Sinapis alba* L.). *Plant Physiology*, **108**, 227–34.

Caldwell, M.M., Robberecht, R. & Flint, S.D. (1983). Internal filters: prospects for UV-accumulation in higher plants. *Physiologia Plantarum*, **58**, 445–50.

Chen, J., Mitchell, D.L. & Britt, A.B. (1994). A light-dependent pathway for the elimination of UV-induced pyrimidine (6–4) pyrimidone photoproducts in *Arabidopsis*. *Plant Cell*, **6**, 1311–17.

Davies, R.J.H. (1995). Ultraviolet radiation damage in DNA. *Biochemical Society Transactions*, **23**, 407–18.

Eastwood, A.C. & McLennan, A.G. (1985). Repair replication in ultraviolet light-irradiated protoplasts of *Daucus carota*. *Biochimica et Biophysica Acta*, **826**, 13–19.

Friedberg, E.C. (1985). *DNA Repair*. W.H. Freeman, San Francisco.

Kim, S.-T., Malhotra, K., Smith, C.A., Taylor, J.-S. & Sancar, A. (1994). Characterisation of (6–4) photoproduct DNA photolyase. *Journal of Biological Chemistry*, **269**, 8535–40.

Langer, B. & Wellmann, E. (1990). Phytochrome induction of photoreactivating enzyme in *Phaseolus vulgaris* L. seedlings. *Photochemistry and Photobiology*, **52**, 861–3.

Leech, R.M. (1984). Chloroplast development in angiosperms: current knowledge and future prospects. In *Topics in Photosynthesis, Vol. 5. Chloroplast Biogenesis*. (Baker, N.R. & Barber, J. eds), pp. 1–21. Elsevier Science Publishers, Amsterdam.

Lehmann, A.R. (1995). Nucleotide excision repair and the link with transcription. *Trends in Biochemical Science*, **20**, 402–5.

Li, J., Ou-Lee, T.-M., Raba, R., Amundson, R.G. & Last, R.L. (1993). *Arabidopsis* flavonoid mutants are hypersensitive to UV-B irradiation. *Plant Cell*, **5**, 171–9.

Lindahl, T. (1976). New class of enzymes acting on damaged DNA. *Nature*, **259**, 64–6.

Lloyd, R.S. & Linn, S. (1993). In *Nucleases* (Lin, S., Lloyd, R.S. & Roberts, R.J., eds) pp. 263–316. Cold Spring Harbor Laboratory Press.

McLennan, A.G. (1987a). DNA damage, repair, and mutagenesis. In *DNA Replication in Plants* (Bryant, J.A. & Dunham, V.E., eds), pp. 135–86. CRC Press, Boca Raton, Florida.

McLennan, A.G. (1987b). The repair of ultraviolet light-induced DNA damage in plant cells. *Mutation Research*, **181**, 1–7.

Malhotra, K., Kim, S.-T., Batschauer, A., Dawut, L. & Sancar, A. (1995). Putative blue-light photoreceptors from *Arabidopsis thaliana* and *Sinapis alba* with a high degree of sequence homology to DNA photolyase cofactors but lack DNA repair activity. *Biochemistry*, **34**, 6892–9.

Markham, K.R. (1982). *Techniques of Flavonoid Identification*. 113 pp. Academic Press, London.

Matsunaga, T., Hatakeyama, Y., Ohta, M., Mori, T. & Nikaido, O. (1993). Establishment and characterisation of a monoclonal antibody recognising the Dewar isomers of (6–4) photoproducts. *Photochemistry and Photobiology*, **57**, 934–40.

Mitchell, D. L. & Karentz, D. (1993). The induction and repair of DNA photodamage in the environment. In *Environmental UV Photobiology* (Young, A. R., ed.). Plenum Press, New York.

Mizuno, T., Matsunaga, T., Ihara, M. & Nikaido, O. (1991). Establishment of a monoclonal antibody recognising cyclobutane-type thymine dimers in DNA: a comparative study with 64M-1 antibody specific for (6–4) photoproducts. *Mutation Research*, **254**, 175–84.

Montecucco, A., Savini, E., Biamonti, G., Stefanini, M., Focher, F. & Ciarrocchi, G. (1995). Late induction of human DNA ligase I after UV-C irradiation. *Nucleic Acids Research*, **23**, 962–6.

Mori, T., Nakane, M., Hattori, T., Matsunaga, T., Ihara, M. & Nikaido, O. (1991). Simultaneous establishment of monoclonal antibodies specific for either cyclobutane pyrimidine dimer or (6,4) photoproduct from the same mouse immunised with ultraviolet-irradiated DNA. *Photochemistry and Photobiology*, **54**, 225–32.

Pang, Q. & Hays, J.B. (1991). UV-B-inducible and temperature-sensitive photoreactivation of cyclobutane pyrimidine dimers in *Arabidopsis thaliana*. *Plant Physiology*, **95**, 536–43.

Paolillo, D.J., Sorrells, M.E. & Keyes, G.J. (1991). Gibberellic acid sensitivity determines the length of the extension zone in wheat leaves. *Annals of Botany*, **67**, 479–85.

Quaite, E.F., Takayanagi, S., Ruffini, J., Sutherland, J.C. & Sutherland, B.M. (1994). DNA damage levels determine cyclobutyl pyrimidine dimer repair mechanisms in alfalfa seedlings. *Plant Cell*, **6**, 1635–41.

Robberecht, R. & Caldwell, M.M. (1978). Leaf transmittance of ultraviolet radiation and its implications for plant sensitivity to ultraviolet-radiation induced injury. *Oceologia*, **32**, 277–87.

Sancar, A. (1994). Structure and function of DNA photolyase. *Biochemistry*, **33**, 2–9.

Sancar, A. & Sancar, G. (1988). DNA repair enzymes. *Annual Reviews of Biochemistry*, **57**, 29–76.

Sancar, A. & Tang, M.-S. (1993). Photobiology school; nucleotide excision repair. *Photochemistry and Photobiology*, **57**, 905–21.

Santerre, A. & Britt, A.B. (1994). Cloning of a 3-methyladenine-DNA glycosylase from *Arabidopsis thaliana*. *Proceedings of the National Academy of Sciences, USA*, **91**, 2240–4.

Seeberg, E., Eide, L. & Bjoras, M. (1995). The base excision repair pathway. *Trends in Biochemical Science*, **20**, 391–7.

Stapleton, A.E. & Walbot, V. (1994). Flavonoids can protect Maize DNA from the induction of ultraviolet radiation damage. *Plant Physiology*, **105**, 881–9.

Strid, A. & Porra, R.J. (1992). Alterations in pigment content in leaves of *Pisum sativum* after exposure to supplementary UV-B. *Plant and Cell Physiology*, **33**, 1015–23.

Talpaert-Borle, M. & Liuzzi, M. (1982). Base excision repair in carrot cells, partial purification and characterisation of Uracil-DNA glycosylase and apurinic/apyrimidinic endodeoxyribonuclease. *European Journal of Biochemistry*, **124**, 435–40.

Taylor, R.M., Nikaido, O., Jordan, B.R., Rosamond, J., Bray, C.M. & Tobin, A.K. (1996). Ultraviolet-B-induced lesions and their removal in wheat (*Triticum aestivum* L.) leaves. *Plant Cell and Environment*, **19**, 171–81.

Taylor, R.M., Tobin, A.K. & Bray, C.M. (1996*b*). Nucleotide sequence of an *Arabidopsis* cDNA At-phrII (Accession No. X99301) encoding a protein with high homology to the class II CPD photolyases present in higher eukaryotes. *Plant Physiology*, **112**, 862 (PGR96-083).

Todo, T., Takemori, H., Ryo, H., Ihara, M., Matsunaga, T., Nikaido, O., Saito, K. & Nomara, T. (1993). A new photoreactivating enzyme that specifically repairs ultraviolet light-induced (6–4) photoproducts. *Nature*, **361**, 371–4.

Trosko, J.E. & Mansour, V.H. (1968). Response of tobacco and *Haplopappus* cells to ultraviolet irradiation after posttreatment with photoreactivating light. *Radiation Research*, **44**, 700–12.

Trosko, J.E. & Mansour, V.H. (1969). Photoreactivation of ultraviolet light-induced pyrimidine dimers in Ginko cells grown *in vitro*. *Mutation Research*, **7**, 120–1.

Tyrrell, R.M. (1993). UV photochemistry and action spectroscopy. In *The Effects of Environmental UV-B Radiation on Health and Ecosystems* (Proceedings of the First European Symposium, Munich, 27–29 October, 1993). Office for Official Publications of the European Community, Luxembourg, pp. 93–6.

Yajima, H., Takao, M., Yasuhira, S., Zhao, J.H., Ishii, C., Inoue, H. & Yasui, A. (1995). A eukaryotic gene encoding an endonuclease that specifically repairs DNA damaged by ultraviolet light. *EMBO Journal*, **14**, 2393–9.

A.B. BRITT

Genetic analysis of DNA repair in plants

Introduction

Predicted increases in solar UV-B radiation have served to focus attention on the toxic effects of UV-B on plants. Nuclear DNA is present in very low copy number, and acts as the template for its own synthesis. For this reason, it is an especially vulnerable target for UV-induced damage. Even a single persisting UV-induced lesion can be a potentially lethal event, particularly in haploid tissues such as pollen grains.

The cyclobutane pyrimidine dimer (CPD) and the pyrimidine (6–4) pyrimidinone dimer (the 6–4 photoproduct) make up the great majority of UV-induced DNA damage products (Fig. 1) (Mitchell & Nairn, 1989). The biological effects of these lesions have been studied extensively in microbial and mammalian systems, where UV-induced DNA damage has been shown to produce two distinct effects: mutagenesis and toxicity. At the molecular level, pyrimidine dimers are known to inhibit the progress of microbial and mammalian DNA polymerases. Because pyrimidine dimers cannot effectively base pair with other nucleotides, they are not *directly* mutagenic, but instead act as blocks to DNA replication. Interestingly, and very significantly in terms of UV-induced toxicity, mammalian RNA polymerases have also been shown to 'stall' at both CPDs and 6–4 photoproducts (Protic-Sabljic & Kraemer, 1986; Mitchell, Vaughan & Nairn, 1989). Thus a single pyrimidine dimer, if left unrepaired, is sufficient to completely eliminate expression of a transcriptional unit. The direct biological effects of UV-induced dimers on transcription and DNA replication have not been well characterised in plants, but are generally assumed to be similar to those observed in the other living kingdoms.

If every pyrimidine dimer acts as a block to transcription and replication, while only a small fraction of dimers result in a mutation (see below), the inhibitory effects of UV on transcription and DNA replication are probably more significant (in terms of plant growth) than its mutagenic effects. Any living tissue, even one in which cell division does not occur, has to be able to either avoid or repair UV-induced

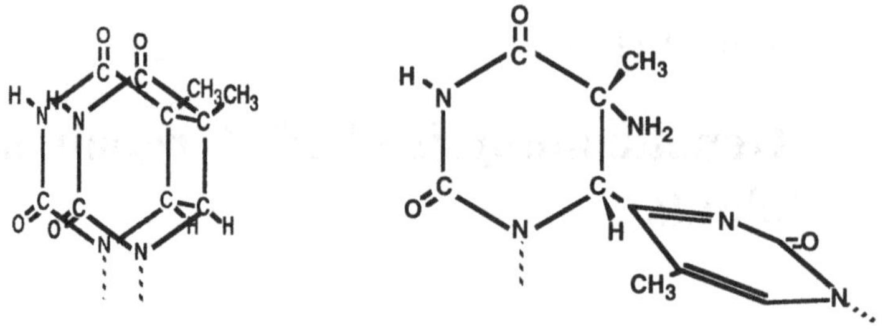

cyclobutane dimer

pyrimidine (6-4)
pyrimidinone dimer

Fig. 1. Cyclobutyl pyrimidine dimers (CPDs) and pyrimidine 6–4 pyrimidinone dimers (6–4 photoproducts) make up the majority of UV-induced DNA damage.

DNA damage if it is to survive. Because no UV-absorbing agent can be 100% effective, plants undoubtedly employ a combination of sunscreens (to reduce the rate of production of DNA damage) and repair mechanisms (to eliminate damage induced by the residual UV) to maintain the integrity of the genome. UV-absorbing pigments are discussed in chapters by Bornman *et al.* and by Jenkins *et al.* (this volume), and a variety of DNA repair mechanisms are discussed in the chapter by Taylor *et al.* (this volume).

Not surprisingly, then, the study of DNA repair in plants is a rapidly growing field. An understanding of the diversity and biological relevance of repair pathways is not only essential to our understanding of UV tolerance, but also provides insights into mutagenic processes required for the creation of genetic diversity. Our current state of knowledge of the various pathways for DNA repair in plants has been recently reviewed (Britt, 1996). This chapter therefore focuses on the rationale and techniques involved in the genetic analysis of DNA repair and UV-sensitivity in *Arabidopsis thaliana*, and will discuss the various methods for the isolation of repair-defective mutants and address the possible biological relevance of these mutants, both in terms of UV-resistance mechanisms and in regard to their possible applications to others fields.

A classical genetic approach can be applied to mechanistic questions such as:

(1)   Which pathways are required for the repair of pyrimidine dimers?

(2)   How many proteins are involved in each of these pathways?

and, perhaps most importantly,

(3)   Do the subtle effects of solar UV on plant growth have anything to do with DNA damage?

Because DNA damage results in the inhibition of plant growth, as well as mutations, we can screen for mutants with enhanced sensitivity to either of these effects. However, mutants expressing either of these 'UV-sensitive phenotypes' might be defective in processes other than DNA repair, including, but not limited to, defects in the production of UV-absorbing pigments, defects in the ability to cope with a variety of stresses, or defects in other UV-sensitive processes, including, perhaps, photosynthesis. Not all of these mutants are useful to an investigator with a focused interest in DNA repair. From the point of view of a plant biologist with an interest in the effects of UV on plant growth, however, the relatively wide net cast by a genetic approach to UV-sensitivity is a point in its favour. By searching for UV-sensitive mutants, rather than directly for DNA repair mutants, we can gain useful insights into the biochemical basis of UV-induced growth inhibition, irrespective of our own initial prejudices in approaching this problem.

Methods for the isolation of UV-sensitive mutants of *Arabidopsis*

Why *Arabidopsis*?

The many qualities of *Arabidopsis thaliana* that make it an excellent model system for the study of a variety of higher-plant processes have been described in several reviews. In brief, *Arabidopsis* is a small, self-pollinating member of the family *Brassicaceae* that has an unusually compact genome and a relatively short generation time (Somerville & Estelle, 1986). These attributes make it a good subject for both classical genetic and molecular studies. The mere fact that it has become a popular model system has enhanced its value as such; an extensive genetic infrastructure has been developed, including an elaborate molecular genetic map (Hauge *et al.*, 1993; Jarvis *et al.*, 1994), and a collection of ordered, contiguous cloned DNAs extending over the majority of the genome. For an investigator with an interest in the repair of UV-induced DNA damage, tiny *Arabidopsis* seedlings have the additional advantage of providing a system in which UV-B induced damage is distributed throughout the seedling with fairly good uniformity, rather than just in the first few cell layers (Chen, Jiang & Britt, 1996). This makes it possible to quantify the induction of damage and the rate of its repair, and also to identify mutants altered in their response to UV, in an intact higher plant. It should be kept in mind, however, that *Arabidopsis* is a winter season, low-light organism, and as such may lack some of the capabilities for UV resistance expressed by high-light summer crops such as maize. Although there are some technical difficulties involved in the study of repair in larger plants, the development of a second model system, particularly a high-light plant, is critical for our understanding of UV resistance mechanisms.

Mutagenesis of *Arabidopsis*

In order to isolate novel mutants, *Arabidopsis* seeds, rather than adult plants, are treated with DNA damaging agents. The mutagenised seeds (termed the M1, or first mutagenised, generation) are both heterozygous and chimaeric for all changes in their DNA sequence; obviously, every cell in the seed will acquire different mutations. However, single *Arabidopsis* flowers are, to a large extent, derived from what was a single cell in the seed (Li & Redei, 1969; Harle, 1972). Because the *Arabidopsis* flowers produce both pollen and ova, self-pollination of these M1 flowers will produce M2 progeny, some of which are homozygous for new mutations. This M2 generation is usually employed in mutant screens as it will contain mutants homozygous (as well as heterozygous) for new mutations. Ethyl methane sulphonate (EMS) is a very effective mutagen for the production of a high density of mutations in *Arabidopsis* seeds; as few as 3000 M2 seeds need to be screened to have a very good probability of finding a mutation that eliminates or reduces the expression of a particular gene of interest (Somerville & Estelle, 1986). The resulting mutants should, however, be back-crossed extensively to their unmutagenised progenitor line, as the newly isolated mutants carry a large number of other mutations. These secondary mutations may confound comparisons to what should be an isogenic 'wild-type' line.

Certain other mutagenic agents produce a much lower density of mutations, and as a result a much larger M2 population must be screened in order to identify a desired mutant. For example, a mutagenic treatment such as T-DNA transfection (Errampalli et al., 1991) or transposon-induced mutagenesis (Long et al., 1993) produces only one, or a few, new mutations per M2 seed. Whether the downside of these mutagenic agents (the larger, by orders of magnitude, M2 populations that must be screened) is balanced by their beneficial aspects (a potentially tagged allele, and the low frequency of secondary mutations) depends on both the overall goals of the investigator and the difficulty of the screen employed to identify UV-sensitive mutants. In addition, it is sometimes desirable to screen for enhanced sensitivity in lines that are already defective in DNA repair or the production of UV-absorbing agents (see below); in this case, neither commercially generated mutant populations nor collections of insertion lines will be available to the investigator. Under these circumstances, the additional burden of generating mutant lines will rest on the individual investigator, making the insertional mutagenesis strategy less efficient, and so less attractive, than chemical mutagenesis.

Identification of UV-sensitive mutants

It is always easier to identify a mutant class by *selecting* for that mutation, that is, by developing a treatment that eliminates every member of the population except the mutants of interest. Unfortunately, none of the few laboratories working in this area have come up with such a scheme, and the original UV-sensitive mutants of *Arabidopsis* (Britt *et al.*, 1993; Harlow *et al.*, 1994; Jenkins *et al.*, 1995) were identified via more labour-intensive 'screens'. Because the desired phenotype was lethality, or growth inhibition, upon exposure to UV, screens were complicated by the fact that the mutant line not only had to be identified, but also rescued from the damaging effects of the screen itself. The Britt group employed a 'root-bending assay' (Fig. 2) to identify M2 families (the progeny of single M2 plants) that uniformly displayed an inhibition of root growth at unusually low doses of UV-B; this screen is rather laborious to set up, but is very precise; mutant families identified using this screen were always homozygous defective for UV-sensitivity, and mutant sibs in the non-UV-treated controls could be rescued and propagated. The Mount group developed a clever screen that enabled them to look for UV-sensitive phenotypes in M2 individuals, rather than M2 families. They placed a drop of foam containing a UV-C absorbing agent on the growing apex of each M2 plant, then UV-C irradiated an entire pot at a time. UV-sensitivity could then be scored as browning or puckering of the exposed outer leaves, and new growth would continue from the protected apical meristem.

Mutant stocks that are screened for defects in dark repair are allowed to recover from their UV treatment either in total darkness, for 3 days, or in the presence of filtered (gold) light. The gold-light treatment enables the seedlings to perform photosynthesis, but not photoreactivation. When screening for mutants defective in light-dependent repair pathways the seedlings are allowed to recover in the presence of full-spectrum light, and a much higher UV test dose is used, as wild-type strains have a much higher level of resistance to UV in the presence of photoreactivating light.

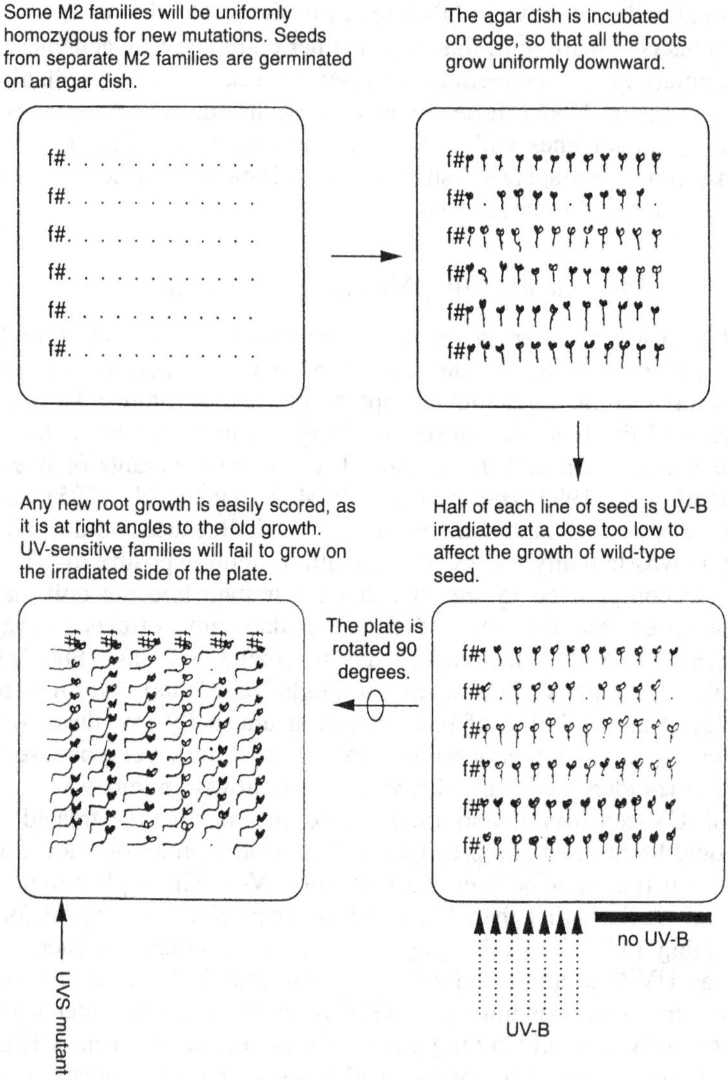

Some M2 families will be uniformly homozygous for new mutations. Seeds from separate M2 families are germinated on an agar dish.

The agar dish is incubated on edge, so that all the roots grow uniformly downward.

Any new root growth is easily scored, as it is at right angles to the old growth. UV-sensitive families will fail to grow on the irradiated side of the plate.

Half of each line of seed is UV-B irradiated at a dose too low to affect the growth of wild-type seed.

The plate is rotated 90 degrees.

no UV-B

UVS mutant

UV-B

Fig. 2. UV-sensitive M2 families of *Arabidopsis* seedlings are identified by a root-bending assay.

The Britt group is currently isolating UV-sensitive mutants using a combination of both approaches. EMS-generated mutant M2 families (that, again, carry a very high density of mutations) are first screened for defects in dark repair using the root bending assay. The majority of families are uniformly UV-resistant. It is possible to screen these same plates of seedlings a second time, this time looking for mutants defective in photoreactivation. After the first screen, the same plates are then laid flat (rather than on edge, as in the root bending assay) and allowed to grow for another week under cool white lamps. They are then subjected to a second, higher UV test dose, in order to screen for defects in light-dependent repair. These seedlings are then allowed to recover under cool white lamps, and after a week the leaves are observed for signs of UV-induced damage. No protecting foam is used. UV-sensitive families display a characteristic phenotype; the outermost edges (especially the distal tips) of the leaves stop growing, pucker and turn yellow or even brown, while the proximal edges and apical tip of the plant, apparently shielded from UV, produce new very dark green growth. Mutants defective in the light-dependent repair of CPDs have been isolated using this screen (Jiang *et al.*, 1997).

When screening large, lightly mutagenised populations (such as T-DNA insertion lines), it is not feasible to screen individual families. In this case, the bulked mutants can simply be grown in flats and screened (without a 'no UV' control) for plants displaying the UV-sensitive phenotype described above. In order to avoid collecting many false positives that are simply mutant individuals with a sickly phenotype that is independent of UV-exposure, the investigators should familiarise themselves with the UV-dependent pattern of necrosis. This can be done simply by irradiating their wild-type strain with higher doses of UV and observing the resulting effects. In any event, the UV-sensitive phenotype can be confirmed by self-pollinating the plant and observing the growth of the resulting progeny with and without UV.

Growth conditions prior to UV-exposure

The manner in which the seedlings are handled prior to, as well as after, UV-irradiation, would be expected to affect the variety of UV-sensitive mutants that will be isolated during a screen. The UV-sensitive mutants of *Arabidopsis* that have been described in the literature were isolated by laboratories primarily interested in generic DNA repair/damage tolerance pathways, rather than specialised pathways for the repair of UV-induced damage, or other mechanisms for UV-resistance. Although prior exposure to light, especially UV-A and blue radiation, has a major impact on plant sensitivity to subsequent UV-exposure (Caldwell, Flint & Searles, 1994; Middleton & Teramura, 1994; Takayanagi et al., 1994), the Arabidopsis seedlings were grown under artificial light (usually Phillips Cool White fluorescent tubes) with no attempt to mimic the solar spectrum. It is especially important to note that the seedlings were not permitted to adapt to UV-B before being subjected to UV-B; if any UV-resistance pathways actually are regulated by UV-B, they would not be expressed in these seedlings. Similarly, both the Mount and Britt groups were careful to maintain their seedlings in the absence of photoreactivating light after UV-irradiation, as they were specifically searching for generic, rather than UV-specific repair defects. Thus mutants defective in photoreactivation of pyrimidine dimers were, of course, not identified.

The test dose

The duration, spectral distribution and intensity of the UV 'test dose' should also be considered both when planning a screen and when interpreting the phenotype of a newly isolated mutant. The published mutants were isolated using a short, intense burst of UV rather than an intensity and spectrum of UV approximating that in the field. One group employed UV-B radiation (a cellulose acetate-filtered DNA transilluminator, λ_{max} = 305 nm) (Britt et al., 1993), and the other UV-C radiation (a germicidal lamp, λ_{max} = 254 nm) (Harlow et al., 1994; Jenkins et al. 1995). In terms of the induction of pyrimidine dimers, this may be an unimportant distinction; the rate of induction of CPDs and 6–4 PDs increases linearly with dose, and the relative rate of induction of CPDs vs 6–4 PDs is similar throughout this range of wavelengths (Rosenstein & Mitchell, 1987). The relatively infrequent induction of cytosine photohydrates also follows this pattern (Mitchell, Jen & Cleaver, 1991). However, it is conceivable that other kinds of damage might be induced exponentially, rather than linearly, with increasing intensities of UV. For example, it is possible that extremely high doses of UV might damage the photosynthetic apparatus, resulting in a burst of oxidative damage. If this were the case, then mutants defective in the repair of these lesions might display a UV-sensitive phenotype due to their inability to repair a lesion that might only rarely, or effectively never, be generated by solar UV. While this particular class of mutants may be interesting and relevant to the study of DNA transactions in plants, and in the study of photoinhibition, it might not represent a pathway that plays a role in resistance to environmental UV.

For these reasons, an investigator with a primary interest in the genetics of UV-resistance should consider screening for mutants in an environment that mimics 'natural conditions' as closely as possible both before and after UV exposure. In fact, it would be wise to screen for mutants in an environment with a constant, realistic, low level of UV-B, UV-A, and blue light. Although these conditions are difficult and expensive to reproduce in the laboratory, it is possible to grow and screen *Arabidopsis* out of doors, provided the temperature can be kept under approximately 30 °C. Robert Last and Laurie Landry, at the Boyce Thompson Institute, have recently isolated several UV-sensitive mutants using a growth chamber that subjects *Arabidopsis* seedlings to continuous UV exposure. Some of these mutants fail to display a UV-repair defect, but at least one mutant is deficient in the photoreactivation of CPDs (Landry *et al.*, 1997).

Whether a pulsed or chronic UV exposure is chosen, the UV test dose required for the screen should be determined empirically by the investigator. Usually a dose of about one-third that required to produce an obvious UV-sensitive reaction in the progenitor line is employed. Higher doses will lead to distracting UV-induced effects in a certain percentage of wild-type plants. In addition, if one is screening M2 individuals, rather than M2 families, valuable mutants with extremely UV-sensitive phenotypes might be killed and lost if the test dose of UV is too high. On the other hand, lower doses might fail to induce an obvious response in mutants with mild UV-sensitive phenotypes.

Selecting a progenitor line

The investigator's specific goals should also be carefully considered when selecting a progenitor line for mutagenesis. For example, researchers interested in the repair of UV-induced damage might want to start with a strain defective in the production of UV-absorbing agents (a *tt5* mutant), simply to avoid collecting novel alleles of genes that have already been identified through mutation analysis. Similarly, an investigator with an interest in damage tolerance might want to use, as a progenitor, a previously isolated mutant that is defective in repair. A single pyrimidine dimer can be processed via a number of different repair and tolerance pathways, and each of these pathways makes its own contribution to UV resistance. In a wild-type background, the fraction of the plant's overall UV-resistance provided, for example, by a dimer bypass mechanism might be very small when compared to that contributed by photoreactivation (Fig. 3). However, in a mutant defective in both excision repair and photoreactivation, dimer bypass might be entirely responsible for what little UV-resistance remains. For this

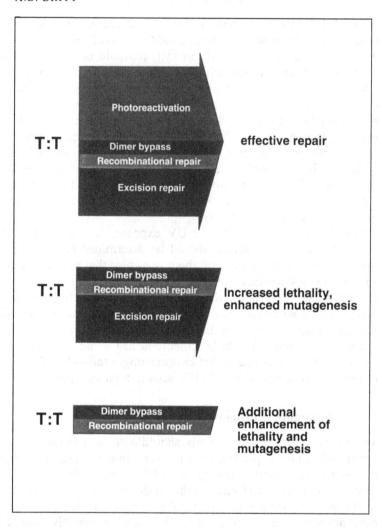

Fig. 3. A dimer may be processed via a variety of repair or tolerance pathways. The UV-sensitive phenotype of a mutant defective in any one pathway will depend on the presence or absence of alternative repair or tolerance mechanisms. The contributions of minor damage tolerance pathways to UV-resistance will be masked by more efficient repair mechanisms.

reason, an investigator interested in identifying genes required for dimer bypass would *start* with a repair defective mutant, treat that with a mutagenic agent, and screen the resulting M2 population for additional enhancement of sensitivity to UV exposure.

Genetic analysis of repair mutants

A large-scale mutant hunt will yield a number of UV-sensitive mutants. Generally only a manageable number of mutant lines are further characterised; the investigator will choose the most UV-sensitive mutants for further analysis. All of the UV-sensitive mutants isolated so far have been recessive to wild-type. All mutant lines should be backcrossed to their progenitor lines to help reduce the number of secondary mutations (which will be very large in a typical EMS-mutagenised population). Unfortunately, secondary mutations mapping nearby will cosegregate with the mutation of interest, and it is always possible that the overall mutant phenotype is due to the combined effects of mutations in two separate genes. However, it is possible to look for UV-resistant revertants of the mutant line, and then determine whether all aspects of the mutant phenotype have reverted to wild type along with the UV-sensitive phenotype. Co-reversion of all aspects of the mutant phenotype suggest that all of these were due to the effects of a single mutation. The search for revertants is relatively painless, as large numbers of seedlings can easily be subjected to killing doses of UV, thereby selecting (rather than screening) for the few revertant survivors.

Mutants should then be crossed to each other for sorting into complementation groups; mutants which, when crossed, yield UV-resistant F1 progeny, are defective in different genes. Mutants defective in the same gene usually produce mutant progeny when hybridised. It is sometimes possible to determine via epistasis analysis (the construction of double homozygous mutant lines) whether the genes produce proteins that participate in the same repair pathway. Theoretically, mutants carrying a single defect in a repair pathway should be no more UV-sensitive than double mutants defective in two different steps of that pathway; in either case, the pathway is defective. In contrast, double mutants defective in two independent pathways will display a synergistic effect that makes the double mutant much more sensitive to UV than either of the original mutants. However, this sort of analysis must be employed with caution when using EMS mutagenesis, as this mutagen produces single base changes, and often results in an incomplete knockout of gene function. In this case, two mutants defective in the same pathway might display a synergistic effect.

Phenotypic characterisation of UV-sensitive mutants

A UV-sensitive phenotype might result from any one of a variety of genetic defects. Mutations known to produce this phenotype are those affecting DNA repair (Britt et al., 1993) and the production of UV-absorbing compounds (Li et al., 1993). Other mutations that might possibly result in UV-sensitivity are defects rendering some other cellular process, such as photosynthesis, particularly susceptible to UV-induced damage, mutations that affect the ability of cells to recognise and degrade UV-damaged cellular components, or mutations resulting in the overproduction, or inappropriate compartmentalisation, of a UV-sensitising agent such as psoralen. Mutations affecting the plant's regulatory response to UV could also be isolated, provided the screen for sensitivity includes a preliminary UV-exposure that allows the plant to adapt to UV radiation.

The first step in the determination of the nature of the genetic defect should be an assay of the rate of induction of UV-induced DNA damage products. This assay indicates whether the mutant has a defect in the production of UV-absorbing agents. The damage should be induced with a short, intense burst of UV, to eliminate any possible influence of repair defects. It is not necessary for the total UV dose to be biologically relevant, as one is simply assaying, essentially, the UV-absorbance of the plant tissue. If the rate of induction of damage is identical in both the mutant and progenitor lines, the mutant must be defective in the processing, rather than the accumulation, of some type of UV-induced damage. This need not be damage to DNA.

Should it be that dimers are induced at a normal rate in the mutant line, the next logical step in characterisation would be the assay of rate of repair of UV-induced lesions in vivo. The rate of repair of 6–4 photoproducts and cyclobutane dimers in total cellular DNA can be determined via radioimmunoassay (Britt et al., 1993; Stapleton & Walbot, 1994); cyclobutane dimers can also be assayed using a lesion-specific glycosylase/endonuclease coupled with some assay for the generation of single-stranded nicks, including alkaline gradient centrifugation (Ganesan, Smith & van Zeeland, 1981), quantitative gel electrophoresis (Quaite, Sutherland & Sutherland., 1994a), or, if sequence-specificity is desired, quantitative PCR (Cannon, Hendrick & Heinhorst, 1995) or Southern blot (Chen et al., 1996).

Because we know so little about the mechanism and specificity of repair pathways for various DNA damage products in plants, the nature of the biochemical defect in each mutant is most convincingly established by the direct assay of repair in vivo, rather than through guesswork based on the characterisation of sensitivity to various DNA damaging agents. However, the demonstration of sensitivity to agents that do not produce pyrimidine dimers obviously proves that the repair, processing or tolerance defect is not specific for pyrimidine dimers.

If the repair of dimers in total cellular DNA is defective, the mutant must be defective in the production of an enzyme or cofactor required for this process, and represents a locus involved in the synthesis or positive regulation of a repair factor. If there is no measurable defect in the overall rate of repair, it might be necessary to investigate the rate of repair of specific subclasses of DNA. We have found no detectable repair (either light-dependent or -independent) of CPDs in the organellar DNAs in young *Arabidopsis* seedlings (Chen *et al.*, 1996), but it is possible that 6–4 photoproducts might be repaired in the organellar genomes, or that repair is more efficient in more mature plants. The transcribed strands of nuclear genes are known to be repaired more efficiently than total cellular DNA in microbes and mammals, and the loss of strand-specific repair results in UV-sensitivity (Hanawalt, 1994). It is possible that plants also possess strand-specific repair. Finally, UV-sensitive mutants that fail to demonstrate a defect in either the production of UV-absorbing agents or the repair of pyrimidine dimers must be defective in the processing or repair of other UV-induced damage products.

Application of DNA repair mutants to the study of UV-resistance

Solar UV has been shown to have a subtle but significant effect on plant growth. These effects include the enhanced expression of pigments, leaf curling, and shortening of internode length (see chapters by Bornman, by Rozema *et al.* and by Björn *et al.*, this volume), and also recently reviewed in Tevini (1994) and Jordan (1996). One of the major thrusts in UV research is the attempt to identify the cause of this change in growth. Is this response the direct effect of UV-induced damage to some cellular component(s), an indirect response to stress, or a programmed, developmental response that optimises plant growth and minimises damage? If this is, even in part, a stress-related response, what is the biochemical basis of this stress? If this is a programmed response to UV-B, what is the UV-B receptor?

The mere fact that mutants defective in repair display a UV-sensitive phenotype does not permit us to conclude that DNA damage is the primary effector of UV-induced inhibition of the growth of *wild-type* plants. It is entirely possible that the DNA repair pathways do their job very effectively, and that UV-induced stress is due to damage to some other cellular component. DNA repair mutants can tell us, however, whether DNA repair pathways are an important component of the battery of plant mechanisms for resistance to UV. In order to determine whether the gene involved represents a significant UV resistance mechanism under natural conditions, it is, of course, critical that the growth of the mutant be characterised under natural conditions. A relatively inexpensive test would involve a comparison of the growth parameters of the mutant and wild-type strains, in the field, under cellulose acetate (which transmits UV-B) vs Mylar polyester (which does not transmit UV-B). Because a significant number of dimers are induced by the UV-A component of solar radiation (Quaite, Sutherland & Sutherland, 1992), one might want to attempt to filter out these wavelengths also. If the mutant displays a *UV-specific* inhibition of growth (when compared to their wild-type progenitors) we can logically conclude that it represents a defect in a gene that plays a significant role in plant resistance to UV. Caution should be employed, however, in extrapolating these results to other plant species. Plants adapted for survival in high light conditions (tropical or summer-season species) may have capacities for UV-repair and tolerance that differ in both efficiency and mechanism from those observed in *Arabidopsis*. Again, the study of the genetics of UV resistance in a model system that is unrelated to *Arabidopsis* would be very helpful. Biochemical studies of repair have recently been performed in soybean (Cannon *et al*, 1995), maize (Stapleton & Walbot, 1994), alfalfa (Quaite *et al*., 1994*b*), and wheat (Taylor *et al*., 1996); genetic studies should also be pursued.

DNA repair mutants, as well as some other classes of UV-sensitive mutants, may also enable us to distinguish between effects on plant growth that are due to a programmed response to environmental influences and effects that are due to a response to generalised stress. If a plant response to UV were mediated *entirely* through a UV-B receptor, then a DNA repair mutant, which accumulates DNA damage at a higher rate than its wild-type progenitor, should have an *unaltered* response to solar UV-B. On the other hand, if the response is actually mediated through the accumulation of pyrimidine dimers, the response will, of course, be exhibited in proportion to accumulation of DNA damage, and the mutant strain will exhibit the entire range of UV-B induced growth effects at a lower UV-B exposure than its wild-type progenitor. It is also possible that some aspects of plant response are 'programmed' while others are the result of DNA damage. In this case, repair mutants will enable us to distinguish between these two classes of UV-induced effects. Thus DNA repair mutants will enable us to clear up some very interesting and important issues in the field of plant response to UV-B.

References

Britt, A.B. (1996). DNA damage and repair in plants. *Annual Review of Plant Physiology and Plant Molecular Biology*, **47**, 75–100.

Britt, A.B., Chen, J.-J., Wykoff, D. & Mitchell, D. (1993). A UV-sensitive mutant of *Arabidopsis* defective in the repair of pyrimidine-pyrimidinone (6–4) dimers. *Science*, **261**, 1571–4.

Caldwell, M.M., Flint, S.D. & Searles, P.S. (1994). Spectral balance and UV-B sensitivity of soybean: a field experiment. *Plant Cell and Environment*, **17**, 267–76.

Cannon, G., Hendrick, L. & Heinhorst, S. (1995). Repair mechanisms of UV-induced DNA damage in soybean chloroplasts. *Plant Molecular Biology*, **29**, 1267–77.

Chen, J-J., Jiang, C-Z. & Britt, A.B. (1996). Little or no repair of cyclobutane pyrimidine dimers is observed in the organellar genomes of the young *Arabidopsis* seedling. *Plant Physiology*, **111**, 19–25.

Errampalli, D., Patton, D., Castle, L., Mickelson, L., Hansen, K., Schnall, J., Feldmann, K. & Meinke, D. (1991). Embryonic lethals and T-DNA insertional mutagenesis in *Arabidopsis*. *Plant Cell*, **3**, 149–57.

Ganesan, A.K., Smith, C.A. & van Zeeland, A.A. (1981). Measurement of the pyrimidine dimer content of DNA in permeabilized bacterial or mammalian cells with endonuclease V of bacteriophage T4. *DNA Repair*, Marcel Dekker, Inc., New York, pp. 89–98.

Hanawalt, P.C. (1994). Transcription-coupled repair and human disease. *Science*, **266**, 1957–8.

Harle, J.R. (1972). A revision of mutation breeding procedures in *Arabidopsis* based on a fresh analysis of the mutant sector problem. *Canadian Journal of Genetics and Cytology*, **14**, 559–72.

Harlow, G.R., Jenkins, M.E., Pittalwala, T.S. & Mount, D.W. (1994). Isolation of *uvh1*, an *Arabidopsis* mutant hypersensitive to ultraviolet light and ionizing radiation. *Plant Cell*, **6**, 227–35.

Hauge, B., Hanley, S., Cartinhour, S., Cherry, J. & Goodman, H. (1993). An integrated genetic/RFLP map of the *Arabidopsis thaliana* genome. *Plant Journal*, **3**, 745–54.

Jarvis, P., Lister, C., Szabo, V. & Dean, C. (1994). Integration of CAPS markers into the RFLP map generated using recombinant inbred lines of *Arabidopsis thaliana*. *Plant Molecular Biology*, **24**, 685–7.

Jenkins, M.E., Harlow, G.R., Liu, Z., Shotwell, M.A., Ma, J. &

Mount, D.W. (1995). Radiation-sensitive mutants of *Arabidopsis thaliana*. *Genetics*, **140**, 725–32.

Jiang, C-Z., Yee, J., Mitchell, D.L. & Britt, A.B. (1997). Photorepair mutants of *Arabidopsis*. *Proceedings of the National Academy of Sciences, USA*, (in press).

Jordan, B. (1996). The effects of ultraviolet-B radiation on plants: a molecular perspective. *Advances in Botanical Research*, **22**, 97–162.

Landry, L.G., Stapleton, A.E., Lim, J., Hoffman, P., Hays, J.B., Walbot, V. & Last, R.L. (1997). An *Arabidopsis* photolyase mutant is hypersensitive to UV-B radiation. *Proceedings of the National Academy of Sciences, USA*, **94**, 328–32.

Li, S.L. & Redei, G.P. (1969). Estimation of mutation rate in autogamous diploids. *Radiation Botany*, **9**, 125–31.

Li, J., Ou-Lee, T.-M., Raba, R., Amundson, R.G. & Last, R.L. (1993). *Arabidopsis* flavonoid mutants are hypersensitive to UV-B radiation. *Plant Cell*, **5**, 171–9.

Long, D., Swinburne, J., Martin, M., Wilson, K., Sundberg, E., Lee, K. & Coupland, G. (1993). Analysis of the frequency of inheritance of transposed Ds elements in *Arabidopsis* after activation by a CaMV 35S promoter fusion to the Ac transposase gene. *Molecular and General Genetics*, **241**, 627–36.

Middleton, E. M. & Teramura, A. H. (1994). Understanding photosynthesis, pigment and growth responses induced by UV-B and UV-A irradiances. *Photochemistry and Photobiology*, **60**, 38–45.

Mitchell, D.L. & Nairn, R.S. (1989). The biology of the (6–4) photoproduct. *Photochemistry and Photobiology*, **49**, 805–19.

Mitchell, D.L., Vaughan, J.E. & Nairn, R.S. (1989). Inhibition of transient gene expression in Chinese hamster ovary cells by cyclobutane dimers and (6–4) photoproducts in transfected ultraviolet-irradiated plasmid DNA. *Plasmid*, **21**, 21–30.

Mitchell, D.L., Jen, J. & Cleaver, J.E. (1991). Relative induction of cyclobutane dimers and cytosine photohydrates in DNA irradiated *in vitro* and *in vivo* with ultraviolet-C and ultraviolet-B light. *Photochemistry and Photobiology*, **54**, 741–6.

Protic-Sabljic, M. & Kraemer, K.H. (1986). One pyrimidine dimer inactivates expression of a transfected gene in *Xeroderma pigmentosum* cells. *Proceedings of the National Academy of Sciences, USA*, **82**, 6622–6.

Quaite, F.E., Sutherland, B. M. & Sutherland, J.C. (1992). Action spectrum for DNA damage in alfalfa lowers predicted impact of ozone depletion. *Nature*, **358**, 576–8.

Quaite, F.E., Sutherland, J.C. & Sutherland, B.M. (1994*a*). Isolation of high molecular weight plant DNA for DNA damage quantitation: relative effects of solar 297 nm UV-B and 365 nm radiation. *Plant Molecular Biology*, **24**, 475–83.

Quaite, F.E., Takayanagi, S., Ruffini, J., Sutherland, J.C. & Sutherland, B.M. (1994*b*). DNA damage levels determine cyclobutane pyrimidine dimer repair mechanisms in alfalfa seedlings. *Plant Cell*, **6**, 1635–41.

Rosenstein, B.S. & Mitchell, D.L. (1987). Action spectra for the induction of pyrimidine(6–4)pyrimidone photoproducts and cyclobutane pyrimidine dimers in normal human skin fibroblasts. *Photochemistry and Photobiology*, **45**, 775–80.

Somerville, C.R. & Estelle, M.A. (1986). The mutants of *Arabidopsis*. *Trends in Genetics*, **16**, 89–93.

Stapleton, A.E. & Walbot, V. (1994). Flavonoids can protect maize DNA from the induction of ultraviolet radiation damage. *Plant Physiology*, **105**, 881–9.

Takayanagi, S., Trunk, J.G., Sutherland, J.C. & Sutherland, B.M. (1994). Alfalfa seedlings grown outdoors are more resistant to UV-induced damage than plants grown in a UV-free environmental chamber. *Photochemistry and Photobiology*, **60**, 363–7.

Taylor, R.M., Nikaido, O., Jordan, B.R., Rosamond, J., Bray, C. M. & Tobin, A.K. (1996). UV-B-induced DNA lesions and their removal in wheat (*Triticum aestivum* L.) leaves. *Plant Cell and Environment*, **19**, 171–81.

Tevini, M. (1994). UV-B effects on terrestrial plants and aquatic organisms. *Progress in Botany*, **55**, 174–90.

N.R. BAKER, S. NOGUÉS and D.J. ALLEN

Photosynthesis and photoinhibition

Introduction

Photosynthesis and photosynthetic productivity in many plant species, although by no means all, can be inhibited by increased exposure to UV-B radiation (Caldwell, Teramura & Tevini, 1989; Tevini & Teramura, 1989; Teramura, Ziska & Sztein, 1991; Tevini, Braun & Fieser, 1991; Middleton & Teramura, 1993; Musil, 1995). To date, there is no consensus for the mechanistic basis of UV-B-induced inhibition of CO_2 assimilation in mature leaves. Decreases in Rubisco activity and stomatal conductance have been implicated as factors limiting CO_2 assimilation in leaves exposed to elevated levels of UV-B. Prolonged exposure to elevated levels of UV-B has been demonstrated to result in decreases in both Rubisco activity and content (Vu, Allen & Garrard, 1984; Strid, Chow & Anderson, 1990; Jordan *et al.*, 1992; He *et al.*, 1993), and is accompanied by large decreases in the mRNA transcripts of both the large and small subunits of Rubisco (Jordan *et al.*, 1992). Such decreases in Rubisco are consistent with the observed decrease in the leaf carboxylation efficiency, determined from the initial slope of the response of CO_2 assimilation to increasing CO_2 concentration, when leaves are given supplemental UV-B radiation (Ziska & Teramura, 1992). Exposure to UV-B can also modify the rates of stomatal opening and closing, and reduce the rate of leaf transpiration (Tevini & Teramura, 1989; Middleton & Teramura, 1993; Day & Vogelmann, 1995). However, numerous studies have demonstrated that photosystem II (PSII) is the most sensitive component of the thylakoid membrane photosynthetic apparatus to increased exposure to UV-B (Noorudeen & Kulandaivelu, 1982; Iwanzik *et al.*, 1983; Renger *et al.*, 1989; Strid, Chow & Anderson, 1990; Melis, Nemson & Harrison, 1992). Consequently PSII damage has often been implicated as the major potential limitation to photosynthesis in UV-B treated leaves (Bornman, 1989; Teramura & Sullivan, 1994; Fiscus & Booker, 1995), as is the case in the photoinhibition of photosynthesis by photosynthetically active radiation (380–700 nm) (Baker & Bowyer, 1994). This chapter considers the role of PSII photoinhibition in the depression of CO_2 assimilation induced by UV-B exposure of leaves.

Photoinhibition of Photosystem II

A schematic model for the PSII complex is shown in Fig. 1. The multi-subunit complex consists of at least 20 proteins, ranging in size from 4

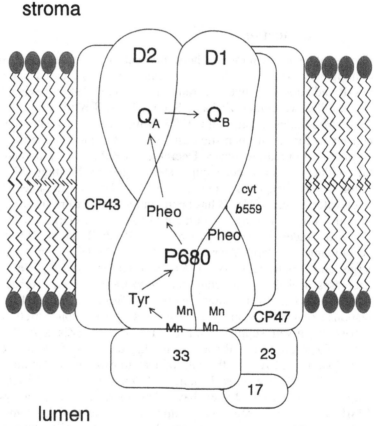

Fig. 1. Schematic model for photosystem II. D1 and D2 are the core proteins of the reaction centre that bind P680, Q_A, Q_B and two pheo-phytin molecules (Pheo). Tyr is a tyrosine residue on the D1 protein that transfers electrons from the manganese-containing complex (4 Mn) to P680. CP43 and CP47 are antennae pigment complexes bind-ing chlorophyll a molecules. Cyt b559 is cytochrome b559. Proteins labelled 17, 23 and 33 (designating molecular mass in kDa) are extrin-sic membrane proteins and are involved in water oxidation.

to 50 kDa, that are encoded by genes located on both the chloroplast and nuclear genomes (for review, see Vermaas, 1993). The largest PSII proteins are the chlorophyll-binding primary antenna proteins CP47 and CP43 (*c.* 47 and 43 kDa), a 33 kDa manganese-stabilising protein and another two chlorophyll-binding proteins of 32 and 34 kDa, known as D1 and D2. Cytochrome b559 is also a component of PSII preparations. The D1 and D2 proteins bind a chlorophyll *a* molecule which when excited will transfer an electron to the adjacent pheophytin. This special chlorophyll is called P680 after its absorption maximum at 680 nm. The D1/D2 heterodimer also binds pheophytin, which is the immediate, metastable electron acceptor from P680, and the two primary, stable quinone electron acceptors, Q_A and Q_B. The immediate electron donor to P680 is a tyrosine residue, Tyr_Z, of the D1 protein.

On transfer of excitation energy from antennae chlorophylls to P680, primary charge separation occurs as the excited P680 transfers an electron to pheophytin. The electron is then transferred from pheophytin to the primary quinone acceptor, Q_A, and then to the secondary quinone acceptor, Q_B. The semiquinone, Q_B^-, is then further reduced by a second turnover of the reaction centre, producing the doubly reduced Q_B^{2-}, which becomes protonated to plastoquinol and dissociates from the reaction centre. Oxidised P680 (P680+) has an extremely high oxidising potential and will oxidise the neighbouring Tyr_Z residue on the D1 protein, which will then be reduced by electron transfer from water via the Mn cluster.

Photoinactivation of the PSII reaction centre can occur by two independent mechanisms, associated with the acceptor and donor sides of PSII respectively, that both result in inhibition of electron transfer through PSII and subsequent degradation of the D1 protein (Styring & Jegerschöld, 1994; Telfer & Barber, 1994). Acceptor side inhibition will occur under high light conditions when the plastoquinone pool is fully reduced, and consequently there is a lack of oxidised plastoquinone to bind to the Q_B site on the D1 protein. In this state Q_A will become doubly reduced on a second turnover of the reaction centre to form Q_A^{2-}, as Q_A^- cannot transfer an electron to Q_B. Q_A^{2-} then becomes protonated to form Q_AH_2 and is released from the Q_A^- binding site on the D1 protein. With the Q_A site vacated, excitation of P680 will result in the formation of the radical pair, P680⁺Pheophytin⁻. Recombination of these radicals will result in the formation of the triplet state of P680, which will react with oxygen to form singlet oxygen. Singlet oxygen is potentially damaging to protein, and is thought to react with the D1 protein thus triggering the degradation of the D1 (Aro, Virgin & Andersson, 1993; Styring & Jegerschöld, 1994; Telfer & Barber, 1994).

Acceptor-side photoinactivation would be expected to occur in leaves when the rate of consumption of the products of electron transport is decreased, as would occur when CO_2 assimilation is restricted, and the processes which quench excitation within the PSII antennae do not have the capacity to dissipate the excess excitation.

Donor-side photoinhibition of PSII will occur when water oxidation is inhibited and the highly reactive $P680^+$ and Tyr_Z^+ are formed. $P680^+$ will oxidise neighbouring molecules. Oxidation of accessory chlorophylls and β-carotene and degradation of D1 have been found to occur under conditions favouring $P680^+$ formation (Telfer & Barber, 1994). Loss of donor-side electron transport and consequent photodamage to the D1 protein has been linked to the generation of a large H^+ electrochemical potential across the thylakoid and the possible associated release of Ca^{2+} from the water-oxidising complex (Ohad et al., 1994). Although donor-side photoinhibition has been frequently observed in in vitro experiments, it has not yet been demonstrated to be a feature of photoinhibition in leaves operating under physiological conditions.

Degradation of the D1 protein is a consequence of photoinactivation of PSII and is generally considered to result from protease activity (Andersson et al., 1994). At present the protease(s) responsible for D1 degradation has not been identified, but it is thought to be associated with the PSII core complex (Andersson et al., 1994). The mechanism by which photodamaged D1 is recognised from native D1 by the proteolytic enzyme is not yet known. The D1 protein is encoded in the chloroplast genome by a gene designated psbA, and its synthesis in the chloroplast is regulated by nuclear-encoded factors that are synthesised in the cytoplasm and imported into the chloroplast (Rochaix, 1992). Consequently, repair of photodamaged PSII complexes requires the synthesis of D1 in the stroma and insertion and integration of the newly synthesised proteins into the damaged PSII complexes.

Kinetics of inhibition of photosynthetic activities induced by UV-B

Analyses of the kinetics of change of a range of photosynthetic activities can be usefully used to probe the intrinsic factors limiting photosynthesis in leaves exposed to UV-B (Nogués & Baker, 1995). UV-B radiation was measured with a Macam SR9910-PC spectroradiometer and converted to a biologically effective dose using the Macam programme for Caldwell's (1971) plant action spectrum. Mature pea leaves were exposed to a UV-B dose of 11.8 μmol m^{-2} s^{-1} (between 280 and 320 nm; equivalent to a biologically effective irradiance of 0.74 W m^{-2}

and a daily biologically effective dosage of 40 kJ m^{-2} d^{-1}, which is about eight times greater than the level detected in midsummer in the UK) during daily photoperiods. After the first 12 h of exposure to UV-B, decreases of about 30% in the light-saturated rate of CO_2 assimilation (A_{sat}) occurred in the absence of any significant changes in the maximum quantum yield of PSII photochemistry, as monitored by Fv/Fm (Fig. 2). The decreases in A_{sat} continued progressively and were accompanied by similar decreases in the quantum efficiencies of CO_2 assimilation (ϕ_{CO_2}) and PSII photochemistry (ϕ_{PSII}) measured at the PPFD under which the leaves had been grown (450 μmol m^{-2} s^{-1}). However, over 93 hours from the onset of exposure to the UV-B only a small decrease in Fv/Fm was observed, which was recoverable within 30 hours of switching off the UV tubes (Fig. 2). The similar decreases in ϕ_{CO_2} and ϕ_{PSII} at a PPFD of 450 μmol m^{-2} s^{-1} in the absence of any similar decrease in Fv/Fm, and the lack of any significant recovery of A_{sat}, ϕ_{CO_2} and ϕ_{PSII} within 30 hours of removing the UV-B irradiation (Fig. 2), indicates that photodamage to PSII cannot account for the decrease in A_{sat} and ϕ_{CO_2}, and also suggests that the decrease in ϕ_{PSII} is attributable to a down-regulation of linear electron transport resulting from a decrease in the rate of operation of the Calvin cycle.

Increasing the UV-B dosage given to the pea leaves to 13.3 μmol m^{-2} s^{-1} (between 280 and 320 nm; equivalent to a biologically effective UV-B irradiance of 0.93 W m^{-2}) induced large changes in Fv/Fm that paralleled the decreases in A_{sat}, ϕ_{CO_2} and ϕ_{PSII} during the 80 hours following the initial exposure to UV-B (Fig. 3). Fv/Fm decreased to about 0.5, and then on removal of the UV-B irradiation recovered back to about 0.7 over 40 hours, whereas recovery of A_{sat}, ϕ_{CO_2} and ϕ_{PSII} was considerably less (Fig. 3). The UV-B-induced decrease in Fv/Fm and its incomplete recovery over 40 hours suggests that the higher dosage of UV-B produces damage to PSII reaction centres. This supposition was confirmed by demonstrating that treatment with 11.8 μmol m^{-2} s^{-1} UV-B did not result in any decrease in the ability of thylakoids isolated from the leaves to bind atrazine, whereas exposure to 13.3 μmol m^{-2} s^{-1} UV-B produced a 40% decrease in atrazine-binding to thylakoids, which was recoverable 40 hours after removing the UV-B treatment (Fig. 4). The atrazine-binding capacity of isolated thylakoids is considered to be a quantitative measure of photoinhibitory damage to the plastoquinone-reductase site of PSII (Kyle, Ohad & Arntzen, 1984), and consequently the decrease observed in atrazine-binding at the higher UV-B dosage confirms the photodamage to PSII reaction centres.

The possibility that UV-B-induced decreases in CO_2 assimilation may

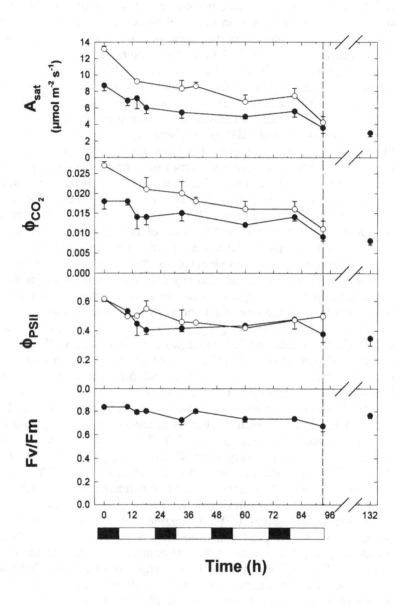

Time (h)

be related to differential effects on photorespiratory and carbon assimilatory metabolism was examined by reducing the O_2 concentration of the atmosphere in which the photosynthetic parameters were measured from 21 to 2% in order to inhibit photorespiration. In control leaves, inhibition of photorespiration resulted in significant increases in A_{sat} and ϕ_{CO_2} but not in ϕ_{PSII} (Figs. 2 and 3), thus demonstrating that the removal of photorespiratory metabolism as a sink for the products of linear electron transport, although increasing the rate of CO_2 assimilation, has no effect on the flux of electrons through PSII under both light-saturated conditions (PPFD 1500 μmol m^{-2} s^{-1}) and at the growth PPFD (450 μmol m^{-2} s^{-1}). The stimulation of CO_2 assimilation by inhibition of photorespiration appeared to be maintained throughout the UV-B treatments (Figs. 2 and 3). These data indicate that effects on photorespiration cannot account for the UV-induced decreases in CO_2 assimilation.

A/c_i responses to UV-B

Analyses of the response of net carbon assimilation (A) to intercellular CO_2 concentration (c_i) allows separation of the relative limitations imposed by stomata, carboxylation efficiency and capacity for regeneration of ribulose 1,5-bisphosphate (RuBP) on leaf photosynthesis (von Caemmerer & Farquhar, 1981). At low c_i values, where A is directly related to c_i, A is limited by the rate of RuBP carboxylation and the *in vivo* maximum rate of RuBP carboxylation by Rubisco ($V_{c,max}$) can be

---

Fig. 2. Changes in the light-saturated rate of CO_2 assimilation (A_{sat}), the quantum yields of CO_2 assimilation (ϕ_{CO_2}) and PSII electron transport (ϕ_{PSII}) of pea leaves at a PPFD of 450 μmol m^{-2} s^{-1}, and Fv/Fm throughout a UV-B treatment of 11.8 μmol m^{-2} s^{-1} (equivalent to a biologically effective UV irradiance of 0.74 W m^{-2} between 280 and 320 nm). Solid (●) symbols indicate that measurements were made on leaves maintained in a normal gas atmosphere (21% O_2, 380 μmol mol^{-1} CO_2); open (○) symbols are for measurements made on leaves maintained in an atmosphere containing 2% O_2 and 380 μmol mol^{-1} CO_2 in *c.* 98% N_2. The open and solid boxes immediately under the abscissa indicate the photoperiods and dark periods, respectively, to which the leaves were exposed; leaves were only exposed to UV during the photoperiods. The vertical dashed line indicates the time at which UV sources were switched off. Data are the means of four replicates and the standard errors are shown when larger than the symbols, and are taken from Nogués and Baker (1995).

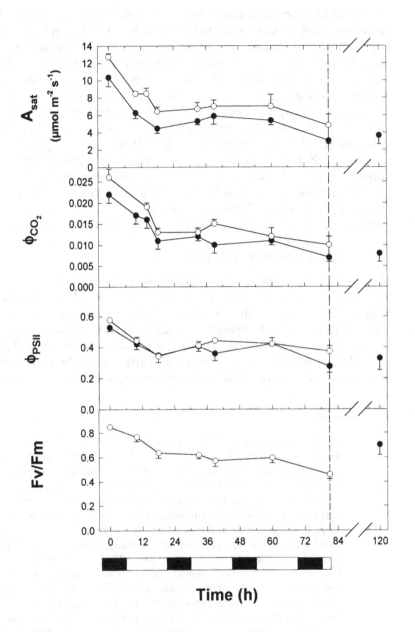

Time (h)

calculated from this relationship (see Fig. 5; McMurtrie & Wang, 1993). With increasing c_i, A approaches a plateau (see Fig. 5) and is limited by the rate of RuBP regeneration by the Calvin cycle, generally assumed to be determined by the rate of supply of reductants and ATP from electron transport, and consequently the maximum rate of non-cyclic electron transport (J_{max}) contributing to CO_2 assimilation can be estimated (McMurtrie & Wang, 1993). Stomatal limitation, the proportion of photosynthesis that is limited by stomatal conductance, is given by $(A_o - A_{sat})/A_o$, where A_o is A at a c_i of 360 μmol mol^{-1} (see Fig. 5; Farquhar & Sharkey, 1982).

Oilseed rape leaves exposed to a UV-B dosage of 10.1 μmol m^{-2} s^{-1} (equivalent to a biologically effective irradiance of 0.63 W m^{-2}) during the daily photoperiod exhibited a continuous decrease in A_{sat} over 5 days (Fig. 6). The decrease in A_{sat} was accompanied by large decreases in both $V_{c,max}$ and J_{max} in the absence of any major decrease in Fv/Fm or changes in stomatal limitation (Fig. 6). Consequently, exposure to UV-B results primarily in an inability to regenerate RuBP and a reduction in carboxylation efficiency. RuBP regeneration could be limited either by an inability to supply reductants and ATP from electron transport or an inactivation or loss of Calvin cycle enzymes other than Rubisco. It is clear, however, that any inability of the thylakoids to supply reductants and ATP cannot be attributable to a reduction in the ability to perform PSII photochemistry, since depressions in J_{max}

Fig. 3. Changes in the light-saturated rate of CO_2 assimilation (A_{sat}), the quantum yields of CO_2 assimilation (ϕ_{CO_2}) and PSII electron transport (ϕ_{PSII}) of pea leaves at a PPFD of 450 μmol m^{-2} s^{-1}, and Fv/Fm throughout a UV-B treatment of 13.3 μmol m^{-2} s^{-1} (equivalent to a biologically effective UV irradiance of 0.93 W m^{-2} between 280 and 320 nm). Solid (●) symbols indicate that measurements were made on leaves maintained in a normal gas atmosphere (21% O_2, 380 μmol mol-1 CO_2); open (○) symbols are for measurements made on leaves maintained in an atmosphere containing 2% O_2 and 380 μmol mol^{-1} CO_2 in c. 98% N_2. The open and solid boxes immediately under the abscissa indicate the photoperiods and dark periods, respectively, to which the leaves were exposed; leaves were only exposed to UV-B during the photoperiods. The vertical dashed line indicates the time at which UV sources were switched off. Data are the means of four replicates and the standard errors are shown when larger than the symbols, and are taken from Nogués and Baker (1995).

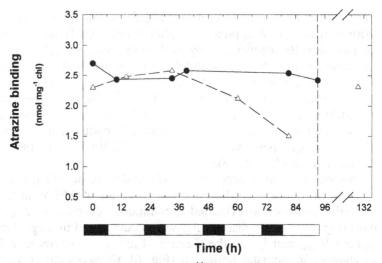

Fig. 4. Changes in the binding of [^{14}C]atrazine to thylakoids isolated from pea leaves throughout the treatments with 11.8 µmol m^{-2} s^{-1} (●) and 13.3 µmol m^{-2} s^{-1} (△) UV-B (see Figs. 2 and 3, respectively). The open and solid boxes immediately under the abscissa indicate the photoperiods and dark periods, respectively, to which the leaves were exposed; leaves were only exposed to UV-B during the photoperiods. The vertical dashed line indicates the time at which UV sources were switched off. Data are the means of four replicates, and are taken from Nogués and Baker (1995); the standard errors of the means were less than 10% of the mean values in all cases.

precede any major photodamage to PSII reaction centres. Decreases in $V_{c,max}$ would result from loss or inactivation of Rubisco.

Rubisco content and activity

After 4 days of the UV-B treatment the Rubisco content of the oilseed rape leaves was *c*. 45% of that of control leaves (Table 1), and was accompanied by 40% and 45% decreases in $V_{c,max}$ and A_{sat}, respectively (Fig. 6). Clearly, the decreases in Rubisco content are consistent with the decreases in these photosynthetic parameters. A 62% decrease was observed in the initial activity of the extracted Rubisco per leaf area from UV-B treated leaves compared to controls, with a 53% decrease in the activity of fully activated Rubisco (Table 1). No significant differences were observed in the specific activity of Rubisco extracted from the UV-B treated leaves and controls (Table 1), consequently the decreases in Rubisco activity can be attributed to UV-B inducing a loss

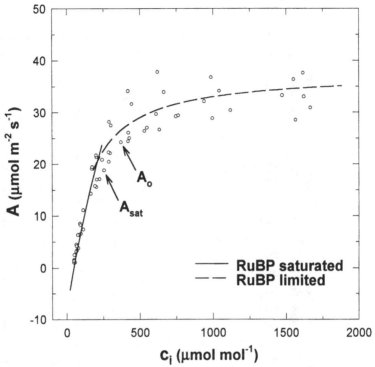

Fig. 5. Typical A/c_i response showing experimental data and the result of this fitted to the McMurtrie & Wang (1993) model. On the initial slope, A is limited by the rate of RuBP carboxylation from which the *in vivo* maximum rate of RuBP carboxylation by Rubisco ($V_{c,max}$) can be calculated. As the A/c_i relationship approaches a plateau, assimilation is limited by the rate of RuBP regeneration by the Calvin cycle, which allows estimation of the maximum rate of non-cyclic electron transport (J_{max}) contributing to CO_2 assimilation. A_o is the A at a c_i of 360 μmol mol^{-1}, and A_{sat} is A at the ambient atmospheric CO_2 concentration of 360 μmol mol^{-1}; stomatal limitation is given by $(A_o - A_{sat})/A_o$.

of the protein (see chapter by Mackerness *et al.*). No data are available to date on the effects of UV-B treatment on other enzymes of the Calvin cycle. If UV-B treatment is producing similar decreases to those observed for Rubisco in the content of other key Calvin cycle enzymes, then this could account for the observed decreases in J_{max}. This being the case, then there would be no need to implicate any UV-B damage to thylakoid membranes in the decrease in J_{max} since the decreased rate

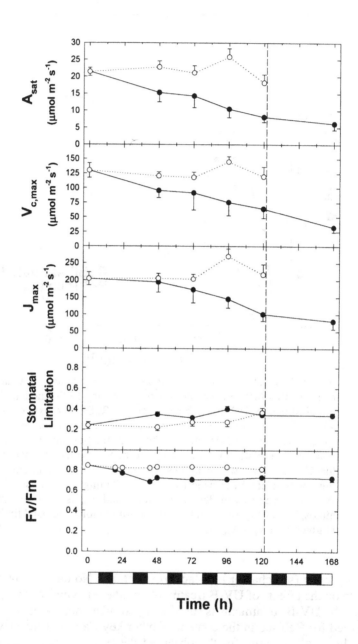

Table 1. *Changes in Rubisco content and activity in mature leaves of oilseed rape (*Brassica napus *cv. Apex) after 4 days of exposure to UV-B dosage of 10.1 μmol m^{-2} s^{-1} (equivalent to a biologically irradiance of 0.63 W m^{-2}) during the daily photoperiod*

Parameter	Treatment	
	UV-B	Control
Rubisco content per leaf area (g m^{-2})	3.66 ± 0.29	6.63 ± 0.27
Initial Rubisco activity (μmol m^{-2} s^{-1})	14.3 ± 0.8	38.2 ± 3.3
Fully-activated rubisco activity (μmol m^{-2} s^{-1})	19.0 ± 0.8	40.2 ± 2.5
Rubisco specific activity (μmol s^{-1} g^{-1})	6.16 ± 0.64	5.30 ± 0.41

Data are the means of five replicates and standard errors are given.

of regeneration of RuBP in UV-B treated leaves could be explained by a decrease in the ability to turnover the Calvin cycle due to reduced enzyme activities associated with component reactions.

Conclusions

As has been frequently reported in the literature, exposure of leaves to elevated levels of UV-B can result in photoinhibition and photodamage to PSII reaction centres. However, analyses of the changes in a range of photosynthetic parameters on exposure of leaves to elevated doses of UV-B have produced compelling evidence that such photoinhibition and photodamage to PSII occurs only after inhibition of CO_2 assimilation characteristics is observed. Consequently, photodamage to

---

Fig. 6. Changes in the light-saturated rate of CO_2 assimilation (A_{sat}), the *in vivo* maximum rate of RuBP carboxylation by Rubisco ($V_{c,max}$) and the maximum rate of non-cyclic electron transport (J_{max}) contributing to CO_2 assimilation of oilseed rape leaves throughout a UV-B treatment of 10.1 μmol m^{-2} s^{-1} (equivalent to a biologically effective UV irradiance between 280 and 320 nm of 0.63 W m^{-2}); parameters shown by ●. Values of these parameters for control leaves not exposed to UV-B are shown by ○. The vertical dashed line indicates the time at which UV sources were switched off. Data are the means of six to eight replicates and the standard errors are shown when larger than the symbols.

PSII can be ruled out as the primary detrimental effect of UV-B on photosynthesis in leaves. Although the primary target site(s) in the photosynthetic apparatus for UV-B has not been resolved to date, it would seem to be associated with factors limiting $V_{c,max}$ and J_{max}. From the data presented here and other literature (e.g. Vu, Allen & Garrard, 1984; Strid, Chow & Anderson, 1990; Jordan *et al.*, 1992; He *et al.*, 1993), loss of Rubisco activity is a prime candidate for the UV-B-induced depressions in $V_{c,max}$, but this cannot account for the depression in J_{max}. An attractive unifying hypothesis to explain simultaneous decreases in $V_{c,max}$ and J_{max} is that elevated levels of UV-B result in destruction of a range of key soluble chloroplast enzymes; the loss of Rubisco would result in a decrease in carboxylation efficiency and account for decreases in $V_{c,max}$, whilst decreases in other Calvin cycle enzyme activities would result in a decrease in the rate of regeneration of RuBP and result in a decrease in J_{max}. Further studies are required to test this hypothesis.

Acknowledgements

We are grateful to I.F. McKee for assistance with estimations of Rubisco activities, and to P.K. Farage and S.P. Long for discussions on A/c_i relationships.

References

Andersson, B., Ponticos, M., Barber, J., Koivuniemi, A., Aro, E.-M, Hagman, Å., Salter, A.H., Dan-Hui, Y. & Lindahl, M. (1994). In *Photoinhibition of Photosynthesis: From Molecular Mechanisms to the Field* (Baker, N.R. & Bowyer, J.R. eds), pp. 143–59. Bios Scientific Publishers, Oxford.

Aro, E.-M., Virgin, I. & Andersson, B. (1993). Photoinhibition of photosystem II. Inactivation, protein damage and turnover. *Biochimica et Biophysica Acta*, **1143**, 113–34.

Baker, N.R. & Bowyer, J.R., eds. (1994). *Photoinhibition of Photosynthesis: From Molecular Mechanisms to the Field*. Bios Scientific Publishers, Oxford.

Bornman, J.F. (1989). Target sites of UV-B radiation in photosynthesis of higher plants. *Journal of Photochemistry and Photobiology*, **4**, 145–58.

Bradbury, M. & Baker, N.R. (1986). The kinetics of photoinhibition of the photosynthetic apparatus in pea chloroplasts. *Plant Cell and Environment*, **9**, 289–97.

Caldwell, M.M. (1971). Solar ultraviolet radiation and the growth and development of higher plants. In *Photophysiology*, Vol. 6 (Giese, A.C., ed.), pp. 131–77. Academic Press, New York.

Caldwell, M.M., Teramura, A.H. & Tevini, M. (1989). The changing solar ultraviolet climate and the ecological consequences for higher plants. *Trends in Ecology and Evolution*, **4**, 363–7.

Day, T.A. & Vogelmann, T.C. (1995). Alterations in photosynthesis and pigment distribution in pea leaves following UV-B exposure. *Physiologia Plantarum*, **94**, 433–40.

Demmig-Adams, B. (1990). Carotenoids and photoprotection in plants: a role for the xanthophyll zeaxanthin. *Biochimica et Biophysica Acta*, **1020**, 1–24.

Farquhar, G.D. & Sharkey, T.D. (1982). Stomatal conductance and photosynthesis. *Annual Review of Plant Physiology*, **33**, 317–45.

Fiscus, E.L. & Booker, F.L. (1995). Is increased UV-B a threat to crop photosynthesis and productivity? *Photosynthesis Research*, **43**, 81–92.

Genty, B., Briantais, J.M. & Baker, N.R. (1989). The relationship between the quantum yield of photosynthetic electron transport and quenching of chlorophyll fluorescence. *Biochimica and Biophysica Acta*, **990**, 87–92.

Groom, Q.J. & Baker, N.R. (1992). Analysis of light-induced depressions of photosynthesis in leaves of a wheat crop during the winter. *Plant Physiology*, **100**, 1217–23.

He, J, Huang, L.-K., Chow, W.S., Whitecross, M.I. & Anderson, J.M. (1993). Effects of supplementary ultraviolet-B radiation on rice and pea plants. *Australian Journal of Plant Physiology*, **20**, 129–42.

Iwanzik, W., Tevini, M., Dohnt, G., Voss, M., Weiss, W., Graber, P. & Renger, G. (1983). Action of UV-B radiation on photosynthetic primary reactions in spinach chloroplasts. *Physiologia Plantarum*, **58**, 401–7.

Jordan, B.R., He, J., Chow, W.S. & Anderson, J.M. (1992). Changes in mRNA levels and polypeptide subunits of ribulose 1,5-bisphosphate carboxylase in response to supplementary ultraviolet-B radiation. *Plant Cell and Environment*, **15**, 91–8.

Kyle, D.J., Ohad, I. & Arntzen, C.J. (1984). Membrane protein damage and repair: selective loss of quinone–protein function in chloroplast membranes. *Proceedings of the National Academy of Sciences USA*, **81**, 4070–4.

McMurtrie, R.E. & Wang, Y.-P. (1993). Mathematical models of the photosynthetic responses of tree stands to rising CO_2 concentrations and temperatures. *Plant Cell and Environment*, **16**, 1–13.

Melis, A., Nemson, J.A. & Harrison, M.A. (1992). Damage to functional components and partial degradation of photosystem II reaction center proteins upon chloroplast exposure to ultraviolet-B radiation. *Biochimica et Biophysica Acta*, **1109**, 312–20.

Middleton, E.M. & Teramura, A.H. (1993). The role of flavonol glycosides and carotenoids in protecting soybean from UV-B damage. *Plant Physiology*, **103**, 741–52.

Musil, C.F. (1995). Differential effects of elevated ultraviolet-B radiation on the photochemical and reproductive performances of dicotyledonous and monocotyledonous arid-environment ephemerals. *Plant Cell and Environment*, **18**, 844–54.

Nogués, S. & Baker, N.R. (1995). Evaluation of the role of damage to photosystem II in the inhibition of CO_2 assimilation in pea leaves on exposure to UV-B. *Plant Cell and Environment*, **18**, 781–7.

Noorudeen, A.M. & Kulandaivelu, G. (1982). On the possible site of inhibition of photosynthetic electron transport by ultraviolet (UV-B) radiation. *Physiologia Plantarum*, **55**, 161–6.

Ohad, I., Keren, N., Zer, H., Gong, H., Mor, T.S., Gal, A., Tal, S. & Domovich, Y. (1994). Light-induced degradation of the photosystem II reaction centre D1 protein *in vivo*: an integrative approach. In *Photoinhibition of Photosynthesis: From Molecular Mechanisms to the Field* (Baker, N. R. & Bowyer, J. R., eds), pp. 161–77. Bios Scientific Publishers, Oxford.

Rackham, O. & Wilson, J. (1968). Integrating sphere for spectral measurements on leaves. In *The Measurement of Environmental Factors in Terrestrial Ecology*. British Ecological Society Symposium No. 8, 259–63.

Renger, G., Volker, M., Eckert, H.J., Fromme, R., Hohm-Veit, S. & Graber, P. (1989). On the mechanism of photosystem II deterioration by UV-B irradiation. *Photochemistry and Photobiology*, **49**, 97–105.

Rochaix, J.-D. (1992). Control of plastid gene expression in *Chlamydomonas reinhardtii*. In *Plant Gene Research: Cell Organelles* (Herrmann, R.G., ed.), pp. 249–74. Springer-Verlag, Vienna.

Rowland, F.S. (1989). Chlorofluorocarbons and the depletion of stratospheric ozone. *American Science*, **77**, 36–45.

Strid, A., Chow, W.S. & Anderson, J.M. (1990). Effects of supplementary ultraviolet-B radiation on photosynthesis in *Pisum sativum*. *Biochimica et Biophysica Acta*, **1020**, 260–68.

Styring, S. & Jegerschöld, C. (1994). Light-induced reactions impairing electron transfer through photosystem II. In *Photoinhibition of Photosynthesis: From Molecular Mechanisms to the Field* (Baker, N.R. & Bowyer, J.R., eds), pp. 51–73. Bios Scientific Publishers, Oxford.

Telfer, A. & Barber, J. (1994). Elucidating the molecular mechanisms of photoinhibition by studying isolated photosystem II reaction centres. In *Photoinhibition of Photosynthesis: From Molecular Mechanisms to the Field* (Baker, N.R. & Bowyer, J.R., eds), pp. 25–49. Bios Scientific Publishers, Oxford.

Teramura, A.H. & Sullivan, J.H. (1994). Effects of UV-B radiation on photosynthesis and growth of terrestrial plants. *Photosynthesis Research*, **39**, 463–73.

Teramura, A.H., Ziska, L.H. & Sztein, A.E. (1991). Changes in growth and photosynthetic capacity of rice with increased UV-B radiation. *Physiologia Plantarum*, **83**, 373–80.

Tevini, M. & Teramura, A.H. (1989). UV-B effects on terrestrial plants. *Photochemistry and Photobiology*, **50**, 479–87.

Tevini, M., Braun, J. & Fieser, G. (1991). The protective function of the epidermal layer of rye seedlings against ultraviolet-B radiation. *Photochemistry and Photobiology*, **53**, 329–33.

Tischer, W. & Strotmann, H. (1977). Relationship between inhibitor binding by chloroplasts and inhibition of photosynthetic electron transport. *Biochimica et Biophysica Acta*, **460**, 113–25.

Vermaas, W. (1993). Molecular-biological approaches to analyze photosystem II structure and function. *Annual Review of Plant Physiology and Plant Molecular Biology*, **44**, 457–81.

von Caemmerer, S. & Farquhar, G.D. (1981). Some relationships between the biochemistry of photosynthesis and the gas exchange of leaves. *Planta*, **153**, 376–87.

Vu, C.V., Allen, L.H. Jr & Garrard, L.A. (1984). Effects of UV-B radiation (280–320 nm) on ribulose-1,5-bisphosphate carboxylase in pea and soybean. *Environmental and Experimental Botany*, **24**, 131–43.

Ziska, L.H. & Teramura, A.H. (1992). CO_2 enhancement of growth and photosynthesis in rice (*Oryza sativa*): modification by increased ultraviolet-B radiation. *Plant Physiology*, **99**, 473–81.

S. A-H. MACKERNESS, B.R. JORDAN
and B. THOMAS

UV-B effects on the expression of genes encoding proteins involved in photosynthesis

Introduction

The depletion of ozone in the stratosphere has been of global environmental concern for two decades. Following the Montreal protocol, established to reduce emissions of man-made ozone-destroying chemicals into the atmosphere, the restriction in the use of chlorofluorocarbons (CFCs) in 1990 has noticeably slowed down the rise in chlorine in the stratosphere (Montzka *et al.*, 1996). However, the total ozone destroying power (including levels of bromine and other minor chemicals) in the atmosphere has only levelled and is not yet decreasing (Montzka *et al.*, 1996; Pyle, this volume). One of the consequences of the depletion of stratospheric ozone is the concomitant increase in the level of UV-B radiation from sunlight reaching the surface of the earth. Increases of 10% in UV-B levels in parts of Europe have been reported over the last decade (see Webb, this volume, for more detail), and this increase is predicted to continue into the next century (Blumthaler & Ambach, 1990). UV-B radiation is potentially damaging to all living organisms, but plants are particularly vulnerable to any changes in UV-B levels as they require sunlight for survival and so are unable to avoid exposure to elevated UV-B levels.

The impact of an increase in UV-B on various physiological parameters and morphological features of plants has been extensively studied and reviewed (Bornman & Teramura, 1993; Tevini & Teramura, 1989). Most notably, exposure leads to a reduction in photosynthesis, resulting in biomass reduction, anatomical changes such as decreases in root-to-shoot ratio and development of smaller, thicker leaves (Teramura, 1980; Krizeck, 1975) and increases in UV-B absorbing flavonoids (Tevini, Braun & Geiser, 1991; Robberecht & Caldwell, 1986; Warner & Caldwell, 1983). Biochemical studies have established the chloroplast as a major site of UV-B damage with impairment of electron transport, phosphorylation and carbon fixation (Bornman, 1989). These changes subsequently lead to the inhibition of photosynthetic function and decreased efficiency of photosynthesis. Recently, a few studies have focused on the molecular aspects of UV-B effects on plants (for review, see Jordan, 1996). Most of those studies have focused specifically on photosynthesis, and molecular mechanisms have been suggested that can account to some degree for the inhibition of photosynthesis. Supplemental UV-B was found to cause a decline in RNA, enzyme activity and protein levels of several key photosynthetic proteins, with nuclear-encoded genes being more sensitive than genes encoded in the chloroplast (Jordan et al., 1991, 1992). The developmental stage of the tissue studied (Jordan et al., 1994) and the irradiance (Jordan et al., 1992; A-H. Mackerness et al., 1996a) were also found to have a profound influence on the relative sensitivity of transcripts to UV-B supplementation.

This chapter discusses the molecular aspects of UV-B radiation effects on photosynthesis.

Effects of UV-B on photosynthetic proteins

UV-B effects on photosynthesis have been studied extensively and multiple sites of functional inhibition identified: ribulose 1,5-bisphosphate carboxylase (Rubisco) and ATPase quantity and activity are reduced (Jordan et al., 1992; Zhang et al., 1994); the quantum efficiency of PS II is reduced (Strid, Chow & Anderson, 1990) with a reduction in the levels of functional D1 polypeptide (Greenberg et al., 1989); chlorophyll content is diminished with effects on the level of chlorophyll a/b-binding proteins (Jordan et al., 1991, 1994; also see section on Cab).

UV-B effects on Rubisco

Rubisco is the major soluble leaf protein and is the primary CO_2-fixing enzyme in C_3 plants, catalysing the initial step in Calvin's reductive pentose phosphate cycle (for reviews, see Gutteridge & Gatenby, 1987; Miziorko, 1983). The pivotal role of this enzyme in photosynthesis and hence its influence on crop productivity, has led to extensive biochemical and more recently molecular studies. The holoenzyme consists of two subunits of c. 55 kDa (LSU: large subunit) and 14.5 kDa (SSU: small subunit). The LSU is encoded in the chloroplast (rbc L) while the SSU is nuclear encoded (Rbc S) and then transported to the chloroplast for holoenzyme assembly (for review, see Gutteridge & Gatenby, 1995).

Rubisco biosynthesis is strongly influenced by the prevailing light environment (for reviews, see Tobin & Silverthorne, 1985: Jordan, Thoman & Partis, 1986), and recent studies have given insight into the effect of UV-B on Rubisco content, activity and gene expression (for review, see Jordan, 1996). UV-B exposure results in a decrease in CO_2 uptake rate (Basiouney, Van & Biggs, 1978; Vu, Allen & Garrard, 1984) and total soluble proteins in leaves, suggesting a possible effect of UV-B radiation on this photosynthetic protein. Recent studies have shown that UV-B exposure results not only in a reduction in Rubisco content (both SSU and LSU) of leaves (Figs. 1b, 2b) but also in a decrease in its activity (note that these levels of UV-B were lower than used by Baker's group, but these experiments were carried out in cabinets without such high-background white light). In pea, Rubisco content was reduced by 35% of the total soluble protein and its activity by 71% after only 3 d of UV-B exposure (Jordan *et al.*, 1992). However, the percentage of activation (the ratio of Rubisco activity obtained promptly upon extraction to that after pre-incubation at 25 °C for 7 min) was increased with supplementary UV-B treatment (Jordan *et al.*, 1992; Strid *et al.*, 1990). Treatment of pea leaves with supplemental UV-B was found to result in a dramatic decrease in the level of *Rbc S* (after 1 d exposure; Fig. 1A) and to a lesser extent *rbc L* mRNA (after 3 d UV-B exposure; Fig. 2A) but had no effect on the *in vitro* functionality of these RNAs. However, this treatment did result in an increase in the rate of degradation of the two polypeptides as determined by inhibitor and ^{35}S feeding experiments (A-H. Mackerness *et al.*, 1996*b*). The decline in SSU polypeptide levels (Fig. 1b) observed due to UV-B treatments was comparable to the loss of *Rbc S* RNA levels (Fig. 1A). Nuclear-encoded genes for chloroplast proteins are frequently regulated at the level of transcription; indeed *Rbc S* has been used as a model for transcriptional regulation (Gilmartin *et al.*, 1990). Thus, although the biosynthesis of Rubisco is slow and the protein is maintained at a relatively stable level in mature leaf tissue (Iwanji, Chua & Siekevitz, 1975; Nikolau & Klessig, 1987) the combination of enhanced degradation of SSU polypeptide and the decline in *Rbc S* transcript levels, in response to UV-B exposure, could account for the SSU pools being depleted at a greater rate than is observed under natural conditions. Therefore, it appears that while a decrease in mRNA abundance in response to UV-B exposure is the primary cause of the reduced levels of the SSU polypeptide, stability of this protein is also an important factor in determining the levels of this protein in response to UV-B irradiation. The effect of UV-B on LSU of Rubisco, however, is more complex. LSU polypeptide levels were reduced by UV-B exposure at the same rate as SSU levels but prior to any effects on *rbc L* RNA levels (compare Fig. 2A and B). It has been shown that chloroplastic genes are primarily controlled at the post-transcriptional level (Gruissem, 1989; Mullet, 1988). The inhi-

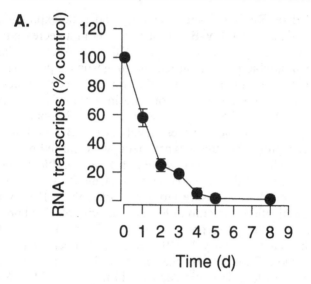

Fig. 1. A. Levels of leaf mRNA after exposure to UV-B (182 mW m^{-2} biologically effective UV-B, based on the generalised plant action spectrum of Caldwell (1971), normalised to unity at 300 nm) given during the 12 h light of a 12 h light:12 h dark lighting schedule. Northern blots, probed with *Rbc S* cDNA probe, were quantified by excision of the appropriate area of the filter and determining the amount of bound radioactive probe by liquid scintillation counting. The effect is defined as the percentage counts obtained from samples from UV-B treated plants compared with counts from samples from control plants, which were treated under the same conditions but in the absence of UV-B, on equivalant days. Error bars indicate the variation in duplicate blots. (Adapted from A-H. Mackerness *et al.*, 1996). B. Western blot analysis of total protein isolated from pea leaves. The proteins were isolated from control plants (c) or from plants exposed to supplemental UV-B (u) and then immunoblotted with an antibody specific for Rubisco.

A.

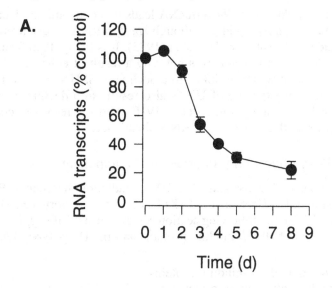

B.

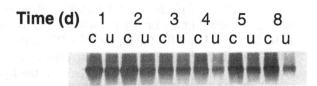

Fig. 2. A. Levels of leaf mRNA after exposure to UV-B (as in Fig. 1). Northern blots, probed with *rbc L* cDNA probe, were quantified by excision of the appropriate area of the filter and determining of the amount of bound radioactive probe by liquid scintillation counting. The effect is defined as the percentage counts obtained from samples from UV-B treated plants compared with counts from samples from control plants, which were treated under the same conditions but in the absence of UV-B, on equivalant days. Error bars indicate the variation in duplicate blots. (Adapted from A-H. Mackerness *et al.*, 1996). B. Western blot analysis of total protein isolated from pea leaves. The proteins were isolated from control plants (c) or from plants exposed to supplemental UV-B (u) and then immunoblotted with an antibody specific for Rubisco.

bition of accumulation of *Rbc S* mRNA leads to a concomitant decrease in SSU and LSU polypeptides, although *rbc L* mRNA levels are maintained (Rodermel, Abbott & Huffaker, 1973). It is likely, therefore, that this initial loss in LSU levels was not only a result of enhanced degradation of LSU but was also linked to a decline in SSU synthesis. However, longer-term exposure of UV-B also results in a decline in *rbc L* mRNA levels, indicating longer-term UV-B exposure results in control also being exerted at the level of RNA abundance.

UV-B effects on photosystem II components

There is a general consensus that UV-B radiation influences PS II activity with little effect on PS I (Kulandaivelu & Noorudeen, 1983; Strid *et al.*, 1990); the following section reviews effects of supplemental UV-B on chlorophyll *a/b*-binding proteins and the D1 polypeptide.

Chlorophyll *a/b*-binding proteins

The chlorophyll *a/b*-binding proteins of the light-harvesting complexes (LHC) are involved in capturing light energy which is then transferred to the reaction centres of photosystems I and II (Paulsen, 1995). The *Cab* proteins are divided into a family consisting of four different types: LHC I, LHC II (types I and II), CP29, CP26, CP24 and the structurally related early light-induced proteins (ELIPs) (White & Green, 1987; Green, Pichersky & Kloppstech, 1991). The units of the light harvesting complexes of PS II are mainly chlorophyll *a/b*-binding proteins consisting of four chl *a*, three chl *b* and one or two xanthophyll molecules per 24–27 kDa polypeptide (Thornber, Markwell & Reinman, 1979). The light-harvesting complexes play an important role with regard to light absorption, thylakoid stacking and distribution of energy and therefore, any damage to these complexes will have multiple effects on the photosynthetic system. It has been suggested that UV-B radiation may result in a functional disconnection of the LHC from the photosystems resulting in impaired energy transfer to the reaction centres (Renger *et al.*, 1986; Tevini *et al.*, 1989), although little work has been carried out to confirm this. Recently, however, the effect of UV-B on the individual components of LHC II has been investigated. In some species, such as pea, supplemental UV-B results in a reduction in total chlorophyll content, with chlorophyll *a* decreasing to a greater extent than chlorophyll *b* (Jordan *et al.*, 1994; Strid *et al.*, 1990; Strid & Porra, 1992). In addition to the decrease in chlorophyll content, there is a change in the relative distribution of chlorophyll with depth in pea leaf tissue (Day & Voglemann, 1995). Chlorophyll biosynthesis does not appear to be inhibited by UV-B irradiation (Strid & Porra, 1992) and therefore it is likely that the reduction in chlorophyll *a* levels is as a result of an increase in degradation. The apparent relative stability in chlorophyll *b* levels may be due to the conversion of chlorophyll *a* or the greater stability of chlorophyll *b* (Brown, Houghton & Henry, 1991). The effect of UV-B on chlorophyll content prompted investigations into the effects of exposure on the chlorophyll *a/b*-binding proteins (Jordan *et al.*, 1991). Treatment of pea leaves with supplemental UV-B was shown to result in a dramatic decrease in the level of *Cab* mRNA (Fig. 3a) with no effect on *in vitro* translatability of the RNA, similar to the response observed for *Rbc S* transcripts. In contrast to the SSU polypeptide, UV-B did not enhance the degradation of the *Cab* proteins and hence, although *Cab* RNA levels were dramatically reduced by UV-B exposure, the effect on protein levels was delayed and protein levels fell at a slower rate than RNA levels (Fig. 3b). However, after 4 d of supplemental UV-B there was a significant reduction in the level of the *Cab* protein (*c.* 55% of levels in control tissue).

Since chlorophyll is normally stably associated with *Cab* proteins in the thylakoid membrane, the reduction in *Cab* proteins on exposure to supplemental UV-B could, therefore, result in the observed increase in chlorophyll degradation. However, further investigations are required to establish the precise mechanisms involved in these chlorophyll changes.

Thus it appears that UV-B does result in damage to or loss of stability of the LHC itself, leading to a reduction in the number of functional units. Therefore, at least part of the inhibition of PS II by supplemental UV-B is a result of inhibition to the light-harvesting potential of the LHC.

D1 polypeptide of photosystem II

The PS II reaction centre is thought to consist of a chlorophyll binding complex composed of three polypeptides: the 32 kDa protein (D1), D2 and cytochrome b_{559}, as well as bound chlorophylls, pheophytins, quinones and non-haeme iron (Nanba & Satoh, 1987; Marder *et al.*, 1987). The 32 kDa protein has received considerable attention because it is the major product of the chloroplast protein synthesising machinery (Mattoo, Marder & Edelman, 1989; Edelman & Reisfeld, 1979). It is rapidly turned over as a function of the visible light intensity (Mattoo *et al.*, 1984) and is the direct target for PS II herbicides such as atrazine and diuron (Pfister *et al.*, 1981; Mattoo *et al.*, 1981).

It has been known for some time that excesses of white light bring about impairment of PS II activity, and this event constitutes the starting point for the physiological phenomenon known as photoinhibition

A.

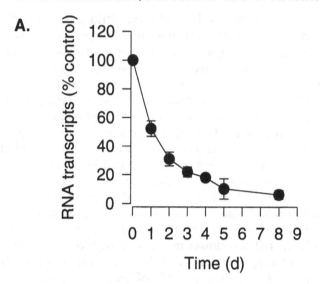

B.

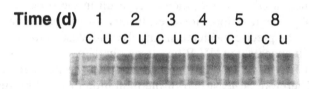

Fig. 3. A. Levels of leaf mRNA after exposure to UV-B (as in Fig. 1). Northern blots, probed with *Cab* cDNA probe, were quantified by excision of the appropriate area of the filter and determining of the amount of bound radioactive probe by liquid scintillation counting. The effect is defined as the percentage counts obtained from samples from UV-B treated plants compared with counts from samples from control plants, which were treated under the same conditions but in the absence of UV-B, on equivalant days. Error bars indicate the variation in duplicate blots. (Adapted from A-H. Mackerness *et al.*, 1996). B. Western blot analysis of total protein isolated from pea leaves. The proteins were isolated from control plants (c) or from plants exposed to supplemental UV-B (u) and then immunoblotted with an antibody specific for *Cab* protein.

(Powles, 1984). One of the main consequences of photoinhibition is the increased turn-over of the D1 protein (Kyle, Ohad & Arntzen, 1984). Our studies have shown that there is a progressive loss of D1 with increasing exposure of pea seedlings to supplemental UV-B (Fig. 4b). In addition, we have found that D1 is considerably more unstable under UV-B conditions than under white light conditions, with higher rates of degradation (Fig. 5) and synthesis (A-H. Mackerness *et al.*, 1997). These results are in agreement with Greenberg *et al.* (1989) who reported that UV-B was very effective at inducing the rapid turn-over of the D1 protein in *Spirodela oligorrhiza*. Spectral evidence has indicated that the major UV-B photosensor for D1 degradation is plastoquinone, although protein components such as tyrosine residues may also be responsible for phototrapping of UV-B energy (Frisco *et al.*, 1994).

The D1 polypeptide is encoded by the chloroplast-located *psb A* gene. In contrast to the four photosynthetic genes discussed above, UV-B exposure did not affect *psb A* mRNA levels substantially until relatively longer exposure times (Fig. 4A). No significant loss of RNA was observed until 5 d of UV-B treatment, at which point levels were at *c*. 70% of control tissue. Jordan *et al.* (1992) also noted a similar insensitivity of *psb A* transcripts to UV-B relative to other photosynthetic proteins. Like other chloroplastic genes, D1 synthesis is regulated at the translational level (Fromm *et al.*, 1985; Staub & Maliga, 1993). It is, therefore, not surprising that UV-B-induced decline in D1 content is not as a result of the down-regulation of *psb A* gene expression but through enhanced turn-over of the polypeptide. The increase in the degradation of the D1 polypeptide is partly offset by increases in synthesis, but the rate of loss exceeds the capacity of the cells to compensate and hence leads to a net decrease in the level of functional D1 protein. This severe sensitivity of the D1 polypeptide is undoubtedly a major contributor to the loss of PS II function on exposure to UV-B radiation. However, although UV-B results in photodamage to the PS II reaction centre, this is not the primary cause of the inhibition of CO_2 assimilation. Thus, other factors, most likely Rubisco, are responsible for the loss of ability to assimilate CO_2 during early stages of irradiation of UV-B (Nogués & Baker, 1995).

Developmental variation in sensitivity to UV-B radiation

The majority of studies carried out on the molecular impact of UV-B on plants have used fully expanded leaves. Recently, however, UV-B sensitivity of *Cab* genes in pea seedlings was found to be dependent on

A.

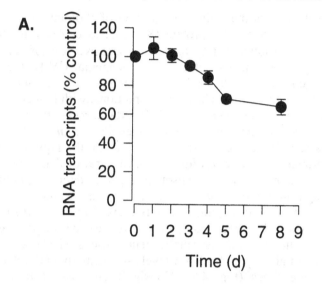

B.

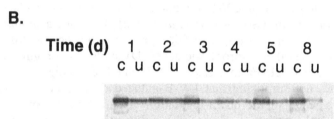

Fig. 4. A. Levels of leaf mRNA after exposure to UV-B (as in Fig. 1). Northern blots, probed with *psb A* cDNA probe, were quantified by excision of the appropriate area of the filter and determining of the amount of bound radioactive probe by liquid scintillation counting. The effect is defined as the percentage counts obtained from samples from UV-B treated plants compared with counts from samples from control plants, which were treated under the same conditions but in the absence of UV-B, on equivalent days. Error bars indicate the variation in duplicate blots. (Adapted from A-H. Mackerness *et al.*, 1996). B. Western blot analysis of total protein isolated from pea leaves. The proteins were isolated from control plants (c) or from plants exposed to supplemental UV-B (u) and then immunoblotted with an antibody specific for the D1 polypeptide.

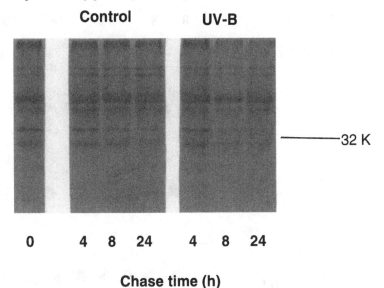

Chase time (h)

Fig. 5. Autoradiograph of *in vivo* translation products obtained from ^{35}S pulse chase experiments, as described in Greenberg *et al.* (1989), using leaf discs, analysed by SDS-PAGE. The proteins were isolated from discs of control plants or those exposed to supplemental UV-B for times indicated.

the developmental stage of the tissue studied (Jordan *et al.*, 1994; A-H. Mackerness *et al.*, unpublished observations). Supplemental UV-B reduced the relative level of *Cab* mRNA transcripts substantially in green leaf tissue and, to a lesser extent, in young bud tissue (Fig. 6a,b). In contrast, *Cab* levels were increased during greening of etiolated tissue exposed to UV-B under the same conditions (Fig. 6c). These results are consistent with previous findings by Jordan *et al.* (1994) in pea, with the exception of those with green bud tissue, where they reported that *Cab* transcripts were more sensitive to UV-B exposure than in fully expanded leaves. This inconsistency is, however, possibly due to the use of different control environment cabinets for the two studies. This could lead to samples at similar chronological ages being at different developmental stages with different sensitivities to UV-B. Our results are consistent with studies in *Arabidopsis* (Lois, 1994; B.R. Jordan, P. James & R.G. Anthony, unpublished observations) where younger tissue was found to be more resistant to UV-B supplementation.

In all plant species studied to date, *Cab* proteins are encoded by

Time (h)   0   8   12   30   78

UV-B Treatment   - +   - +   - +   - +

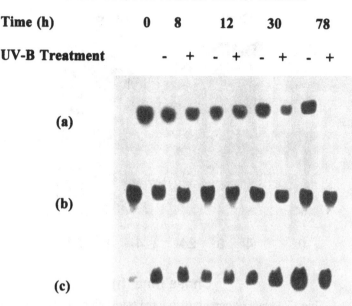

(a)

(b)

(c)

Fig. 6. Autoradiographs of Northern blot analysis of total RNA (20 μg) from a) pea leaves, b) green bud and c) etiolated bud. The RNA was isolated from plants treated with (+) or without (−) UV-B radiation and then hybridised with ^{32}P-labelled *Cab* cDNA fragment.

nuclear multigene families. In pea there are seven known *Cab* genes encoding the LHC II polypeptides (summarised by White *et al.*, 1992). Individual members of the *Cab* gene family have been shown to be sensitive to different light qualities and independently regulated at different developmental stages (Chang & Walling, 1992; White *et al.*, 1992, 1995). We have determined the expression of the individual *Cab* genes in etiolated bud and green leaf tissue of pea seedlings, after exposure to supplemental UV-B, by using a highly sensitive quantitative PCR method developed by White *et al.* (1992). Exposure to UV-B resulted in the reduction of transcript levels for all seven *Cab* genes in green leaves and an increase in the levels of these transcripts in etiolated buds (A-H. Mackerness, L.S. Liu, B.R. Jordan, B. Thomas, A.J. Thompson & M.J. White, unpublished observations). These results indicate that the differing sensitivity of etiolated bud and leaf tissue to UV-B is not due to the differential expression of individual members of the *Cab* family. However, although the effects of UV-B were shown to be consistent for all the seven *Cab* genes, resulting in either inhibition or increase in green leaves and etiolated bud, respectively, not all the genes shared a common pattern of expression. Two distinct groups of genes could be identified with respect to their sensitivity to UV-B. *Cab-9*, *AB66* and *AB80* were found to be more sensitive to UV-B exposure than *Cab-8*, *Cab-215*, *AB96* and *Cab-315*. The same groups have been identified with respect to their sensitivity to red and blue light and their relative levels of expression in bud and leaf tissue (White *et al.*, 1995). The four genes, *Cab-8*, *Cab-215*, *AB96* and *Cab-315*, showed relatively strong stimulatory responses to blue and red light pulses, while *Cab-9*, *AB66* and *AB80* showed little or no response under the same conditions. It appears, therefore, that the former group of genes, although showing a stronger response to red and blue light, are in fact less sensitive to UV-B radiation in leaves. Our results, therefore, corroborate previous studies suggesting that the control of individual *Cab* genes is sensitive to light quality (White *et al.*, 1992; 1995; Marrs & Kaufman, 1991; Karlin-Neumann, Sun & Tobin, 1988; Sun & Tobin, 1990) and also indicate that the response of the two *Cab* groups to red, blue and UV-B light during bud de-etiolation may be mediated through a common regulatory or signal transduction pathway, most likely phytochrome.

Interaction of UV-B and PAR

It has been recognised for some time that longer wavelength radiation (PAR: 400–700 nm) can minimise UV-B induced damage. This protection is effective at the physiological, biochemical and also at the molecular level (Cen & Bornman, 1990; Flint, Jordan & Caldwell, 1985; Jordan *et al.*, 1992). Increased irradiance, both before and during exposure to UV-B irradiation, has been shown to ameliorate the inhibitory effects of UV-B on photosynthesis. The pre-treatment of plants with high PAR results in thicker leaves and increased levels of flavonoids as compared to pre-treatment under low light conditions (Teramura, 1980; Cen & Bornman, 1990; Tevini, Braun & Geiser, 1991), and this tends to lessen the sensitivity of plants to UV-B exposure (Warner & Caldwell, 1983; Murali & Teramura, 1985). These changes cannot, however, explain why plants grown under low PAR and then exposed to UV-B under high PAR are less sensitive to UV-B-induced damage than those maintained at the lower PAR during the UV-B treatment (Kramer, Krizek & Mirecki, 1992; Mirecki & Teramura, 1984). In addition, protection of photosynthetic transcripts against UV-B radiation, by high PAR, can be demonstrated after only a few hours and so is unlikely to be caused by changes in leaf morphology or pigment content (Jordan *et al.*, 1992; A-H. Mackerness *et al.*, 1996). Under these conditions, protection by higher PAR has been generally attributed to increases in the activity of the DNA repair enzyme, photolyase (Pang & Hays, 1991). This enzyme uses energy in the range of 300 to 500 nm to reverse UV-B-induced photoproducts, cyclobutane pyrimidine dimers (Sutherland, 1981). However, photorepair mechanisms are normally saturated at relatively low irradiances (Taylor *et al.*, 1996) and are therefore unlikely to account for the entire effect of high PAR. Alternatively, protection may be afforded through photosynthesis, as high irradiance is known to alter the thylakoid membrane appression (Davis, Chow & Jordan, 1986*a*), photosynthetic function (Davis *et al.*, 1986*b*) and the biochemical composition of the chloroplast (Davis *et al.*, 1987, 1991). These changes could influence the response to UV-B by, for instance, providing additional substrate through increases in photosynthesis, for the repair or the replacement of damaged organelles or tissues. This was elegantly demonstrated by Adamse & Britz (1992), who used different concentrations of CO_2 to modify the photosynthetic potential, and showed that increased CO_2 could counteract the effect of UV-B on plant growth. Similar observations were made in studies on the influence of high PAR on gene expression (A-H. Mackerness *et al.*, 1996). Using a combination of approaches, including high CO_2, low pressure sodium lights and inhibitors, photosynthesis was manipulated without altering the activity of photolyase, and effects on photosynthetic transcripts followed. It was concluded that the protection of gene expression was related to photosynthetic activity and not photorepair mechanisms. The results also suggested that there is a requirement for ATP provided by photophosphorylation. These studies, therefore, suggest that the likely mechanism of this type of protection, under high light conditions is through increases in photosynthesis and not through increases in the activity of DNA repair enzymes.

Discussion

Recent investigations have begun to allow us to understand the molecular mechanisms involved in the effects of UV-B on plant development and productivity. The mechanism by which UV-B influences gene expression in higher plants is at present unclear. UV-B can damage DNA directly, and this consequence of UV-B exposure has been suggested as the mechanism of UV-B effects on the photosynthetic transcripts. However, there is strong evidence that the UV-B-induced down-regulation of gene expression is not simply a consequence of non-specific damage to DNA. These are summarised as follows:

- In parallel with the down-regulation of photosynthetic genes, other genes encoding components of plant defence mechanisms, such as chalcone synthase and glutathione reductase, are up-regulated. Figure 7 shows that the level of leaf mRNA coding for chalcone synthase increases after exposure.
- The developmental variation in the response to UV-B also indicates a more specific response. Photosynthetic genes in etiolated tissue, which have lower levels of protective pigments than green bud tissue, are not down-regulated after UV-B exposure. The penetration of UV-B in these tissues and hence the potential for DNA damage would appear to be greater, yet the less-developed tissue is less responsive to UV-B at the molecular level.
- Finally, protection from UV-B damage under high light

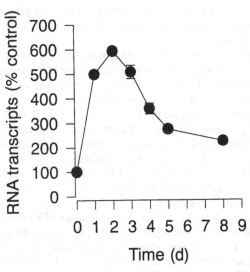

Fig. 7. Levels of leaf mRNA after exposure to UV-B (as in Fig. 1). Northern blots, probed with *chs* cDNA probe, were quantified by excision of the appropriate area of the filter and determining of the amount of bound radioactive probe by liquid scintillation counting. The effect is defined as the percentage counts obtained from samples from UV-B-treated plants compared with counts from samples from control plants, which were treated under the same conditions but in the absence of UV-B, on equivalent days. Error bars indicate the variation in duplicate blots. (Adapted from A-H. Mackerness *et al.*, 1996).

conditions does not appear to be as a result of increased photolyase activity, indicating that DNA damage is not the major cause of the down-regulation of photosynthetic genes.

It is obvious, therefore, that the UV-B-induced down-regulation of gene expression is not simply a consequence of reduced transcription rates caused by non-specific DNA damage and that there is a selective and specific regulation of gene expression by UV-B radiation.

The molecular mechanisms involved in UV-B responses remain to be fully characterised. These mechanistic studies, therefore, require carefully controlled growth environments to accurately determine the responses. Understanding these processes will eventually lead to a greater understanding of how plants overcome environmental stress in more natural environments.

References

Adamse, P. & Britz, S.J. (1992). Amelioration of UV-B damage under high irradiance: role of photosynthesis. *Photochemistry and Photobiology*, **56**, 645–50.

A-H. Mackerness, S., Butt, P.J., Jordan, B.R. & Thomas, B. (1996). Amelioration of UV-B down-regulation of mRNA transcripts for chloroplast proteins, by high irradiance, is mediated by photosynthesis. *Journal of Plant Physiology*, **148**, 100–6.

A-H. Mackerness, S., J., Jordan, B. R. & Thomas, B. (1997). The effect of supplementary UV-B radiation on mRNA transcripts and stability of chloroplast proteins and pigment formation in *Pisum sativum*. *Journal of Experimental Botany*, **48**.

Basiouney, F.M., Van, T.K. & Biggs, R.H. (1978). Some morphological and biochemical characteristics of C3 and C4 plants irradiated with UV-B. *Plant Physiology*, **42**, 29–32.

Blumthaler, M. & Ambach, W. (1990). Indications of increasing solar UV-B radiation flux in the alpine regions. *Science*, **248**, 206–8.

Bornman, J.F. (1989). Target sites of UV-B radiation in photosynthesis of higher plants. *Journal of Photochemistry and Photobiology*, **4**, 145–58.

Bornman, J.F. & Teramura, A.H. (1993). In *Environmental Photobiology*. (Young, A.R., Bjorn, L.O., Moan, J. & Nultsch, W., eds). Plenum Press, New York.

Brown, S.B., Houghton, J.D. & Henry, G.A.F. (1991). Chlorophyll breakdown. In *Chlorophylls* (Scheer, H., ed.), pp. 465–89. CRC Press, Florida.

Caldwell, M.M. (1971). Solar UV-radiation and the growth and development of higher plants. In *Photophysiology*, vol. 6. Academic Press, New York.

Cen, Y.P. & Bornman, J.F. (1990). The response of bean plants to UV-B radiation under different irradiances of background visible light. *Journal of Experimental Botany*, **41**, 1489–95.

Chang, Y.C. & Walling, L.L. (1992). Chlorophyll *a/b*-binding protein genes are differentially expressed during soybean development. *Plant Molecular Biology*, **19**, 217–30.

Davis, E.C., Chow, W.S. & Jordan, B.R. (1986a). A study of the factors which regulate membrane appression of lettuce thylakoids in relation to irradiance. *Photosynthesis Research*, **9**, 359–70.

Davis, E.C., Chow, W.S., Le Fay, J.M. & Jordan, B.R. (1986b). Acclimation of tomato leaves to changes in light intensity; effects on the function of the thylakoid membrane. *Journal of Experimental Botany*, **37**, 211–20.

Davis, E.C., Jordan, B.R., Partis, M.D. & Chow, W.S. (1987). Immunological investigation of thylakoid coupling factor protein during photosynthetic acclimation to irradiances. *Journal of Experimental Botany*, **38**, 1517–27.

Davies, H., Jordan, B.R. & Harwood, J.L. (1991). Changes in acyl-lipid, protein and chlorophyll content of lettuce and tomato chloroplast during acclimation to changes in irradiance. *Plant Physiology and Biochemistry*, **29**, 281–88.

Day, T.A. & Vogelmann, T.C. (1995). Alterations in photosynthesis and pigment distributions in pea leaves following UV-B exposure. *Physiologia Plantarum*, **94**, 433–40.

Edelman, M. & Reisfeld, A. (1979). In *Genome Organisation and Expression in Plants*. (Leaver, C.J., ed.), pp. 353–62. Plenum Press, New York.

Flint, S.D., Jordan, P.W. & Caldwell, M.M. (1985). Plant protective response to enhanced UV-B radiation under field conditions: leaf optical properties and photosynthesis. *Photochemistry and Photobiology*, **41**, 95–9.

Frisco, G., Spetea, C., Giacometti, G.M., Vass, I. & Barbato, R. (1994). Degradation of photosystem II reaction centre D1 protein induced by UV-B radiation in isolated thylakoids. Identification and characterisation of C- and N-terminal breakdown products. *Biochimica et Biophysica Acta*, **1184**, 78–84.

Fromm, H., Devic, M., Flur, R. & Edelman, M. (1985). Control of *psb A* gene expression in mature *Spirodela* chloroplasts; light regulation of 32 kDa protein synthesis is independent of transcript level. *EMBO Journal*, **4**, 291–5.

Gilmartin, P.M., Sarokin, L., Memlink, J. & Chua, N.-H. (1990). Molecular light switches for plant genes. *The Plant Cell*, **2**, 369–78.

Green, R. & Flur, R. (1995). UV-B-induced PR-1 accumulation is mediated by active oxygen species. *The Plant Cell*, **7**, 203–12.

Green, B. R., Pichersky, E. & Kloppstech, K. (1991). Chlorophyll *a/b*-binding proteins: an extended family. *Trends in Biochemical Science*, **16**, 181–6.

Greenberg, B.M., Bruce, M., Gaba, V., Cannaani, O., Malkins, S., Matto, A. & Edelman, M. (1989). Separate photosensitiser mediated degradation of the 32 kDa PS II reaction centre protein in visible light and UV spectra regions. *Proceedings of the National Academy of Sciences, USA,* **86**, 6617–20.

Gruissem, W. (1989). Chloroplast gene expression: how plants turn their plastids on. *Cell*, **56**, 161–70.

Gutteridge, S. & Gatenby, A.A. (1987). The molecular analysis of the assembly, structure and function of Rubisco. In *Oxford Surveys of Plant Molecular and Cell Biology, Vol. 4*. (Miflin, B.J., ed.), pp. 95–135. Oxford University Press, Oxford.

Gutteridge, S. & Gatenby, A.A. (1995). Rubisco synthesis, assembly, mechanism and regulation. *Plant Cell*, **7**, 809–19.

Iwanij, V., Chua, N.-H. & Siekevitz, P. (1975). Synthesis and turnover of ribulose bisphosphate carboxylase and of its subunits during the cell cycle of *Chlamydomonas reinhardtii*. *Journal of Cell Biology*, **64**, 572–85

Jordan, B.R. (1996). The effects of UV-B radiation on plants: a molecular perspective. In *Advances in Botanical Research*. (Callow, J.A., ed.). Academic Press, Oxford.

Jordan, B.R., Thomas, B. & Partis, M. (1986). The biology and molecular biology of phytochrome. In *Oxford Surveys of Plant Molecular and Cell Biology, Vol. 3* (Miflin, B.J., ed.), pp. 315–62. Oxford University Press, Oxford.

Jordan, B.R., Chow, W.S., Strid, A. & Anderson, J.M. (1991). Reduction in *Cab* and *psb A* RNA transcripts in response to supplemental UV-B radiation. *FEBS Letters*, **284**, 5–8.

Jordan, B.R., He, J., Chow, W.S. & Anderson, J.M. (1992). Changes in mRNA levels and polypeptide subunits of ribulose 1,5-bisphosphate carboxylase in response to supplemental UV-B radiation. *Plant Cell and Environment*, **15**, 91–8.

Jordan, B.R., James, P., Strid, A. & Anthony, R.G. (1994). The effect of supplemental UV-B radiation on gene expression and pigment composition in etiolated and green pea leaf tissue: UV-B induced changes are gene specific and dependant on the developmental stage. *Plant Cell and Environment*, **17**, 45–54.

Karlin-Neumann, G.A., Sun, L. & Tobin, E.M. (1988). Expression of light harvesting chlorophyll *a/b*-binding protein genes is phytochrome regulated in etiolated *Arabidopsis thaliana* seedlings. *Plant Physiology*, **88**, 1323–31.

Kramer, G.F., Krizek, D.T. & Mirecki, R.M. (1992). Influence of photosynthetically active radiation and spectral quality on UV-B induced polyamine accumulation in soybean. *Photochemistry*, **31**, 1119–25.

Krizeck, D.T. (1975). Influence of ultraviolet radiation on germination and early seedling growth. *Physiologia Plantarum*, **34**, 182–6.

Kulandaivelu, G. & Noorudeen, A.M. (1983). Comparative study of the action of ultraviolet C and ultraviolet B radiation on photosynthetic electron transport. *Physiologia Plantarum*, **58**, 389–94.

Kyle, D.J., Ohad, I. & Arntzen, C.J. (1984). Membrane protein damge and repair: Selective loss of a quinone-protein function in chloroplast membranes. *Proceedings of the National Academy of Sciences, USA*, **81**, 4070–4.

Landry, L.G., Chapple, C.C.S. & Last, R.L. (1995). *Arabidopsis* mutants lacking phenolic sunscreens exhibit enhanced UV-B injury and oxidative damage. *Plant Physiology*, **109**, 1159–66.

Li, J., Ou-Lee, T.-M., Raba, R., Amundson, R.G. & Last, R.L. (1993). *Arabidopsis* flavonoid mutants are hypersensitive to UV-B irradiation. *The Plant Cell*, **5**, 171–9.

Lois, R. (1994). Accumulation of UV-absorbing flavonoids induced by UV-B radiation in *Arabidopsis thaliana* L. *Planta*, **194**, 498–503.

Lois, R. & Buchanan, B.B. (1994). Severe sensitivity to UV-B radiation in an *Arabidopsis* mutant deficient in flavonoid accumulation. II. Mechanism of UV-B resistance in *Arabidopsis*. *Planta*, **194**, 504–9.

Marder, J.B., Chapmann, D.J., Telfer, A., Nixon, P.J. & Barber, J. (1987). Identification of *Psb A* and *Psb D* gene products, D1 and D2, as reaction centre proteins of photosystem 2. *Plant Molecular Biology*, **9**, 325–33.

Marrs, K.A. & Kaufman, L.S. (1991). Rapid translational regulation of the *Cab* and pEA207 gene families in peas by blue light in the absence of cytoplasmic protein synthesis. *Planta*, **183**, 327–33.

Mattoo, A.K., Pick, U., Hoffman-Falk, H. & Edelman, M. (1981). The rapidly metabolized 32,000-dalton polypeptide of the chloroplast is the 'proteinaceous shield' regulating photosystem electron transport and mediating diuron herbicide sensitivity. *Proceedings of the National Academy of Sciences, USA*, **78**, 1572–6.

Mattoo, A.K., Hoffman-Falk, H., Marder, J.B. & Edelman, M. (1984). Regulation of protein metabolism: coupling of photosynthetic electron transport to in vivo degradation of the rapidly metabolized 32-kilodalton protein of the chloroplast membranes. *Proceedings of the National Academy of Sciences, USA*, **81**, 1380–4.

Mattoo, A.K., Marder, J.B. & Edelman, M. (1989). Dynamics of the photosystem II reaction centre. *Cell*, **56**, 241–6.

Mirecki, R.M. & Teramura, A.H. (1984). Effects of UV-B radiation on soybean. 5. The dependence of plant-sensitivity on the photosynthetic photon flux-density and after leaf expansion. *Plant Physiology*, **74**, 475–80.

Miziorko, H.M. (1983). Ribulose 1,5-bisphosphate carboxylase-oxygenase. *Annual Review of Biochemistry*, **52**, 507–35.

Montzka, S.A., Butler, J.H., Myers, R.C., Thompson, T.H., Clerke, A.D., Lock, L.T. & Elkiss, J.W. (1996). Decline in the tropospheric abundance of halogen from halocarbons: Implications for stratospheric ozone depletion. *Science*, **272**, 1318–22.

Mullet, J.E. (1988). Chloroplast development and gene expression. *Annual Review of Plant Physiology*, **39**, 475–502.

Murali, N.S. & Teramura, A.H. (1985). Effects of UV-B irradiation on soybean. Influence of phosphorus nutrition on growth and flavonoid content. *Plant Physiology*, **95**, 536–43.

Nanba, O. & Satoh, K. (1987). Isolation of a photosystem II reaction center consisting of D1 and D2 polypeptides and cytochrome b-$_{559}$. *Proceedings of the National Academy of Sciences, USA*, **84**, 109–12.

Nikolau, B.J. & Klessig, D.F. (1987). Coordinate, organ specific and developmental regulation of ribulose 1,5-bisphosphate carboxylase gene expression in *Amaranthus hypochondriacus*. *Plant Physiology*, **85**, 167–73.

Nogués, S. & Baker, N.R. (1995). Evaluation of the role of damage to photosystem II in the inhibition of CO_2 assimilation in pea leaves on exposure to UV-B radiation. *Plant Cell and Environment*, **18**, 781–7.

Pang, Q. & Hays, J.B. (1991). UV-B-inducible and temperature sensitive photoreactivation of cyclopyrimidine dimers in *Arabidopsis thaliana*. *Plant Physiology*, **95**, 536–43.

Paulsen, H. (1995). Chlorophyll *a/b*-binding proteins. *Photochemistry and Photobiology*, **62**, 367–82.

Pfister, K., Steinbeck, K.E., Gardner, G. & Arntzen, C.J. (1981). Photoaffinity labeling of a herbicide receptor protein in chloroplast membranes. *Proceedings of the National Academy of Sciences, USA*, **78**, 981–5.

Powles, S.B. (1984). Photoinhibition of photosynthesis induced by visible light. *Annual Review of Plant Physiology*, **35**, 15–44.

Renger, G., Voss, M., Graber, P. & Schuize, A. (1986). The effects of UV-B irradiation on different partial reactions of the primary processes of photosynthesis. In *Stratospheric Ozone Reduction, Solar Ultraviolet Radiation and Plant Life*. (Caldwell, M.M., ed.), pp. 171–84. Springer-Verlag, Berlin.

Robberecht, R. & Caldwell, M.M. (1986). Leaf optical properties of *Rumex patientia* L. and *Rumex obtusifolius* L. in regard to a protection mechanism against solar UV-B radiation injury. In *Stratospheric Ozone Reduction, Solar Ultraviolet Radiation and Plant*

Life. (Worrest, R.C. & Caldwell, M.M., eds), pp. 251–59. Springer-Verlag, Berlin.

Rodermel, S.R., Abbott, M.S. & Huffaker, R.C. (1973). Evidence for the lack of turn over of RuDP carboxylase in barley leaves. *Plant Physiology*, **51**, 1042–5.

Staub, J.M. & Maliga, P. (1993). Accumulation of D1 polypeptide in tobacco plastids is regulated via the untranslated region of the *psb A* mRNA. *EMBO Journal*, **12**, 601–6.

Strid, A. (1993). Alterations in expression of defence genes in *Pisum sativum* after exposure to supplementary UV-B radiation. *Plant Cell and Environment*, **34**, 949–53.

Strid, A., Chow, W.S. & Anderson, J.M. (1990). Effects of supplementary UV-B radiation on photosynthesis of *Pisum sativum*. *Biochemica et Biophysica Acta*, **1020**, 260–8.

Strid, A. & Porra, R.J. (1992). Alterations in pigment content in leaves of *Pisum sativum* on exposure to supplemental UV-B. *Plant Cell Physiology*, **33**, 1015–23.

Sun, L. & Tobin, E.M. (1990). Phytochrome regulated expression of genes encoding light-harvesting chlorophyll *a/b*-protein in two long hypocotyl mutants and wild type plants of *Arabidopsis thaliana*. *Photochemistry and Photobiology*, **52**, 51–6.

Sutherland, B.M. (1981). Photoreactivating enzymes. In *The Enzymes* (Boyer, P.D., ed.). Academic Press, New York.

Taylor, R.M., Nikaid, O., Jordan, B.R., Rosamond, A., Bray, C.M. & Tobin, A.K. (1996). Ultraviolet-B-induced DNA lesions and their removal in wheat (*Triticum aestivum* L.) leaves. *Plant Cell and Environment*, **19**, 171–81.

Teramura, A.H. (1980). Effects of UV-B radiation on soybean. *Physiologia Plantarum*, **48**, 333–9.

Tevini, M. & Teramura, A.H. (1989). UV-B effects on terrestrial plants. *Photochemistry and Photobiology*, **50**, 479–87.

Tevini, M., Mark, V., Fieser, G. & Saile, M. (1989). Effects of enhanced solar UV-B radiation on growth, function and composition of crop plant seedlings. In Final report, GSF, Munchen, Part A, pp. 1–100.

Tevini, M., Braun, J. & Gieser, G. (1991). The protective function of the epidermal layer of rye seedlings against UV-B radiation. *Physiologia Plantarum*, **53**, 329–33.

Thornber, J.P., Markwell, J.P. & Reinman, S. (1979). Plant chlorophyll-protein complexes: recent advances. *Photochemistry and Photobiology*, **29**, 1205–16.

Tobin, E.M. & Silverthorne, J. (1985). Light regulation of gene expression in higher plants. *Annual Review of Plant Physiology*, **36**, 569–93.

Vu, C.V., Allen, L.H. & Garrard, L.A. (1984). Effects of enhanced UV-B radiation (280–320 nm) on ribulose-1,5-bisphosphate carboxylase in pea and soybean. *Environmental and Experimental Botany*, **24**, 131–46.

Warner, C.W. & Caldwell, M.M. (1983). Influence of photon flux density in the 400–700 nm waveband on inhibition of photosynthesis by UV-B irradiation in soybean leaves: separate and indirect and immediate effects. *Photochemistry and Photobiology*, **38**, 341–5.

White, M.J. & Green, B.R. (1987). Antibodies to the photosystem I chlorophyll *a* + *b* antenna cross react with polypeptides of CP29 and LHCII. *European Journal of Biochemistry*, **163**, 545–51.

White, M.J., Fristensky, B., Falconet, D., Childs, L.C., Watson, J.C., Alexander, L., Roe, B.A. & Thompson, W.F. (1992). Expression of the chlorophyll *a/b*-binding protein multigene family in pea (*Pisum sativum* L.): evidence for distinct developmental responses. *Planta*, **188**, 190–8.

White, M.J., Kaufman, L.S., Horwitz, B.A., Briggs, W.R. & Thompson, W.F. (1995). Individual members of the *Cab* gene family differ widely in fluence response. *Plant Physiology*, **107**, 161–5.

Zhang, J.E., Hu, X., Henkow, L., Jordan, B.R. & Strid, A. (1994). The effects of UV-B radiation on CF_0F_1-ATPase. *Biochemica et Biophysica Acta*, **1185**, 295–302.

G.I. JENKINS, G. FUGLEVAND
and J.M. CHRISTIE

UV-B perception and signal transduction

Introduction

Plant growth and development are regulated by aspects of the environment throughout the life cycle. Plants continually gather information about a wide range of environmental variables which elicit particular responses. Such environmental factors function as cues to initiate developmental transitions, for instance, in the photoperiodic induction of flowering observed in many plant species. However, environmental signals may also be potentially harmful and the plant's response is then to minimise damage. Examples of abiotic environmental stresses include drought, extremes of temperature and UV-B irradiation, the subject of this volume.

Although responses to external signals, such as UV-B radiation, may be studied at the whole plant level, the signals are perceived and responded to at the cellular level. In addition to external signals, plant cells respond to a range of endogenous signals, such as the concentrations of metabolites and growth regulators. Thus, individual plant cells have the capacity to respond to a range of signals, which regulate their growth and differentiation and affect their survival. A key point is that cells must have mechanisms which allow signals to be detected and acted upon to give rise to particular responses. The detection of signals in many cases is likely to involve specific cellular components termed receptors. Reception is coupled to the terminal response by signal transduction mechanisms. The signal transduction process often serves to amplify the initial signal and in some cases may store it for periods of time. Information about plant signal perception and transduction is now accumulating rapidly as a consequence of the application of new cell physiological techniques (for example, Knight & Knight, 1995; Maathuis & Sanders, 1995; Miller, 1995) in conjunction with molecular and genetic approaches (Bowler & Chua, 1994).

UV-B irradiation (280–320 nm) has well-documented effects on plants, which are discussed in detail in other chapters and in several recent reviews (Tevini & Teramura, 1989; Stapleton, 1992). Many of its effects are damaging to cellular components such as DNA and other macromolecules. To a large extent these changes are likely to be caused by direct absorption of UV-B wavelengths by the molecules in question (Mitchell & Karentz, 1993; see chapter by Taylor *et al.*, this volume). The energy of the radiation is sufficient to cause photochemical changes. Clearly, such UV-B damage processes do not involve specific cellular receptors and signal transduction components in the accepted sense. Further damaging effects of UV-B may result indirectly from the generation of reactive oxygen species (Bowler, van Montagu & Inzé, 1992; Green & Fluhr, 1995).

However, it is evident that not all the effects of UV-B on plants involve macromolecular damage. Indeed, plants exhibit a wide range of responses to UV-B (Tevini & Teramura, 1989; Stapleton, 1992), including physiological responses which help to protect them from damaging UV-B wavelengths. In particular, the synthesis of UV-absorbing protective molecules, such as flavonoids, hydroxycinnamic acids and related compounds (see below) is not a damage response and involves the stimulation of expression of particular genes. In these cases the responses are likely to involve specific UV-B photoreceptors and signal transduction processes which lead to the regulation of transcription. This chapter discusses what is presently known about these cellular processes, with particular reference to the regulation of genes concerned with production of the protective flavonoid pigments.

UV-B stimulates the synthesis of protective pigments in plants

Accumulation of UV-absorbing pigments

It is well established that exposure of plants to UV-B stimulates the accumulation of UV-absorbing pigments which provide a protective screen (Tevini & Teramura, 1989; Stapleton, 1992; Beggs & Wellmann, 1994; and chapter by Bornman *et al.*, this volume). Accumulation of the protective pigments is primarily in epidermal layers (Schmelzer, Jahnen & Hahlbrock, 1988; Tevini, Braun & Fieser, 1991; Day, Martin & Voglemann, 1993) and the epidermis can prevent the transmission of all but a few per cent of the incident UV radiation (Robberecht & Caldwell, 1978). The UV-absorbing molecules include hydroxycinnamic acid esters, flavonoids, particularly flavonols and flavones, and related molecules. The absorption spectra of the pigments differ. Flavonoids have absorbtion maxima at 270 and 345 nm, whereas derivatives of hydroxycinnamic acid have an absorbtion maximum at

320 nm (Schnabl *et al.*, 1989), so both classes may protect against damaging UV wavelengths. Flavonoids have been shown to protect against DNA damage *in vitro* (Kootstra, 1994). The anthocyanins, which visibly accumulate in many plants following UV-B exposure, do not absorb strongly in the UV region and are unlikely to be key UV-protectants.

There is now strong evidence that flavonoids and hydroxycinnamic acid derivatives protect against UV damage *in vivo*. In addition to the strong correlative evidence provided by many researchers (Tevini & Teramura, 1989; Stapleton, 1992; Lois, 1994), recent genetic studies demonstrate directly that particular classes of molecules provide UV protection. Mutants that fail to make these compounds are more sensitive to UV-B damage. One of the first uses of mutants to investigate responses to UV-B was reported by Li *et al.* (1993). They studied two mutants of *Arabidopsis*, *tt4* and *tt5* which, respectively, are deficient in chalcone synthase (CHS) and chalcone isomerase (CHI), the initial steps in the flavonoid biosynthesis pathway (Fig. 1). Li *et al.* (1993) showed that these mutants were unable to produce the spectrum of UV-absorbing molecules that wild-type *Arabidopsis* produces following exposure to UV-B. The lack of production of these molecules was correlated with a much increased sensitivity to UV-B. Similarly, Stapleton and Walbot (1994) reported that a flavonoid-deficient maize line had increased UV-induced DNA damage, and Lois and Buchanan (1994) found that a mutant with increased sensitivity to UV-B lacked a particular sub-class of flavonoids.

Of particular interest is recent research (Landry, Chapple & Last, 1995) with an *Arabidopsis* mutant, *fah1* (previously *sin1*; Chapple *et al.*, 1992), that is deficient in the synthesis of sinapic acid esters (Fig. 1), a class of hydroxycinnamic acid esters. This mutant is more sensitive to UV-B than *tt4* and *tt5*, suggesting that sinapic acid esters may be more important UV protectants than flavonoids in *Arabidopsis*. Hence the use of mutants, in conjunction with biochemical analysis, is a powerful means of evaluating the significance of particular classes of compounds as UV protectants. The work of Landry, Chapple and Last (1995) suggests that the relative importance of flavonoids and hydroxycinnamic acid derivatives as protectants should be re-examined in other species.

Regulation of phenylpropanoid and flavonoid biosynthesis gene expression by UV-B

The production of flavonoids is from a branch of the phenylpropanoid pathway (Fig. 1; Dixon & Paiva, 1995). Phenylalanine ammonia-lyase

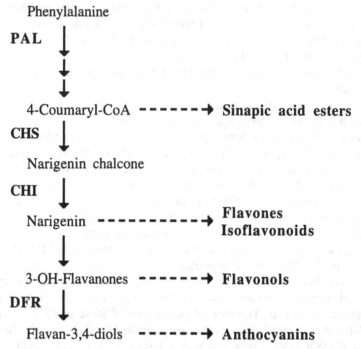

Fig. 1. The phenylpropanoid and flavonoid biosynthesis pathways illustrating reactions catalysed by the enzymes phenylalanine ammonia-lyase (PAL), chalcone synthase (CHS) and dihydroflavonol reductase (DFR).

(PAL) is the first step in the phenylpropanoid pathway and CHS is the key point which commits the pathway to flavonoid biosynthesis. Subsequently the pathway branches to synthesise the different classes of flavonoids and anthocyanins.

It is well established that exposure of plants to UV-B stimulates expression of genes encoding PAL, CHS and several other phenyl-propanoid and flavonoid biosynthesis enzymes in a range of species (for reviews see Hahlbrock & Schell, 1989; Dangl, Hahlbrock, & Schell, 1989; van der Meer, Stuitje & Mol, 1993; Beggs & Wellmann, 1994). For example, in parsley cell cultures, UV-B induces genes encoding PAL and other phenylpropanoid pathway enzymes including the *CHS* genes (Hahlbrock & Schell, 1989). In *Arabidopsis*, UV-B exposure elevates transcript levels of *PAL1*, *CHS*, *CHI* and dihydroflavonol reductase (*DFR*) genes (Kubasek *et al.*, 1992; Li *et al.*, 1993). Action spectra of *PAL* and *CHS* transcript accumulation in carrot cell cultures indicate the involvement of a UV-B photoreceptor (Takeda, Obi & Yoshida, 1994).

The well-studied *PAL* and *CHS* genes are in fact regulated by a range of environmental and endogenous signals (Dangl *et al.*, 1989; van der Meer *et al.*, 1993; Mol *et al.*, 1996). Depending on the species, they may be regulated, for example, by light absorbed by phytochrome and UV-A/blue photoreceptors, by several abiotic stresses, pathogenesis, metabolites and plant growth regulators. In several cases, information on the signal transduction events which lead to the regulation of transcription of specific phenylpropanoid and flavonoid biosynthesis genes has been obtained (Mol *et al.*, 1996). Furthermore, the transcription factor targets of the signal transduction pathways have, in some instances, been defined (see below).

In common with other aspects of their regulation, the stimulation of *PAL* and *CHS* genes by UV-B is therefore likely to involve cellular signal perception and transduction processes resulting in the stimulation of transcription. Hence these genes provide ideal subjects for investigation of the cellular and molecular mechanisms involved.

UV-B photoreception

Possible mechanisms of UV-B perception

The mechanisms by which higher plants perceive UV-B wavelengths and initiate physiological responses, in particular the regulation of gene expression, are not fully understood. There are several possible mechanisms for the specific detection of UV-B. First, direct absorption of UV-B by DNA in the nucleus could, in principle, result in the generation of some type of 'signal' which stimulates the rate of transcription of specific genes. While there is no experimental evidence to support this conjecture, it is premature to dismiss it out of hand. Secondly, UV-B could be detected via its ability to generate reactive oxygen species, such as singlet oxygen. In this case, the increases in transcription observed following exposure to UV-B would be oxidative stress responses rather than photo-responses to UV-B wavelengths *per se*. In support of this hypothesis, Green and Fluhr (1995) demonstrated that the accumulation of a pathogenesis-related protein, PR-1, and its transcripts, in response to UV-B in tobacco leaves was mediated by the production of reactive oxygen species. They found that the UV-B response was greatly diminished in the presence of antioxidants and that the generation of singlet oxygen could substitute for UV-B irradiation in inducing PR-1 accumulation. Thirdly, UV-B could be detected by photoreceptor molecules analogous to other photoreception systems in higher plants. The existence of specific UV/blue photoreceptors and of chromophores that could absorb UV-B wavelengths makes this an attractive possibility, and the evidence for this hypothesis is discussed further below. Of course, the above mechanisms are not mutually exclusive, and it is quite feasible that UV-B regulates gene expression by parallel photoreceptor-mediated and oxidative stress-mediated signalling processes. It is now a priority to establish the relative importance of these mechanisms for a range of genes.

Predictions about the likely properties of a specific UV-B photoreceptor can best be made from examination of other known UV/blue light-absorbing proteins in plants and other organisms. Studies of the action spectra of various UV/blue photoresponses, and experiments with compounds that interact with flavins, indicate that plant UV/blue photoreceptors are likely to be proteins with a bound flavin chromophore (Briggs & Iino, 1983; Galland & Senger, 1988a; Short & Briggs, 1994). Pterins are also strong candidates as chromophores for UV/blue photoreceptors (Galland & Senger, 1988b; Schmidt et al., 1990). The possibility that the chromophore could be a carotenoid was considered (see Briggs & Iino, 1983) but has little experimental support. As discussed below, the UV/blue-absorbing DNA photolyase enzymes in a range of organisms contain a bound flavin and pterin. In addition, a higher plant UV-A/blue photoreceptor has recently been characterised and is also reported to bind a flavin and pterin when expressed in heterologous systems (Lin et al., 1995a; Malhotra et al., 1995; see below).

DNA photolyases

DNA photolyases are enzymes which catalyse the light-dependent repair of DNA (Sancar, 1994; see chapter by Taylor et al., this volume). They have been studied principally in bacteria, fungi, yeast and animal cells. They absorb UV/blue light which provides the energy for catalysis. The chromophore of the photolyase is either a pterin or deazaflavin and is bound to the N-terminal domain of the protein. Following light absorption this chromophore transfers energy to a reduced flavin adenine dinucleotide ($FADH_2$) cofactor in the C-terminal region to initiate DNA repair. The microbial DNA photolyases have little amino acid sequence homology to animal photolyases even though they have the same pterin and flavin chromophores. These enzymes therefore provide interesting comparative models in the study of plant UV/blue photoreceptors.

UV-A/blue photoreceptors in plants

Only recently has molecular information on a plant UV/blue photo-receptor been obtained (Ahmad & Cashmore, 1993; Lin *et al.*, 1995*a,b*; Malhotra *et al.*, 1995). This photoreceptor, termed CRY1 (for *cry*ptochrome), mediates several responses to UV-A/blue light in *Arabidopsis*. The function of CRY1 is known from the phenotypic character-isation of the *hy4* mutant alleles, which are deficient in the photorecep-tor. The original *hy4* mutant, *hy4–2.23N* (Koornneef, Rolff & Spruit, 1980) appears to be a null mutant (C. Lin & A.R. Cashmore, personal communication). The *hy4–2.23N* mutant has longer hypocotyls than the wild-type in blue, UV-A and green light (Koornneef *et al.*, 1980; Ahmad & Cashmore, 1993; Lin *et al.*, 1995*a,b*; Jackson & Jenkins, 1995). Furthermore, it is altered in several other extension growth responses and has reduced blue light-induction of *CHS*, *CHI* and *DFR* transcripts (Jackson & Jenkins, 1995). Anthocyanin induction by blue light is also much reduced (Jackson & Jenkins, 1995; Ahmad, Lin & Cashmore, 1995). However, we have recently found that *hy4–2.23N* is not altered in the UV-B induction of *CHS* transcripts (Fuglevand, Jack-son & Jenkins, 1996), providing genetic evidence that separate UV-B and UV-A/blue photoreceptors regulate expression of the *Arabidopsis CHS* gene.

The gene encoding the CRY1 photoreceptor was cloned through the availability of a *hy4* mutant allele tagged with a T-DNA insertion (Ahmad & Cashmore, 1993). Examination of the derived CRY1 amino acid sequence revealed considerable homology with microbial DNA photolyases in the N-terminal region of the protein. However, the CRY1 protein, produced by expression in *E. coli*, does not have photolyase activity (Lin *et al.*, 1995*a*; Malhotra *et al.*, 1995). The C-terminal domain of the protein has sequence similarity to rat smooth muscle tropomyosin A, and could be involved in initiating signal transduction rather than photoreception. The amino acid sequence homology between CRY1 and the DNA photolyases suggested that CRY1 might bind $FADH_2$ and deazaflavin chromophores. Lin *et al.* (1995*a*) expressed the CRY1 protein in insect cells and showed that it bound an oxidised FAD chromophore which could be reduced to a flavosemiqui-none that absorbed UV-A, blue and green wavelengths, consistent with the expected properties of the photoreceptor in mediating responses *in vivo*. The CRY1 protein, excluding the C-terminal domain, expressed in *E. coli* was reported to bind flavin (FAD) and pterin (methenyltetrahydrofolate) chromophores (Malhotra *et al.*, 1995).

Experiments are now required to determine which chromophores are associated with CRY1 in *Arabidopsis* plants.

The possible nature of a UV-B photoreceptor

It is evident from the above discussion that a UV-B photoreceptor would probably be a protein with flavin and/or pterin chromophores. Pterins in particular are strong candidates for the chromophore of a UV-B photoreceptor. The absorption spectra of pterins are dependent on their redox state and a reduced pterin would be able to absorb in the UV-B region (Galland & Senger, 1988*b*). Khare and Guruprasad (1993) reported that the induction of anthocyanin synthesis in maize by UV-B was inhibited by compounds that inhibit flavin photoreactions, but such compounds are also known to interact with pterins (Warpeha, Kaufman & Briggs, 1992).

No information is available on the likely nature of the UV-B photoreceptor apoprotein. Ahmad and Cashmore (1993) reported that a CRY1 DNA probe hybridised to several fragments on Southern blots of *Arabidopsis* genomic DNA, and it is therefore likely that one or more genes related to CRY1 are present in the genome. Recently, a CRY2 sequence has been reported which resembles CRY1 in the chromophore-binding domain but not the C-terminal domain (Ahmad & Cashmore, 1996). The function of this putative photoreceptor is unknown but, given its similarity to CRY1, may be another UV-A/blue photoreceptor. In *Sinapis*, a protein with sequence similarity to microbial DNA photolyases has been cloned (Batschauer, 1993) which does not have significant photolyase activity (Malhotra *et al.*, 1995) and is a candidate UV/blue photoreceptor. This protein, expressed in *E. coli* cells binds the same flavin and pterin as CRY1 (Malhotra *et al.*, 1995). Photophysiological, biochemical and genetic data indicate that there are likely to be a number of different UV/blue photoreceptors in plants (Kaufman, 1993; Short & Briggs, 1994; Jenkins *et al.*, 1995). Hence the fact that further sequences related to photolyases and CRY1 have not yet been isolated may suggest that there is significant sequence divergence among the photoreceptors. Although cloning strategies based on possible sequence homologies may be successful, the isolation of a mutant altered specifically in UV-B photoreception will enable a function to be assigned unequivocally to a putative UV-B photoreceptor.

UV-B signal transduction

Information on signal transduction in plants is now starting to accumulate by a combination of cell physiological, biochemical, molecular and genetic approaches. At present there is no published information specifically on UV-B signal transduction. However, some information is available on blue light signal transduction in various responses (Kaufman, 1993; Short & Briggs, 1994; Jenkins *et al.*, 1995; Christie & Jenkins, 1996). With regard to the stimulation of *CHS* expression, information has been published on phytochrome signal transduction (Neuhaus *et al.*, 1993; Bowler *et al.*, 1994*a,b*) and transduction of other signals (Mol *et al.*, 1996). Studies of *PAL* signal transduction have primarily focused on its induction by pathogen attack (Lamb *et al.*, 1989; Dixon & Lamb, 1990). The phytochrome signal transduction pathway which regulates *CHS* expression has been studied by microinjection of putative signal transduction components, inhibitors or agonists into tomato hypocotyl cells and by pharmacological experiments in cultured soybean cells (Neuhaus *et al.*, 1993; Bowler *et al.*, 1994*a,b*). The pathway involves a G-protein as an early component and cGMP as a second messenger rather than calcium. Calcium functions in other signal transduction pathways initiated by phytochrome (Millar, McGrath & Chua, 1994).

Pharmacological experiments in cultured cells

A particularly promising approach to obtain information on the signal transduction processes involved in the regulation of expression of *CHS* and other genes by UV-B is to carry out pharmacological experiments in cellular systems. We (Christie & Jenkins, 1996) have used an *Arabidopsis* cell suspension culture for this purpose. These cells show the same responses to different light qualities in the regulation of *CHS* expression as mature *Arabidopsis* leaf tissue. That is, *CHS* transcript levels are induced by UV-B and UV-A/blue light rather than red and far-red light. Thus there appears to be little response mediated by phytochrome. Following exposure to UV-B and UV-A/blue light *CHS* transcript levels increase within a few hours in the cells, which facilitates biochemical studies of the signal transduction processes. We have introduced inhibitors of known cellular signal transduction components into the cells and have started to define which components are involved in the UV-B and UV-A/blue signal transduction pathways. In contrast to the phytochrome signal transduction pathways described in tomato and soybean, the UV-B and UV-A/blue induction of *CHS* in *Arabidopsis* cells is likely to involve calcium because it is inhibited by known calcium channel blockers. Moreover, there is no induction by cGMP. Both UV-B and UV-A/blue induction are likely to involve reversible protein phosphorylation because the responses are inhibited by kinase and phosphatase inhibitors. The only difference we have detected between the UV-B and UV-A/blue pathways to date is that calmodulin is involved in the UV-B pathway but not the UV-A/blue pathway. The evidence for this is that *CHS* expression in response to UV-B is inhibited by the calmodulin antagonist W-7, whereas the UV-A/blue induction is not. These experiments provide the first information on the cellular processes involved specifically in UV-B signal transduction.

Synergistic interaction between UV-B, UV-A and blue light signal transduction pathways

Further experiments indicate that the signal transduction pathway responsible for the induction of *CHS* expression by UV-B in plants does not function independently of other signal transduction pathways. The response to UV-B is enhanced synergistically by both UV-A and blue light (Fuglevand *et al.*, 1996). In these experiments transgenic *Arabidopsis* plants containing the β-glucuronidase reporter gene fused to a *Sinapis alba CHS* gene promoter (Batschauer, Ehmann & Schäfer, 1991) were used. The promoter drives expression of the GUS gene product and this can easily be assayed in plant extracts. Hence GUS activity is a measure of *CHS* transcription. In mature leaves of the transgenic *Arabidopsis* plants, *CHS–GUS* expression is stimulated by UV-B and blue light, but not by red light (Jackson *et al.*, 1995). We have found that the increase in *CHS–GUS* expression, and endogenous *CHS* transcripts, is enhanced several fold if blue light or UV-A is given simultaneously with UV-B (G. Fuglevand and G.I. Jenkins, unpublished observations). The effect is more than additive of the separate effects of these light qualities and is therefore synergistic. The synergistic response with blue light is observed if blue light precedes UV-B but not vice versa. This indicates that blue light produces a relatively stable signal which enhances the subsequent response to UV-B. In contrast, synergism with UV-A is only observed when it is applied together with UV-B, suggesting that the UV-A signal is transient. When blue light is given, followed by UV-A plus UV-B, the fold induction is approximately twice as great as with either blue plus UV-B or UV-A plus UV-B, indicating that the two synergistic interactions work together in an essentially additive manner to give maximal expression. Previous experiments in parsley cells (Ohl, Hahlbrock & Schäfer, 1989) have also provided evidence for a stable signal produced by blue light which potentiates a subsequent increase in response to UV-B in the regulation of *CHS* expression, but in this case the effect of UV-A was not investigated.

If the above observations are considered in relation to the natural environment, it is evident that the synergistic interactions described would be the norm, since plants are not exposed to UV-B in the absence of UV-A or blue light. This will promote maximal, rapid induction of protective flavonoids. There may, however, be instances where plants receive significant UV-A and blue light in the absence of UV-B. In this case blue light will potentiate the plant to produce high levels of protective flavonoids rapidly when subsequently exposed to UV-B. The synergistic interactions may therefore have adaptive significance. In addition, in many species, including *Arabidopsis*, phytochrome acts in young seedlings to stimulate *CHS* expression and flavonoid biosynthesis (Batschauer *et al.*, 1991; Frohnmeyer *et al.*, 1992; Kaiser *et al.*, 1995). Hence different photoreceptors, acting in some cases at different stages in development, work together to protect the plant against damaging UV-B wavelengths.

Coupling signal transduction to transcription

The stimulation of transcription of *CHS* and other genes is a terminal response of UV-B signal transduction. It is therefore pertinent to consider how the signal transduction pathway is coupled to the regulation of transcription. *CHS*, *PAL* and other UV-regulated genes have been studied in several species and information has been obtained on DNA sequence elements in the promoters of these genes which are concerned with UV-B control. Probably the best studied gene in this respect is the parsley *CHS* gene (Weisshaar *et al.*, 1991*a*). Studies with a parsley cell culture and parsley plants demonstrate that UV-B and blue light are the principal light qualities regulating *CHS* transcription (Hahlbrock & Schell, 1989; Ohl *et al.*, 1989; Frohnmeyer *et al.*, 1992).

Transient expression assays of *CHS–GUS* promoter–reporter fusions in protoplasts obtained from parsley cells enabled functional dissection of the promoter (Schulze-Lefert *et al.*, 1989). The promoter region contains two light regulatory units (LRUs), LRU1 and 2, each of which can mediate UV light regulation and which together give high levels of UV-induction (Fig. 2; Weisshaar *et al.*, 1991*a*). Each LRU contains binding sites, termed boxes, for two transcription factors. Thus LRU1 contains boxes I and II and LRU2 contains boxes III and IV (Fig. 2). Mutation of specific nucleotides within these elements leads to loss of function. Studies of other UV-regulated *CHS* promoters, for example, in *Sinapis* (Batschauer *et al.*, 1991) and *Arabidopsis* (B. Weisshaar, personal communication), reveal a similar structure, at least in the presence of a light response unit comparable to LRU1 in parsley. The boxes

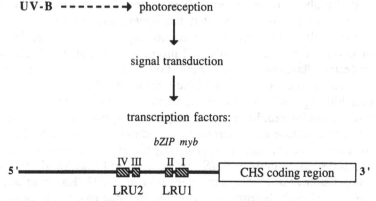

Fig. 2. The promoter region of the parsley *CHS* gene showing DNA sequence elements involved in binding transcription factors concerned with UV-induction.

in the parsley *CHS* promoter bind transcription factors. Boxes II and III bind a basic leucine zipper class of transcription factor which has been cloned (Weisshaar *et al.*, 1991*b*). Box I appears to bind a *myb*-like transcription factor (B. Weisshaar, personal communication) whereas the factor binding to box IV is unknown.

The *Sinapis CHS* promoter region equivalent to the parsley LRU1 is sufficient for induction by UV-B in transgenic tobacco seedlings (Kaiser *et al.*, 1995). Similarly, the *Arabidopsis CHS* LRU region is sufficient for the induction of transcription specifically by UV-B, as well as UV-A/blue light, in transient expression experiments with protoplasts derived from an *Arabidopsis* cell culture (J.M. Christie, U. Hartmann, B. Weisshaar & G. I. Jenkins, unpublished observation).

We can therefore conclude that UV-B signal transduction terminates with an effect on the transcription factors which bind to the LRU region. In principle, UV-B could regulate biogenesis and/or activation of relevant transcription factors. Present evidence suggests that both types of regulation may be involved. UV-containing white light stimulates an increase in transcripts encoding parsley CPRF1, a transcription factor which interacts with the box II and III sequence elements, which is more rapid than the increase in *CHS* transcripts (Weisshaar *et al.*, 1991*b*; Feldbrügge *et al.*, 1994). There is also evidence that transcription factors which interact with box II are regulated by phosphorylation (Harter *et al.*, 1994). Harter *et al.* (1994) report that transcription factors that interact with box II are present in the cytosol and become activated and transported to the nucleus following illumination in reactions that involve protein phosphorylation. Hence the molecular mechanisms of *CHS* transcriptional regulation by UV-B are likely to be complex and much remains to be learnt about the processes through which cellular signal transduction ultimately causes an increase in transcription. The availability of functionally defined promoter elements, cloned transcription factors and cellular systems in which biochemical and molecular experimentation can be combined will facilitate progress.

The genetic approach

Use of mutants in studies of UV-B responses

In recent years the application of a genetic approach, involving the isolation and characterisation of mutants, has provided new information on plant responses to UV-B. Some of this work is discussed in the section on 'Accumulation of UV-absorbing pigments'. *Arabidopsis thaliana* has been chosen for much of this research because of its well documented advantages for genetic experimentation combined with its small genome size, which facilitates gene isolation. The value of the genetic approach is several-fold. Firstly, the examination of mutant phenotypes provides valuable insights into gene function. Secondly, genetic analysis, through the construction and phenotypic characterisation of double mutants, can be used to obtain information on the functional interactions between genes. Thirdly, in *Arabidopsis* it is possible to isolate the genes identified by mutation using several approaches.

The prospect of isolating UV-B perception or signal transduction mutants

Is it possible to use the genetic approach to identify components concerned specifically with UV-B perception and signal transduction? The answer is surely yes, although no mutants altered specifically in these processes have yet been reported. It might be argued that one way to isolate such mutants is to screen for increased sensitivity to UV-B. Mutants isolated in this way may be unable to produce protective molecules through a failure to detect or respond to a UV-B signal. In principle, this is a valid approach, but the problem is that several other classes of mutants will be obtained at probably greater frequency. For example, mutants deficient in key enzymes involved in flavonoid or sinapic acid ester biosynthesis will have increased sensitivity to UV-B as shown by Li *et al.* (1993), Lois and Buchanan (1994) and Landry *et al.* (1995). Furthermore, mutants impaired in the ability to repair or tolerate UV-B damage may also be obtained. In fact several workers have isolated mutants by screening for increased sensitivity to UV-B and these do fall into the above categories rather than being altered in UV-B perception and signal transduction. The *uvr1* mutant isolated by Britt *et al.* (1993) and *uvh1* isolated by Harlow *et al.* (1994) both show increased UV-induced DNA damage and are likely to be altered in DNA repair mechanisms.

An alternative approach to isolate mutants altered in UV-B perception and signal transduction is to use a transgene expression screen. The strategy is to look for mutants which are altered in the stimulation of expression of specific genes by UV-B. Such mutants will not be altered in DNA repair or in biosynthetic enzymes, but will be altered in the ability to detect UV-B and transduce the signal to stimulate gene expression. The basis of the screen is to produce transgenic plants in which the expression of a suitable reporter gene is driven by a promoter which is responsive to UV-B. The *CHS* promoter is well suited to this purpose, but other promoters could be used. The 'wild-type' transgenic seedlings will show elevated expression of the reporter in response to UV-B and so M2 plants, derived from a selected transgenic line which is isogenic and homozygous for the transgene, can be screened for altered expression of the reporter gene. The choice of reporter gene is critical. Strong selectable markers such as antibiotic resistance genes (Susek *et al.*, 1993), alcohol dehydrogenase (Severin, Wagner & Schoff, 1995), the *tms2* gene of *Agrobacterium* T-DNA (Karlin-Neumann, Brusslan & Tobin, 1991) or cytosine deaminase (Perera, Linard & Signer, 1993) could be used. These provide powerful approaches to isolate mutants that either fail to stimulate *CHS* promoter activity or switch on the promoter under non-inductive conditions. The alternatives to selectable markers are screenable markers such as GUS (Takahashi, Naito, Komeda, 1992; Jackson *et al.*, 1995) or luciferase (Millar *et al.*, 1995). These allow mutants to be isolated upon a background of expression, i.e. an altered quantitative response to the stimulus, but are more laborious to perform. Plants expressing combinations of transgenes increase the applicability of the approach. They also enable spurious mutants to be eliminated which arise from mutation of the transgene itself rather than in a component which regulates transgene expression: the probability of two transgenes being mutated in the same plant is very low.

In principle, with transgene expression screens, it is feasible to isolate mutants with either decreased or increased gene expression in response to UV-B. It is therefore possible to identify components which are essential for the response to UV-B and also those which limit the extent of the response. Examples of 'negative regulators', i.e. components which constrain the response under particular conditions, have been identified by mutation in other systems, and appear to be a feature of plant signal transduction (Bowler & Chua, 1994). Mutants in such components would not so easily be identified in conventional screens for altered UV-B sensitivity. The feasibility of isolating mutants with impaired UV-B photoreception has been established by the characterisation of the UV-A/blue photoreceptor CRY1, identified by the *hy4* mutant. Moreover, there are examples of mutants which appear to be altered in light signal transduction (Whitelam *et al.*, 1993) so, in principle, UV-B signal transduction mutants could be obtained. Such mutants are likely to be defective in components specific to the UV-B response, because components involved in a range of cellular signal transduction processes may give lethal mutant phenotypes.

Several workers have reported the successful application of transgene expression screens of various types (Susek *et al.*, 1993; Li *et al.*, 1993, 1995; Jackson *et al.*, 1995; Millar *et al.*, 1995). The only screen to employ a *CHS* promoter was reported by Jackson *et al.* (1995). In this case, transgenic *Arabidopsis* plants were produced in which GUS was fused to a *Sinapis alba CHS* promoter. The plants showed a large increase in GUS expression in response to UV-B, blue light or an increased fluence rate of white light. A screen was undertaken based on the fluorimetric assay for GUS in which leaves from individual M2 transgenic seedlings were assayed in microtitre plates. Putative mutants with either increased or decreased *CHS–GUS* expression in white light were obtained. The critical test for a regulatory mutant is whether expression of the endogenous *CHS* gene is altered identically to the transgene. The mutants with low *CHS–GUS* expression had normal white light-stimulation of *CHS* transcript levels and so were not regulatory mutants. It appeared that transgene inactivation had occurred in these individuals. In contrast, mutants with elevated transgene expression were similarly altered in *CHS* transcript levels. Jackson *et al.* (1995) showed that one mutant, designated *icx1* (increased chalcone synthase expression), had enhanced white light stimulation of *CHS*, *CHI* and *DFR* transcript levels and elevated light induction of anthocyanin. This mutant therefore appears to be altered in a negative regulator of *CHS* transcription. As expected, since the screen was undertaken in white light, *icx1* is not altered in its response specifically to UV-B (H.K. Wade, G. Fuglevand & G.I. Jenkins, unpublished data). Nevertheless, this work demonstrates the feasibility of isolating mutants which are specifically altered in response to UV-B. Therefore, further screens have since been undertaken in UV-B to identify mutants with altered *CHS* expression (G. Fuglevand & G.I. Jenkins, unpublished data). Several more mutants have been isolated with increased or decreased expression of endogenous *CHS* transcripts in UV-B and further research will establish the specificity of these mutants with respect to UV-B and other wavelengths.

Summary and perspectives

Recent research supports the hypothesis that plant cells have UV-B photoreception-transduction pathways that are responsible for the stimulation of transcription of genes concerned with the production of UV-protective molecules. This is an important conclusion because it counters the notion that UV-B has only damaging effects on cells in which the radiation is absorbed directly by the affected macromolecules. However, very little information is yet available on the nature of the putative UV-B photoreceptor(s), the UV-B signal transduction processes and the molecular mechanisms through which UV-B signal transduction is coupled to transcription. Moreover, the extent to which UV-B responses are the consequence of oxidative stress has still to be established.

It is likely that the most rapid progress in dissecting these important cellular processes will be made through a combination of complementary experimental approaches. Systems are now available in which biochemical studies of UV-B signal transduction coupled to the transcription of specific genes can be undertaken. Cultured cells in which specific transcripts or promoter–reporter fusions can be assayed in response to UV-B provide the basis for such research. Experiments in cellular or cell-free sytems in which defined promoter elements are used enable the transcription factor targets of the signal transduction pathways to be defined and their mechanisms of activation to be investigated. The application of a genetic approach should lead to the isolation of mutants which are altered specifically in UV-B photoreception and signal transduction. Transgene expression screens offer the best approach for isolating such mutants because screens for UV-B sensitivity alone generate additional classes of mutants. The genetic approach will, in due course, result in the isolation of key UV-B phototransduction components which can then be studied using the systems developed for biochemical and molecular experiments. Hence there is every prospect that considerable progress will be made in understanding these important processes within the next five years. This research will lead to an understanding of the genetic basis of variation in susceptibility to UV-B in natural populations and provide the means of manipulating crop plants to increase their UV-B tolerance.

Acknowledgements

GIJ thanks the BBSRC for a research grant in the UV-B section of the Biological Adaptation to Global Environmental Change Programme which supports GF. We are grateful to the Gatsby Charitable Foundation for further financial support and the award of a Sainsbury PhD Studentship to JMC. Thanks are due to other members of the laboratory and to Dr Bernd Weisshaar and colleagues at the Max-Planck-Institut, Köln, for discussions of this research.

References

Ahmad, M. & Cashmore, A.R. (1993). *HY4* gene of *A. thaliana* encodes a protein with characteristics of a blue-light photoreceptor. *Nature*, **366**, 162–6.

Ahmad, M. & Cashmore, A.R. (1996). Seeing blue: the discovery of cryptochrome. *Plant Molecular Biology*, **30**, 851–61.

Ahmad, M., Lin, C. & Cashmore, A.R. (1995). Mutations throughout an *Arabidopsis* blue-light photoreceptor impair blue-light-responsive anthocyanin accumulation and inhibition of hypocotyl elongation. *Plant Journal*, **8**, 653–8.

Batschauer, A. (1993). A plant gene for photolyase: an enzyme catalyzing the repair of UV-light-induced DNA damage. *Plant Journal*, **4**, 705–9.

Batschauer, A., Ehmann, B. & Schäfer, E. (1991). Cloning and characterisation of a chalcone synthase gene from mustard and its light-dependent expression. *Plant Molecular Biology*, **16**, 175–85.

Beggs, C.J. & Wellmann, E. (1994). Photocontrol of flavonoid biosynthesis. In *Photomorphogenesis in Plants*, 2nd edn (Kendrick, R.E. & Kronenberg, G.H.M. eds), pp. 733–751. Kluwer, Dordrecht.

Bowler, C. & Chua, N-H. (1994). Emerging themes of plant signal transduction. *Plant Cell*, **6**, 1529–41.

Bowler, C., Neuhaus, G., Yamagata, H. & Chua, N-H. (1994*a*). Cyclic GMP and calcium mediate phytochrome phototransduction. *Cell*, **77**, 73–81.

Bowler, C., van Montagu, M. & Inzé, D. (1992). Superoxide dismutase and stress tolerance. *Annual Review of Plant Physiology and Plant Molecular Biology*, **43**, 83–116.

Bowler, C., Yamagata, H., Neuhaus, G. & Chua, N-H. (1994*b*). Phytochrome signal transduction pathways are regulated by reciprocal control mechanisms. *Genes and Development*, **8**, 2188–202.

Briggs, W.R. & Iino, M. (1983). Blue-light-absorbing photoreceptors in plants. *Philosophical Transactions of the Royal Society, London, B*, **303**, 347–59.

Britt, A.B., Chen, J-J., Wykoff, D. & Mitchell, D. (1993). A UV-sensitive mutant of *Arabidopsis* defective in the repair of pyrimidine-pyrimidinone (6–4) dimers. *Science*, **261**, 1571–4.

Chapple, C.S., Vogt, T., Ellis, B.E. & Somerville, C.R. (1992). An *Arabidopsis* mutant defective in the general phenylpropanoid pathway. *Plant Cell*, **4**, 1413–24.

Christie, J.M. & Jenkins, G.I. (1996). Distinct UV-B and UV-A/blue light signal transduction pathways induce chalcone syntase gene expression in *Arabidopsis* cells. *Plant Cell*, **8**, 1555–67.

Dangl, J.L., Hahlbrock, K. & Schell, J. (1989). Regulation and structure of chalcone synthase genes. In *Plant Nuclear Genes and their Expression, Vol. 6*, (I.K. Vasil & Schell, J. eds), pp. 155–73. Academic Press, New York.

Day, T.A., Martin, G. & Vogelmann, T.C. (1993). Penetration of UV-B radiation in foliage: evidence that the epidermis behaves as a non-uniform filter. *Plant Cell and Environment*, **16**, 735–41.

Dixon, R.A. & Lamb, C.J. (1990). Molecular communication in interactions between plants and microbial pathogens. *Annual Review of Plant Physiology and Plant Molecular Biology*, **41**, 339–67.

Dixon, R.A. & Paiva, N.L. (1995). Stress-induced phenylpropanoid metabolism. *Plant Cell*, **7**, 1085–97.

Feldbrügge, M., Sprenger, M., Dinkelbach, M., Yazaki, K., Harter, K. & Weisshaar, B. (1994). Functional analysis of a light-responsive plant bZIP transcriptional regulator. *Plant Cell*, **6**, 1607–21.

Frohnmeyer, H., Ehmann, B., Kretsch, T., Rocholl, M., Harter, K., Nagatani, A., Furuya, M., Batschauer, A., Hahlbrock, K. & Schäfer, E. (1992). Differential usage of photoreceptors for chalcone synthase gene expression during plant development. *Plant Journal*, **2**, 899–906.

Fuglevand, G., Jackson, J.A. & Jenkins, G.I. (1996). UV-B, UV-A, and blue light signal transduction pathways interact synergistically to regulate chalcone synthase gene expression in *Arabidopsis*. *Plant Cell*, **8**, 2347–57.

Galland, P. & Senger, H. (1988*a*). The role of flavins as photoreceptors. *Journal of Photochemistry and Photobiology B*, **1**, 277–94.

Galland, P. & Senger, H. (1988*b*). The role of pterins in the photoreception and metabolism of plants. *Photochemistry and Photobiology*, **48**, 811–20.

Green, R. & Fluhr, R. (1995). UV-B-induced PR-1 accumulation is mediated by active oxygen species. *Plant Cell*, **7**, 203–12.

Hahlbrock, K. & Schell, D. (1989). Physiology and molecular biology of phenyl-propanoid metabolism. *Annual Review of Plant Physiology and Plant Molecular Biology*, **40**, 347–69.

Harlow, G.R., Jenkins, M.E., Pittalwala, T.S. & Mount, D.W. (1994). Isolation of *uvh-1*, an *Arabidopsis* mutant hypersensitive to ultraviolet light and ionizing radiation. *Plant Cell*, **6**, 227–35.

Harter, K., Kircher, S., Frohmeyer, Krenz, M., Nagy, F. & Schäfer, E. (1994). Light-regulated modification and nuclear translocation of cytosolic G-box binding factors in parsley. *Plant Cell*, 6, 545–59.

Jackson, J.A. & Jenkins, G.I. (1995). Extension growth responses and flavonoid biosynthesis gene expression in the *Arabidopsis hy4* mutant. *Planta*, 197, 233–9.

Jackson, J.A., Fuglevand, G., Brown, B.A., Shaw, M.J. & Jenkins, G.I. (1995). Isolation of *Arabidopsis* mutants altered in the light-regulation of chalcone synthase gene expression using a transgenic screening approach. *Plant Journal*, 8, 369–80.

Jenkins, G.I., Christie, J.M., Fuglevand, G., Long, J.C. & Jackson, J.A. (1995). Plant responses to UV and blue light: biochemical and genetic approaches. *Plant Science*, 112, 117–38.

Kaiser, T., Emmler, K., Kretsch, T., Weisshaar, B., Schäfer, E. & Batschauer, A. (1995). Promoter elements of the mustard *CHS1* gene are sufficient for light regulation in transgenic plants. *Plant Molecular Biology*, 28, 219–29.

Karlin-Neumann, G.A., Brusslan, J.A. & Tobin, E.M. (1991). Phytochrome control of the *tms2* gene in transgenic *Arabidopsis*: a strategy for selecting mutants in the signal transduction pathway. *Plant Cell*, 3, 573–82.

Kaufman, L.S. (1993). Transduction of blue light signals. *Plant Physiology*, 102, 333–7.

Khare, M. & Guruprasad, K.N. (1993). UV-B-induced anthocyanin synthesis in maize regulated by FMN and inhibitors of FMN photoreactions. *Plant Science*, 91, 1–5.

Knight, H. & Knight, M.R. (1995). Recombinant aequorin methods for intracellular calcium measurement in plants. In *Methods in Cell Biology* Vol. 49, (Galbraith, D.W., Bohnert, H.J. & Bourque, D.P. eds), pp. 201–16. Academic Press, London.

Koornneef, M., Rolff, E. & Spruit, C.J.P. (1980). Genetic control of light-inhibited hypocotyl elongation in *Arabidopsis thaliana*. *Zeitschrift für Pflanzenphysiologie*, 100, 147–60.

Kootstra, A. (1994). Protection from UV-B-induced DNA damage by flavonoids. *Plant Molecular Biology*, 26, 771–4.

Kubasek, W.L., Shirley, B.W., McKillop, A., Goodman, H.M., Briggs, W. & Ausubel, F.M. (1992). Regulation of flavonoid biosynthetic genes in germinating *Arabidopsis* seedlings. *Plant Cell*, 4, 1229–36.

Lamb, C.J., Lawton, M.A., Dron, M. & Dixon, R.A. (1989). Signals and transduction mechanisms for activaton of plant defence against microbial attack. *Cell*, 56, 215–24.

Landry, L.G., Chapple, C.C.S. & Last, R.L. (1995). *Arabidopsis* mutants lacking phenolic sunscreens exhibit enhanced ultraviolet-B injury and oxidative damage. *Plant Physiology*, 109, 1159–66.

Li, J., Ou-Lee, T.-M., Raba, R., Amundson, R.G. & Last, R.L. (1993).

Arabidopsis flavonoid mutants are hypersensitive to UV-B irradiation. *Plant Cell*, **5**, 171–9.

Lin, C., Robertson, D.E., Ahmad, M., Raibekas, A.A., Jorns, M.S., Dutton, P.L. & Cashmore, A.R. (1995*a*). Association of flavin adenine dinucleotide with the *Arabidopsis* blue light receptor CRY1. *Science*, **269**, 968–70.

Lin, C., Ahmad, M., Gordon, D. & Cashmore, A.R. (1995*b*). Expression of an *Arabidopsis* cryptochrome gene in transgenic tobacco results in hypersensitivity to blue, UV-A and green light. *Proceedings of the National Academy of Sciences, USA*, **92**, 8423–7.

Lois, R. (1994). Accumulation of UV-absorbing flavonoids induced by UV-B radiation in *Arabidopsis thaliana* L. *Planta*, **194**, 498–503.

Lois, R. & Buchanan, B.B. (1994). Severe sensitivity to ultraviolet radiation in an *Arabidopsis* mutant deficient in flavonoid accumulation. *Planta*, **194**, 504–9.

Maathuis, F.J.M. & Sanders, D. (1995). Patch-clamping plant cells. In *Methods in Cell Biology*, Volume 49 (Galbraith, D.W., Bohnert, H.J. & Bourque, D.P. eds.), pp. 293–302. Academic Press, London.

Malhotra, K., Kim, S.-T., Batschauer, A., Dawut, L. & Sancar, A. (1995). Putative blue-light photoreceptors from *Arabidopsis thaliana* and *Sinapis alba* with a high degree of sequence homology to DNA photolyase contain the two photolyase cofactors but lack DNA repair activity. *Biochemistry*, **34**, 6892–9.

Millar, A.J., McGrath, R.B. & Chua, N.-H. (1994). Phytochrome phototransduction pathways. *Annual Review of Genetics*, **28**, 325–49.

Millar, A.J., Carre, I.A., Strayer, C.A., Chua, N.-H. & Kay, S.A. (1995). Circadian clock mutants in *Arabidopsis* identified by luciferase imaging. *Science*, **267**, 1161–3.

Miller, A.J. (1995). Ion-selective microelectrodes for measurement of intracellular ion concentrations. In *Methods in Cell Biology Volume 49*, (Galbraith, D.W., Bohnert, H.J. & Bourque, D.P. eds), pp. 275–91. Academic Press, London.

Mitchell, D.L. & Karentz, D. (1993). The induction and repair of DNA photodamage in the environment. In *Environmental UV Photobiology* (Young, A.R., ed.), pp. 345–77. Plenum Press, New York.

Mol, J., Jenkins, G.I., Schäfer, E. & Weiss, D. (1996). Signal perception, transduction and gene expression involved in anthocyanin biosynthesis. *Critical Reviews in Plant Science* (in press).

Neuhaus, G., Bowler, C., Kern, R. & Chua, N.-H. (1993). Calcium/calmodulin-dependent and -independent phytochrome signal transduction pathways. *Cell*, **73**, 937–52.

Ohl, S., Hahlbrock, K. & Schäfer, E. (1989). A stable blue-light-derived signal modulates ultraviolet-light-induced activation of the chalcone synthase gene in cultured parsley cells. *Planta*, **177**, 228–36.

Perera, R.J., Linard, G.C. & Signer, E.R. (1993). Cytosine deaminase as a negative selective marker for *Arabidopsis*. *Plant Molecular Biology*, **23**, 793–9.

Robberecht, R. & Caldwell, M.M. (1978). Leaf epidermal transmittance of ultraviolet radiation and its implication for plant sensitivity to ultraviolet-radiation induced injury. *Oecologia*, **32**, 277–87.

Sancar, A. (1994). Structure and function of DNA photolyase. *Biochemistry*, **33**, 2–9.

Schmelzer, E., Jahnen, W. & Hahlbrock, K. (1988). *In situ* localisation of light-induced chalcone synthase mRNA, chalcone synthase, and flavonoid end products in epidermal cells of parsley leaves. *Proceedings of the National Academy of Sciences, USA*, **85**, 2989–93.

Schmidt, W., Galland, P., Senger, H. & Furuya, M. (1990). Microspectrophotometry of *Euglena gracilis*. Pterin- and flavin-like fluorescence in the paraflagellar body. *Planta*, **182**, 375–81.

Schnabl, H., Weissenböck, G., Sachs, G. & Scharf, H. (1989). Cellular distribution of UV-absorbing compounds in guard and subsidiary cells of *Zea mays* L. *Plant Physiology*, **135**, 249–52.

Schulze-Lefert, P., Becker-Andre, M., Schulz, W., Hahlbrock, K. & Dangl, J.L. (1989). Functional architecture of the light-responsive chalcone synthase promoter from parsley. *Plant Cell*, **1**, 707–14.

Severin, K., Wagner, A. & Schoff, F. (1995). A heat-inducible ADH gene as a reporter gene for a negative selection in transgenic *Arabidopsis*. *Transgenic Research*, **4**, 163–72.

Short, T.W. & Briggs, W.R. (1994). The transduction of blue light signals in higher plants. *Annual Review of Plant Physiology and Plant Molecular Biology*, **45**, 143–71.

Stapleton, A.E. (1989). Ultraviolet radiation and plants: burning questions. *Plant Cell*, **4**, 1353–8.

Stapleton, A.E. & Walbot, V. (1994). Flavonoids can protect maize DNA from the induction of ultraviolet radiation damage. *Plant Physiology*, **105**, 881–9.

Susek, R.E., Ausubel, F.M. & Chory, J. (1993). Signal transduction mutants of *Arabidopsis thaliana* uncouple nuclear *CAB* and *RBCS* gene expression from chloroplast development. *Cell*, **74**, 787–99.

Takahashi, T., Naito, S. & Komeda, Y. (1992). The *Arabidopsis* HSP18.2 promoter/GUS fusion in transgenic *Arabidopsis* plants: a powerful tool for the isolation of regulatory mutants of the heat shock response. *Plant Journal*, **2**, 751–61.

Takeda, J., Obi, I. & Yoshida, K. (1994). Action spectra of phenylalanine ammonia-lyase and chalcone synthase expression in carrot cells in suspension. *Physiologia Plantarum*, **91**, 517–521.

Tevini, M. & Teramura, A.H. (1989). UV-B effects on terrestrial plants. *Photochemistry and Photobiology*, **50**, 479–87.

Tevini, M., Braun, J. & Fieser, G. (1991). The protective function of the epidermal layer of rye seedlings against ultraviolet-B radiation. *Photochemistry and Photobiology*, **53**, 329–33.

van der Meer, I.M., Stuitje, A.R. & Mol, J.N.M. (1993). Regulation of general phenylpropanoid and flavonoid gene expression. In *Control of Plant Gene Expression* (Verma, D.P.S., ed.), pp. 125–55. CRC Press, Boca Raton, Florida.

Warpeha, K.M.F., Kaufman, L.S. & Briggs, W.R. (1992). A flavoprotein may mediate the blue light-activated binding of guanosine 5'-triphosphate to isolated plasma membranes of *Pisum sativum* L. *Photochemistry and Photobiology*, **55**, 595–603.

Weisshaar, B., Block, A., Armstrong, G.A., Herrmann, A., Schulze-Lefert, P. & Hahlbrock, K. (1991*a*). Regulatory elements required for light-mediated expression of the *Petroselinum crispum* chalcone synthase gene. In *Molecular Biology of Plant Development, Symp. Soc. Exp. Biol. XLV.* (Jenkins, G.I. & Schuch, W., eds), pp. 191–210. Company of Biologists, Cambridge.

Weisshaar, B., Armstrong, G.A., Block, A., da Costa e Silva, O. & Hahlbrock, K. (1991*b*). Light-inducible and constitutively expressed DNA-binding proteins recognizing a plant promoter element with functional relevance in light-responsiveness. *EMBO Journal*, **10**, 1777–86.

Whitelam, G.C., Johnson, E., Peng, J., Carol, P., Anderson, M.L., Cowl, J.S. & Harberd, N.P. (1993). Phytochrome A null mutants of *Arabidopsis* display a wild-type phenotype in white light. *Plant Cell*, **5**, 757–68.

J.F. BORNMAN, S. REUBER, Y-P. CEN
and G. WEISSENBÖCK

Ultraviolet radiation as a stress factor and the role of protective pigments

Introduction

Ozone-depleting substances as well as global climatic changes (for example, increasing 'greenhouse' gases) may alter ozone chemistry by creating conditions that facilitate an increased ozone degradation. It is also of concern that ozone reductions are not confined to the Antarctic, but extend to mid-latitudes in both hemispheres (see chapter by Pyle, this volume). A reduced stratospheric ozone layer will result in a selective increase at the earth's surface of ultraviolet radiation in the spectral region 280–320 nm (UV-B). The wavelength specificity is due to the absorption coefficient of ozone, which decreases sharply in the short wave UV-B region. Enhanced levels of UV-B radiation may affect the regulatory mechanisms of many plant processes, causing changes in the internal photosynthetic light micro-environment of the leaf, inducing ultrastructural changes, altering morphology and compounding photo-inhibition by visible radiation, to name but a few.

The tolerance and avoidance of UV-B stress is manifested in a range of plant strategies, with increased UV-screening compounds being one of the most widely occurring responses. In addition, plants growing in environments of naturally high UV-B irradiances tend to be more tolerant than plants from low irradiance areas (Robberecht & Caldwell, 1986; Ziska, Teramura & Sullivan, 1992). The importance of the distribution within leaves of UV-absorbing pigments can be shown by comparing internal radiation gradients with the pigment concentrations and absorption curves. Visible radiation, UV-A (320–400 nm) and UV-B influence the induction and quantitative and qualitative pattern of these UV-screening phenylpropanoid compounds. Phenylalanine ammonia-lyase (PAL) and chalcone synthase (CHS) are key regulatory enzymes in the phenylpropanoid and flavonoid pathway, respectively, and together with the epidermis-specific flavonoids may be useful as markers for assessing UV-adaptability of leaves (Reuber, Bornman & Weissenböck, 1996a,b). The different kinds of phenolic compounds occurring in epidermal and mesophyll cells reflect UV response by specific tissue types with epidermal flavonoids playing a major role. The protective role of flavonoids can be most strikingly demonstrated by comparison of flavonoid mutants and their corresponding mother variety or wildtype, or by a metabolic reduction using an efficient inhibitor of PAL (Reuber *et al.*, 1993). This chapter will attempt to demonstrate the role played by protective plant pigments in the expression of tolerance to UV-B radiation, and also discuss certain correlations between different plant responses.

Relationships between UV-B levels and protective pigments

Two main approaches can be used to study relationships between increased UV-B levels and pigment accumulation. These are a) the induction of phenolic compounds and their enzymatic regulation in plants grown under an enhanced UV-B environment, and b) the behaviour of mutants deficient in flavonoid compounds. Apart from the UV-absorbing phenylpropanoid compounds, a discussion of protective plant pigments should also mention the carotenoids. What then are some of the roles of these compounds which contribute to UV-B radiation tolerance?

The photoprotective role of carotenoids is expressed by the quenching of excess excitation energy of triplet chlorophyll (*Chl), thus preventing *Chl from passing on the energy to oxygen, and consequently decreasing the probability of formation of singlet oxygen ($^1O_2$). Carotenoids may also quench $^1O_2$ directly, thus preventing reactions culminating in peroxidation of membrane lipids. The interaction of carotenoids in this way results in ground-state molecules of both Chl and oxygen. The carotenoids involved in the xanthophyll cycle also play an important role in maintaining energy balance. This cycle has been shown to contribute to photoprotection under visible light by the thermal dissipation of excess excitation energy (Demmig-Adams, 1990). The xanthophyll cycle may be a target of UV-B radiation whereby the de-epoxidation of violaxanthin to zeaxanthin is inhibited (Pfündel, Pan & Dilley, 1992). This was shown for *in vitro*-irradiated chloroplasts as well as for chloroplasts isolated from UV-B treated plants. At the same time, studies done on leaves of *Brassica napus* cv. Paroll of photoinhibition by visible light together with enhanced levels of UV-B radiation tend to show that the xanthophyll cycle activity is increased, as reflected in a greater non-photochemical quenching compared to plants not exposed to photoinhibition and UV-B radiation (Bornman & Sundby-Emanuelsson, 1995). The level of visible radiation applied together with UV-B radiation no doubt influences the cycle in different ways.

Apart from the UV-filtering role of polyphenol compounds, they can function as reducing agents, metal chelators and $^1O_2$ quenchers as well as hydrogen-donating antioxidants (Rice-Evans, 1995). Photobleaching of carotenoids can also be prevented by flavonoids such as kaempferol and quercetin (Takahama, 1982). Of interest is the fact that flavonoids with several hydroxyl groups may confer antioxidant activity (Harborne, 1986; Larson, 1988). This role may be particularly important with regard to plant exposure to UV-B radiation.

Differential flavonoid response

Although for many plant species it has been demonstrated that UV-stimulated flavonoid metabolism is a general response (for a review, see Bornman & Teramura, 1993), it has also been found that UV-B changes the amounts of certain major flavonoid compounds, favouring those with additional hydroxyl groups on ring B of the flavonoid skeleton. This was demonstrated in leaves of *Brassica napus* L. and *Hordeum vulgare* L. (Cen, Weissenböck & Bornman, 1993; Liu, Gitz III & McClure, 1995; Reuber, Bornman & Weissenböck, 1996b). Of the two major compounds in leaves of *B. napus*, namely, quercetin-glycoside and kaempferol-glycoside, the former increased markedly compared to kaempferol. Similarly in *H. vulgare*, where lutonarin and saponarin are the major compounds, lutonarin has been found to increase between 100 and 500% after UV-B irradiation (Liu *et al*, Gitz III & McClure, 1995; Reuber, Bornman & Weissenböck, 1996b). In both *B. napus* and *H. vulgare*, the over-induced compounds possess an additional hydroxyl group on ring B of the flavonoid skeleton as compared to the second major flavonoid. This suggests an increase in the potential antioxidant activity of the plants. However, although flavonoids have been proposed to function as antioxidants in several different ways, for example, through inhibition of lipid peroxidation, via their metal chelating potential and their interaction with ascorbate (Moru *et al.*, 1990; Afanas'ev *et al.*, 1989; Bors, Michel & Schikora, 1995, respectively), these interactions are highly dependent on the redox potentials of the participating compounds (Bors, Michel & Schikora, 1995). Consequently, while flavonoids may be effective radical scavengers, many flavonols have quite high redox potentials which may result in oxidation rather than reduction reactions (Bors *et al.*, 1995).

An example of the HPLC profiles for *B. napus* grown with or without enhanced levels of UV-B radiation, which shows the increase in quercetin under elevated UV-B, is given in Fig. 1. Thus apart from the obvious UV-filtering function of phenolic compounds, other roles may also contribute to increasing the overall protective mechanisms of the plant.

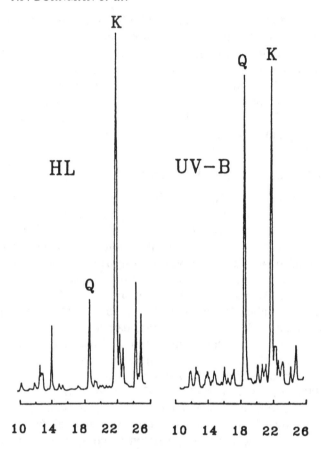

Retention time, min

Fig. 1. HPLC profiles obtained after acid hydrolysis of the crude extracts of leaves of *Brassica napus* grown for 16 days under a photosynthetically active radiation (PAR, HL) of 750 μmol m^{-2} s^{-1} with or without biologically effective UV-B radiation (8.9 kJ m^{-2} day^{-1} from UVB-313 Q-PANEL lamps provided with cellulose diacetate and FBL.2458 Plexiglas to remove wavelengths below 280 nm). Relative absorption was at 335 nm. K, kaempferol; Q, quercetin. (From Cen, Weissenböck & Bornman, 1993.)

Localisation in leaves

In addition to the roles mentioned above, another feature of importance for an effective protection in plants is the pigment distribution across leaves. A study of the localisation of UV-absorbing pigments in leaves shows that certain types of compounds have a specific tissue localisation. Using the primary leaf of rye (*Secale cereale* L.) as an example, different sets of secondary phenolic substances have been localised in the epidermal (hydroxycinnamic acid esters, isovitexin glycosides) and mesophyll tissues (luteolin glucuronides, cyanidin glucosides) (see Anhalt & Weissenböck, 1992). This tissue-specific distribution was studied in both previously etiolated seedlings, exposed to 6 h of UV-B radiation and low visible light, as well as in plants grown under UV-B and relatively high visible light levels (Fig. 2). Of interest is the fact that the hydroxycinnamic acid (HCA) compounds showed little change with irradiation treatments, whereas epidermal flavonoids showed a

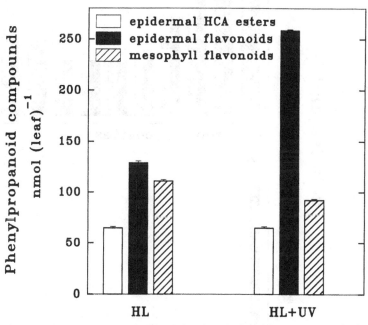

Fig. 2. Tissue-specific distribution of UV-absorbing compounds in leaves of previously etiolated seedlings of *Secale cereale*. Plants were grown under relatively high visible light levels (photosynthetically active radiation (PAR) of 750 μmol m^{-2} s^{-1}) with or without 13 kJ m^{-2} day^{-1} biologically effective UV-B radiation (UVB-313 Q-PANEL lamps provided with cellulose diacetate and FBL .2458 Plexiglas to remove wavelengths below 280 nm). Analyses were done after 7 days of light treatment. Extraction of phenylpropanoids and HPLC analysis was according to Reuber *et al.* (1993). (Figure from Reuber, Bornman & Weissenböck, 1996a; slightly modified.)

162   J.F. BORNMAN *et al.*

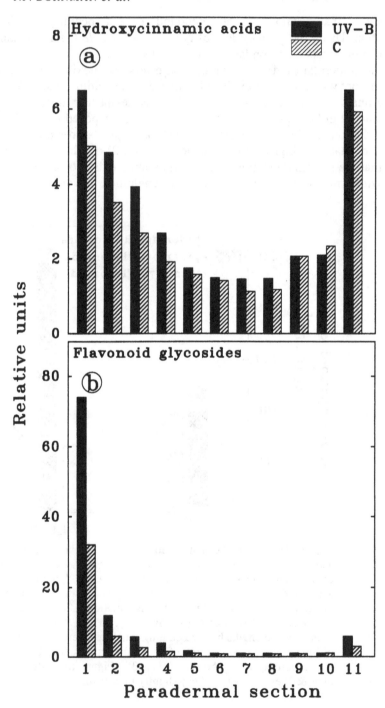

two-fold increase compared with plants not exposed to UV-B radiation. The light sensitivity of the mesophyll flavonoids was characteristic only for etiolated leaves, where UV penetration would be significant, since their synthesis was not stimulated in light-grown material. Thus the HCA compounds are probably under endogenous control, forming a constitutive shield against UV-B radiation, in contrast to the light-inducible epidermal flavonoids (Reuber *et al.*, 1996*a*). Evidence for this was also seen in a preliminary study of *B. napus* (Cen, Weissenböck & Bornman, 1993), where analysis of HCA derivatives from paradermal leaf sections showed a much smaller stimulation by UV-B radiation (Fig. 3a) in contrast to the large induction of flavonoid glycosides (Fig. 3b).

In certain species, leaf trichomes may also play a significant role as a UV-filter, since several species have been reported to have UV-absorbing trichomes (Karabourniotis *et al.*, 1992). The protective effect was demonstrated in *Olea europaea* for which it was shown that the photosynthetic capacity, as measured by chlorophyll fluorescence, declined in dehaired leaves exposed to UV-B radiation (Karabourniotis, Kyparissis & Manetas, 1993). It also has been suggested that, in some species, notably xerophytes, the protective filtering effect of the trichomes may outweigh their role in water conservation (Grammatikopoulos *et al.*, 1994).

Significance of correlations with internal radiation gradients

In order to evaluate the consequences of acclimation to an increased UV-B level and to attempt to link tissue-specific phenylpropanoid compounds with plant tolerance, the distribution of UV-B radiation can be followed through the leaf. This is most effectively done using quartz optical fibres. The fine tip of these microprobes can be inserted into leaf tissue, and by coupling the other end of the fibre to a spectroradiometer,

---

Fig. 3. Distribution of flavonoid glycosides (a) and hydroxycinnamic acid derivatives (b) from paradermal leaf sections of *Brassica napus* (sections 1 to 11 were taken from the adaxial to the abaxial surface, respectively) after 16 days of radiation treatment (see Fig. 1 legend). Relative values were taken from integrated peak areas after detection at 335 nm and calculated on the same leaf area basis. (From Cen, Weissenböck & Bornman, 1993; slightly modified.)

gradients of UV radiation can be collected (Bornman & Vogelmann, 1988, 1991; Cen & Bornman, 1993; Ålenius, Vogelmann & Bornman, 1995; Reuber, Bornman & Weissenböck, 1996*a,b*). Figure 4 shows the penetration of UV-B radiation (310 nm) into primary leaves of 7-day-old rye grown under relatively high PAR and an enhanced UV-B level, with the inset emphasising the difference in the subepidermal, mesophyll layer; penetration of radiation was significantly reduced in plants which had been grown under supplementary UV-B. Using a three-dimensional visualisation of data, good correlation was shown between

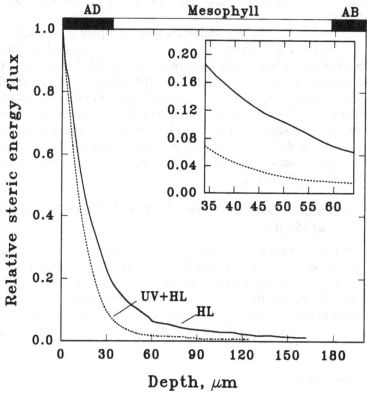

Fig. 4. Penetration of UV-B (310 nm) radiation from the adaxial leaf surface of 7-day-old *Secale cereale*. Growth and irradiation conditions as in Fig. 2. The insert, with the *x*-axis expanded, serves to emphasise the penetration of 310 nm radiation in the subepidermal cell layer. AD, adaxial epidermis; AB, abaxial epidermis. (Figure from Reuber, Bornman & Weissenböck, 1996*a*.)

absorption by bulk UV-screening pigments and the penetration of UV-B radiation in *B. napus* (Ålenius, Vogelmann & Bornman, 1995).

These correlations of pigment distribution and UV penetration are also reflected in photosynthetic response, which is particularly evident with the use of mutants deficient in flavonoid compounds. Photosynthetic capacity in a barley mutant, estimated from the ratio of variable to maximum chlorophyll fluorescence showed a corresponding decrease of 76% compared to the mother variety (Reuber, Bornman & Weissenböck, 1996*b*). However, not all photosynthetic parameters may respond similarly, since Middleton and Teramura (1993) reported that gas exchange in soybean mutants was not correlated with UV-absorbing pigment content.

Conclusions

The modifying influence on plants of an increased UV-B climate remains of concern, since UV-B radiation has been shown to alter the balance of many plant systems, both at the biochemical and whole plant level. With regard to the whole plant level, earlier work on changes in morphology and shifts in competitive balance among different species serve to illustrate the range of UV-B effects (Caldwell, 1977; Gold & Caldwell, 1983; Barnes, Flint & Caldwell, 1990). One response to UV-B radiation is often an increase in various defence mechanisms in plants. An obvious effective way is the reduction in penetration of UV-B radiation by increasing UV-absorbing polyphenol compounds. However, it has been shown that UV-B not only influences the amount of these pigments, but also their composition. Overall, many beneficial functions of polyphenols besides UV-screening appear to be enhanced, and this will lead to protection of an increased number of sensitive processes.

Acknowledgements

We acknowledge financial support from the Natural Science Research Council to J.F.B. and to G.W. (We 630/11–1) from the Deutsche Forschungsgemeinschaft, Bonn, Germany.

References

Afanas'ev, I.B., Dorozhko, A.J., Brodskii, A.V., Kostyuk, V.A. & Potapovitch, A.J. (1989). Chelating and free radical scavenging mechanisms of inhibitory action of rutin and quercetin in lipid peroxidation. *Biochemical Pharmacology*, **38**, 1763–9.

166 J.F. BORNMAN *et al.*

Ålenius, C.M. Vogelmann, T.C. & Bornman, J.F. (1995). A three-dimensional representation of the relationship between penetration of UV-B radiation and UV-screening pigments in leaves of *Brassica napus*. *New Phytologist*, **131**, 297–302.

Anhalt, S. & Weissenböck, G. (1992). Subcellular localization of luteolin glucuronides and related enzymes in rye mesophyll. *Planta*, **187**, 83–8.

Barnes, P.W., Flint, S.D. & Caldwell, M.M. (1990). Morphological responses to crop and weed species of different growth forms to ultraviolet-B radiation. *American Journal of Botany*, **77**, 1354–60.

Bornman, J.F. & Sundby-Emanuelsson, C. (1995). Response of plants to UV-B radiation: Some biochemical and physiological effects. In *Environment and Plant Metabolism: Flexibility and Acclimation. Environmental Plant Biology Series*. (Smirnoff, N., ed.), pp. 245–62. BIOS Science Publications, Oxford, UK.

Bornman, J.F. & Teramura, A.H. (1993). Effects of UV-B radiation on terrestrial plants. In *Environmental UV Photobiology*. (Young, A.R., Björn, L.O., Moan, J. & Nultsch, W., eds), pp. 427–71. Plenum Publ. Co., New York.

Bornman, J.F. & Vogelmann, T.C. (1988). Penetration of blue and UV radiation measured by fiber optics in spruce and fir needles. *Physiologia Plantarum*, **72**, 699–705.

Bornman, J.F. & Vogelmann, T.C. (1991). The effect of UV-B radiation on leaf optical properties measured with fibre optics. *Journal of Experimental Botany*, **42**, 547–54.

Bors, W., Michel, C. & Schikora, S. (1995). Interaction of flavonoids with ascorbate and determination of their univalent redox potentials: a pulse radiolysis study. *Free Radical Biology and Medicine*, **19**, 45–52.

Caldwell, M.M. (1977). The effects of solar UV-B (280–315 nm) on higher plants: Implications of stratospheric ozone reduction. In *Research in Photobiology*. (Castellani, A., ed.), pp. 597–607. Plenum Publishing Co., New York.

Cen, Y-P. & Bornman, J.F. (1993). The effect of exposure to enhanced UV-B radiation on the penetration of monochromatic and polychromatic UV-B radiation in leaves of *Brassica napus*. *Physiologia Plantarum*, **87**, 249–55.

Cen, Y-P., Weissenböck, G. & Bornman, J.F. (1993). The effects of UV-B radiation on phenolic compounds and photosynthesis in leaves of *Brassica napus*. In Physical, Biochemical and Physiological Effects of Ultraviolet Radiation on *Brassica napus* and *Phaseolus vulgaris*. PhD thesis, Lund, Sweden.

Demmig-Adams, B. (1990). Carotenoids and photoprotection in plants: a role for the xanthophyll zeaxanthin. *Biochimica et Biophysica Acta*, **1020**, 1–24.

Gold, W.G. & Caldwell, M.M. (1983). The effects of ultraviolet-B radiation on plant competition in terrestrial ecosystems. *Physiologia Plantarum*, **58**, 435–44.

Grammatikopoulos, G., Karabourniotis, G., Kyparissis, A., Petropoulou, Y. & Manetas, Y. (1994). Leaf hairs of olive (*Olea europaea*) prevent stomatal closure by ultraviolet-B radiation. *Australian Journal of Plant Physiology*, **21**, 293–301.

Harborne, J.B. (1986). Nature, distribution and function of plant flavonoids. In *Plant Flavonoids in Biology and Medicine*. (Cody, V. Middleton, E. & Harborne, J.B. eds), Vol. 213, pp. 15–24. Alan R. Liss, New York.

Karabourniotis, G., Papadopoulos, K., Papamarkou, M. & Manetas, Y. (1992). Ultraviolet-B radiation absorbing capacity of leaf hairs. *Physiologia Plantarum*, **86**, 414–18.

Karabourniotis, G., Kyparissis, A. & Manetas, Y. (1993). Leaf hairs of olive (*Olea europaea*) protect underlying tissues against ultraviolet-B radiation damage. *Environmental and Experimental Botany*, **33**, 341–5.

Larson, R.A. (1988). The antioxidants of higher plants. *Phytochemistry*, **27**, 969–78.

Liu, L., Gitz III, D. C. & McClure, J. W. (1995). Effects of UV-B on flavonoids, ferulic acid, growth and photosynthesis in barley primary leaves. *Physiologia Plantarum*, **93**, 725–33.

Middleton, E.M. & Teramura, A.H. (1993). The role of flavonol glycosides and carotenoids in protecting soybean from ultraviolet-B damage. *Plant Physiology*, **103**, 741–52.

Moru, A., Paya, M., Rios, J.L. & Alcaraz, M.J. (1990). Structure-activity relationships of polymethoxyflavones and other flavonoids as inhibitors of nonenzymic lipid peroxidation. *Biochemical Pharmacology*, **40**, 451–8.

Pfündel, E.E., Pan, R.-S. & Dilley, R.A. (1992). Inhibition of violaxanthin deepoxidation by ultraviolet-B radiation in isolated chloroplasts and intact leaves. *Plant Physiology*, **98**, 1372–80.

Reuber, S., Leitsch, J., Krause, G. H. & Weissenböck, G. (1993). Metabolic reduction of phenylpropanoid compounds in primary leaves of rye (*Secale cereale* L.) leads to increased UV-B sensitivity of photosynthesis. *Zeitschrift für Naturforschung*, **48c**, 749–56.

Reuber, S., Bornman, J.F. & Weissenböck, G. (1996a). Phenylpropanoid compounds in primary leaves of rye (*Secale cereale*) – light regulation of their biosynthesis and the possible role in UV-B protection. *Physiologia Plantarum*, **97**, 160–8.

Reuber, S., Bornman, J.F. & Weissenböck, G. (1996b). A flavonoid mutant of barley (*Hordeum vulgare* L.) exhibits increased sensitivity to UV radiation in the primary leaf. *Plant Cell Environment*, **19**, 593–601.

Rice-Evans, C. (1995). Plant polyphenols: free radical scavengers or chain-breaking antioxidants? In *Free Radicals and Oxidative*

Stress: Environment, Drugs and Food Additives. (Rice-Evans, C., Halliwell, B. & Lunt, G.G., eds), Vol. 61, pp. 103–16. Portland Press, London.

Robberecht, R. & Caldwell, M.M. (1986). Leaf UV optical properties of *Rumex patientia* L. and *Rumex obtusifolius* L. in regard to a protective mechanism against solar UV-B radiation injury. In *Stratospheric Ozone Reduction, Solar Ultraviolet Radiation and Plant Life.* (Worrest, R.C. & Caldwell, M.M. eds), Vol. G8, pp. 251–9. NATO ASI Series. Springer-Verlag, Berlin.

Takahama, U. (1982). Suppression of carotenoid photobleaching by kaempferol in isolated chloroplasts. *Plant Cell Physiology*, **23**, 859–64.

Ziska, L.H., Teramura, A.H. & Sullivan, J.H. (1992). Physiological sensitivity of plants along an elevational gradient to UV-B radiation. *American Journal of Botany*, **79**, 863–71.

PART III

Effects of UV-B at the whole
plant and community level

D-P. HÄDER

Effects of solar UV-B radiation on aquatic ecosystems

Introduction

Ozone depletion caused by the production and release of anthropogenic trace gases has been documented to be responsible for the increase of solar ultraviolet radiation in the UV-B wavelength range (280–320 nm) over polar and mid-latitudes, while in the tropics little or no increase has been measured (Rowland, 1989; Wei, 1991). Due to the long half-lives of the chlorinated fluorocarbons (CFCs) and the fact that their production and emission will only be gradually phased out, it is likely that the peak of ozone depletion is still to come. The current predictions envisage a peak of ozone depletion around the year 2000; decreased ozone values and increased solar UV-B radiation will therefore be with us well through the next century (Madronich et al., 1995). In addition to the Antarctic ozone hole, abnormal high chlorine concentrations and concomitant decreases in the ozone layer have been found over the Northern Hemisphere (Madronich et al., 1995; Blumthaler & Ambach, 1990). The resulting increased solar UV-B radiation has been suspected to affect aquatic ecosystems which are responsible for half of the biomass production of the planet. In order to determine whether or not aquatic ecosystems are at risk, a number of key questions need to be investigated.

- What is the penetration of solar UV radiation into the water column?
- What is the vertical distribution of ecologically major biomass producers?
- What is the biological sensitivity of aquatic organisms?
- What is the effectiveness of mitigating and protective measures in aquatic ecosystems?

Penetration of solar UV-B radiation into the water column

The penetration of solar radiation depends on the wavelength (Jerlov, 1970); both shorter (UV) and longer (red and infra-red) wavelengths are absorbed more than blue and green. Another factor which dramatically affects the transmission is the amount of dissolved (Gelbstoff) and particulate substances. In coastal waters with high turbidity and large concentrations of Gelbstoff, UV-B penetrates only a few decimetres or metres into the column (Piazena & Häder, 1994) while, in clear oceanic waters, UV-B has been measured to penetrate to depths of dozens of metres (Baker & Smith, 1982). Record penetration was observed in Antarctic waters, where UV-B radiation was determined with a new instrument (LUVSS) which has a resolution of 0.2 nm from 250 to 350 nm and a 0.8 nm resolution from 350 to 700 nm (Smith et al., 1992). The instrument is installed on a remotely operated vehicle which measures data while moving freely in the water column. Recent measurements clearly indicated when the ozone hole vortex was over the measurement site.

Accurate spectral measurements in the UV-B range can only be performed with a double monochromator spectroradiometer (Piazena & Häder, 1994). Short wavelength radiation in particular undergoes multiple (Rayleigh) scattering, so that even after a short distance from the surface a considerable fraction of upwelling (from below) radiation can be measured. In order to minimise this problem, we have developed a spherical detector which receives radiation from a 4π geometry.

Vertical distribution of biomass producers in the water column

Macroalgae constitute a large share of the primary biomass producers. With only a few exceptions, these plants are sessile and cannot actively move vertically in the water column. However, they follow a distinct vertical distribution which is believed to be dominated by light (Lüning, 1985). Some algae are adapted to the unattenuated solar radiation near, or even above, the water surface while others are restricted to greater depths or crevices protected from direct sunshine.

In contrast, phytoplankton, which even outcompete macroalgae in terms of biomass productivity, are free to move vertically in the water column. Phytoplankton depend on the availability of sunlight for energy acquisition and consequently need to be close to the water surface (Häder, 1991a; Häder & Worrest, 1991; Häder et al., 1995). On the other hand, most phytoplankton do not tolerate the unattenuated solar radiation at the surface. Consequently, the organisms tend to move to a specific depth, and in fact, phytoplankton form typical vertical distribution patterns both in freshwater and in marine habitats (Cabrera & Montecino, 1987; Häder, 1995b). Depending on the transparency of the water, the maximum depth varies between a few decimetres (in turbid coastal waters and eutrophic freshwater lakes) and tens of metres (in clear open oceanic waters), although this typical vertical distribution pattern of phytoplankton is disturbed by passive mixing due to high wind and waves (Ignatiades, 1990). The consumers follow the primary producers and are thus exposed to the same radiation regime as the phytoplankton (Brown & Cochrane, 1991).

Rather than defining a physical depth, it is biologically more useful to define a euphotic zone, the lower limit of which is the depth where the surface photosynthetically active radiation (PAR) has decreased to 0.1%. This concept is meaningful since, at that irradiance, photosynthesis is balanced by respiration in most organisms.

Inhibition of phytoplankton by solar UV-B radiation

Most phytoplankton are unicellular and thus lack an epidermal layer to protect them from short wavelength irradiation. Even at current solar radiation levels, phytoplankton seem to be under considerable UV-B stress as indicated by two observations. On a global basis, phytoplankton are not uniformly distributed in the oceans, but rather are concentrated in the circumpolar regions, while in equatorial waters the concentrations are 100 to 1000 times smaller (Lohrenz *et al.*, 1988). In addition to other environmental factors, such as non-permissive temperatures and lack of nutrients, solar UV-B levels may be responsible for the pattern of distribution. UV-B levels are much higher in the tropics and subtropics than at higher latitudes (Smith, 1989). The second indication for a high UV-B sensitivity is the fact that, at mid-latitudes, the large phytoplankton blooms occur in spring and disappear when the level of solar ultraviolet irradiation increases during the summer. There is often a second, smaller algal bloom in autumn, depending on permissive temperatures and sufficient nutrients. Thus, many aquatic ecosystems appear to be under considerable UV-B stress even at current radiation levels (Maske, 1984).

Numerous studies have substantiated the pronounced sensitivity to solar UV-B radiation in phytoplankton (Cullen & Lesser, 1991; Raven, 1991; Karentz, 1991; Häder, 1995*a*). Most earlier research has been carried out using artificial UV sources, the emission spectrum of which strongly deviated from that of solar radiation. Since the action spectra of inhibition responses in phytoplankton differ between species and even between individual physiological responses, results from exposure to different artificial radiation sources are difficult to interpret. One of the alternatives is exclusion studies: the organisms are exposed to ambient solar radiation, and shorter wavelengths are removed by appropriate cut-off filters (e.g. Schott WG filters) or filter foils. Alternatively, the experimental site can be moved from mid latitudes toward the equator or up on a mountain, since at both locations higher ambient UV-B radiation is recorded (Gerber & Häder, 1993).

Solar UV-B effects on motility and orientation

Phytoplankton display a distinct vertical distribution in the water column to optimise their light input for growth and survival as discussed above. Some organisms are capable of active movement powered by flagella or cilia. Others modify their buoyancy by producing gas vacuoles (Walsby, Kinsman & George, 1992) or oil droplets (Gosink, Irgens & Staley, 1993) to perform vertical migrations in the water column. Many phytoplankton move toward the surface before sunrise and during the morning and move into deeper waters at night. Some dinoflagellates have been found to migrate up to 15 m up and down in the water column (Burns & Rosa, 1980).

In addition to endogenous rhythms, vertical migrations are controlled by external physical and chemical factors such as light and gravity (Häder, 1991a,b). Figure 1 shows the vertical distribution of the dinoflagellate *Prorocentrum micans* over a 38-h period (Eggersdorfer & Häder, 1991a,b). The cells move toward the surface before sunrise and populate the layer near the surface before noon. At times of excessive

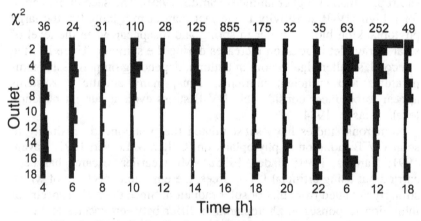

Fig. 1. Vertical distribution of the marine dinoflagellate *Prorocentrum micans* in a 3 m water column. (Modified after Eggersdorfer & Häder, 1991b).

irradiation, the cells dive to a deeper layer and come back later in the afternoon.

Many actively motile phytoplankton use positive phototaxis to move to the surface. This behaviour is augmented by negative gravitaxis (Fig. 2a) which helps the cells to find the way to the surface even in muddy waters or in darkness (Häder, 1987). The resulting upward motility is counterbalanced by a downward movement guided by negative phototaxis in high irradiances (Fig. 2b). These two antagonistic responses lead to the vertical distribution discussed above (Häder *et al.*, 1981). Both phototaxis and gravitaxis have been found to be affected by solar radiation (Fig. 3) (Häder *et al.*, 1990*a,b*). In addition to the white light and UV-A components, the UV-B wavelengths in particular have detrimental effects as indicated by inhibition with artificial UV radiation.

As well as the orientation mechanisms, both the percentage of motile cells and the swimming velocity of the still motile cells decrease dramatically when exposed to unfiltered radiation (Fig. 4). Since the position in the water column is of considerable ecological significance, the

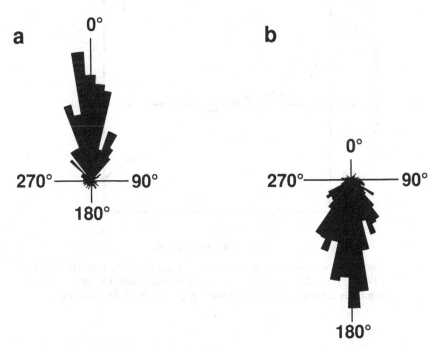

Fig. 2. Circular histograms for negative gravitaxis (a) and negative phototaxis (b) in the flagellate *Euglena gracilis*.

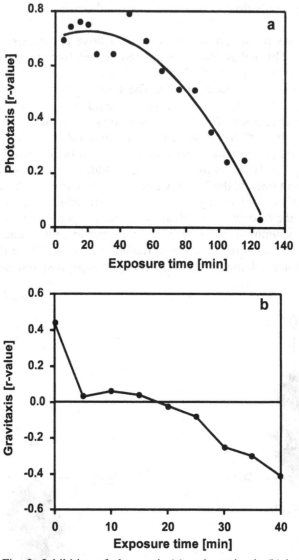

Fig. 3. Inhibition of phototaxis (a) and gravitaxis (b) in the dinoflagellate *Gymnodinium* sp. (Y100) by solar radiation. Ordinate: r-value,
which is a statistical measure for the precision of orientation.

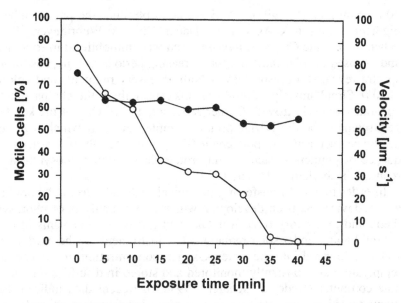

Fig. 4. Inhibition of motility (percentage of motile cells, open circles, and swimming velocity, closed circles) in *Gymnodinium* sp. (Y100) by solar radiation.

inhibition of motility and orientation by increased UV-B radiation would diminish the chances for growth and survival. Most organisms studied so far do not orientate with respect to solar UV-B radiation, but rather use UV-A and visible radiation for orientation. Consequently, they cannot move to greater depth to protect themselves from the UV stress.

Biochemical studies of *Euglena gracilis* indicated that UV-B radiation affects the proteins of the paraflagellar body (PFB), the putative photoreceptor organelle (Häder & Brodhun, 1991). A comparison of the protein patterns from UV-B exposed cells with those from control cells clearly indicated the loss of specific PFB proteins (Häder & Brodhun, 1991). In addition, the chromophoric groups, pterins and flavins (Galland *et al.*, 1990), were bleached by excessive radiation.

Metabolism

Solar UV-B radiation affects many physiological and biochemical reactions in phytoplankton and thus impairs growth and development. Recent studies in Antarctica have indicated that, under the ozone 'hole', photosynthesis is decreased by as much as 6–12% in the top 10 to

20 m (Smith *et al.*, 1992). UV-B radiation bleaches the photosynthetic pigments (Nultsch & Agel, 1986; Häder, Rhiel & Wehrmeyer, 1988). When the marine *Cryptomonas maculata* was immobilised in solid agar and exposed to solar radiation for increasing periods of time, absorption spectra revealed a drastic loss of both accessory pigments and chlorophyll *a*. Simultanously, several proteins especially of the photosynthetic apparatus were destroyed (Gerber & Häder, 1992). One of the key targets seems to be the D1/D2 protein complex in PS II which regulates the linear electron transport chain (Renger *et al.*, 1989). As a consequence of pigment bleaching and protein destruction photosynthesis is reduced (Häberlein & Häder, 1992).

In order to avoid transferring biological material into the laboratory, a new device has been developed, which allows analysis of photosynthetic and respiratory oxygen uptake and release in organisms in their natural habitat under solar radiation. The instrument can be used above water or lowered into the water. Oxygen concentration, irradiance and temperature are constantly monitored and stored in disc files on a portable computer (Häder & Schäfer, 1994a,b). Recent data indicate that many organisms operate only in a limited irradiance window for net photosynthetic oxygen production. At too low irradiances, respiration exceeds photosynthesis and at too high irradiances photosynthesis is shut down by the process of photoinhibition.

Another valuable method for the determination of the status of the photosynthetic apparatus is pulse amplitude modulation (PAM) fluorescence (Schreiber, Schliwa & Bilger, 1986; Krause & Weis, 1991). Fluorescence quenching analysis is based on the hypothesis that two processes with different time kinetics can decrease the maximal fluorescence yield Fm: the fast photochemical quenching and the slower non-photochemical quenching which is thought to be mainly based on the energisation of the thylakoid membrane (Schreiber *et al.*, 1995; Krause & Weis, 1991). Empirical expressions for the quantum yield are based on the fluorescence parameters measured during quenching analysis (Weis & Berry, 1987). The validity of these expressions has been supported by concomitant gas exchange measurements (Schreiber, Bilger & Neubauer, 1994). PAM fluorescence techniques were first used in higher plants; later on, these techniques were successfully applied to the measurements in unicellular algae and macroalgae. Several algal groups show a qualitatively different behaviour than higher plants which is believed to be based on different regulatory mechanisms in various algal taxonomic groups (Büchel & Wilhelm, 1993).

Effects of UV-B on cyanobacteria

Cyanobacteria developed during a time when neither oxygen nor a protecting ozone layer was present in the primitive atmosphere. Because of the resulting high levels of UV radiation it has been assumed that these organisms must have developed at a considerable depth of water, which would have absorbed most of the short wavelength radiation (Häder, 1996). This assumption is supported by the fact that many cyanobacteria are adapted to very low PAR levels.

In contrast to eukaryotes, the prokaryotic cyanobacteria (and some other bacteria) are capable of utilising atmospheric nitrogen and thus occupy a central position in nutrient cycling. They use the enzyme nitrogenase to convert N_2 into ammonium (NH_4). This nitrogen is also available to eukaryotes, and it has been calculated that cyanobacteria fix over 35 million tonnes of nitrogen annually (Häder, Worrest & Kumar, 1989). Cyanobacteria are also responsible for making available a significant share of the nitrogen consumed by eukaryotic phytoplankton algae in the oceans (Carpenter & Romans, 1991) and freshwater habitats (Storch, Saunders & Ostrofsky, 1990). The role of nitrogen fixing cyanobacteria as providing natural biofertiliser has been particularly demonstrated in rice paddy fields (Kumar & Kumar, 1988; Sinha & Kumar, 1992; Huang & Chow, 1988; Padhy, 1985), where up to 20–30 kg of biologically fixed nitrogen could be provided per hectare of rice crop per season. The effects of artificial UV-B irradiation on growth, survival, pigmentation, nitrate reductase, glutamine synthetase (GS) and total protein profile have been studied in a number of N_2-fixing cyanobacterial strains isolated from rice paddy fields in India (Sinha *et al.*, 1995). NR activity was found to increase, while GS activity decreased following exposure to UV-B in all tested organisms. Simultaneously, the protein content decreased with increasing UV-B exposure time.

Filamentous, gliding cyanobacteria also orient in their habitat by light-controlled movement (Häder & Hoiczyk, 1992). Thus, the ecological consequences of decreased motility due to solar UV are similar to those in other motile organisms. Both the percentage of motile filaments and their linear velocity have been found to decrease upon excessive solar radiation (Häder, Watanabe & Furuya, 1986; Häder & Häder, 1990*a*; Donkor & Häder, 1991). Exposure experiments in Ghana have shown that cyanobacteria exposed to tropical solar radiation were affected within minutes (Donkor, Amewowor & Häder, 1993*a,b*), and that the damage encountered could only be partially repaired and only if the exposures were short term.

Different organisms show different UV-B sensitivity in terms of growth and survival. *Anabaena* sp. and *Nostoc carmium* were killed after 2 h of UV-B exposure, whereas *Nostoc commune* and *Scytonema* sp. tolerated longer exposure times. Pigment content, particularly phycocyanin, severely decreased following UV-B irradiation. Spectroscopic analysis of the photosynthetic pigments indicated that the phycobilins are bleached first, followed by the carotenoids, and finally chlorophyll *a* is affected. Fluorescence studies during UV-B exposure indicate that, as a first effect, the excitation energy cannot be passed to the reaction centres and consequently is lost in the form of fluorescence (Fig. 5). Later, the fluorescence yield decreases again, indicating a gradual loss in biliproteins. In addition, a shift to shorter wavelengths is encountered which has been interpreted to reflect a disassembly of the phycobilisomes first into hexamers, then trimers and finally (α,β) monomers.

Effects of UV-B on macroalgae and seagrasses

In recent years there has been an increased interest in UV-B effects on macroalgae and seagrasses. As indicated above, most macrophytes are

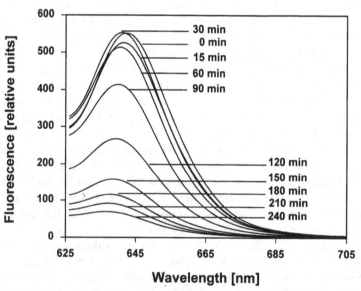

Fig. 5. Fluorescence emission spectra (excited at 620 nm) of the lowest fraction from a sucrose gradient obtained after solubilisation of cytoplasm from the cyanobacterium *Anabaena* sp. before and after increasing exposure times to artificial UV radiation from a transilluminator. (Modified from Sinha *et al.*, 1995*b*).

attached to their growing site and are thus restricted to certain depth zones above, below or within the tidal zone. If the UV-B/PAR ratio increases, the organisms will be exposed to enhanced short wavelength radiation they may not be adapted to. Indeed, many red, brown and green benthic algae have been found to be affected by solar UV-B radiation (Larkum & Wood, 1993). Using the recently developed instrumentation described above, photosynthetic oxygen production was determined in several macroalgae at various depths (Häder *et al.*, 1996) (Fig. 6). In most cases, optimal photosynthesis was found at the depth where the algae grow in their natural habitat and exposing them close to the surface for increasing periods of time resulted in rapid photoinhibition (Fig. 7).

When using PAM (pulse amplitude modulation) fluorescence measurements, deep-water benthic algae were most sensitive, while intertidal algae were least sensitive. As in oxygen measurements, photoinhibition could be induced in most algae even by short exposure to solar radiation at, or near, the surface (Fig. 8). However, it is interesting to note that many algae showed some degree of photoinhibition even at their natural growth site at high solar angles (Fig. 9).

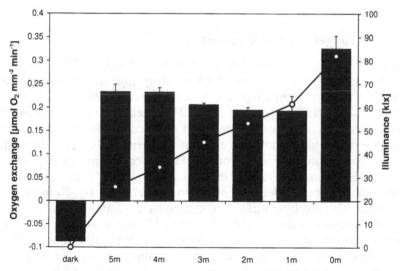

Fig. 6. Photosynthetic oxygen production in the red macroalga *Laurencia obtusa* as a function of the depth in the water column (bars), and corresponding light intensity (circles). The algae were harvested from a rock pool (0 m), and photosynthetic activity then measured at the depths indicated.

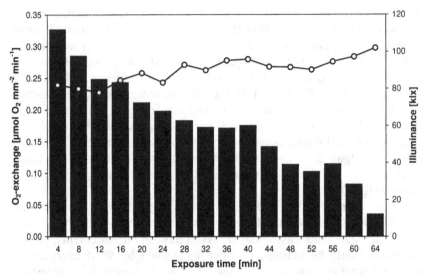

Fig. 7. Inhibition of photosynthetic oxygen production in the red macroalga *Laurencia obtusa* induced by exposure to solar radiation at the surface. Algae were harvested from a rock pool (0 m): symbols as in Fig. 6.

Targets of UV-B radiation

In order to determine the molecular targets of solar UV-B radiation, action spectra have been measured in many microorganisms and for very different responses (Cullen, Neale & Lesser, 1992; Häder, Worrest & Kumar, 1991; Häder *et al.*, 1994). These action spectra can be monochromatic, that is they are based on fluence rate response curves measured at individual wavelengths. Alternatively, polychromatic action spectra are measured. For this purpose organisms are exposed to the whole spectral range, and increasing wavelength bands are cut off from the short wavelength end (also see chapter by Holmes, this volume). The resulting difference in activity is plotted versus the dose which is removed by the cut off. Figure 10 shows the polychromatic action spectrum for the inhibition of motility in the flagellate *Euglena gracilis* (Gerber, Biggs & Häder, 1996). In this instance, the data were first plotted in the form of dose–response curves, from which the linear regression was calculated, and the inverse values plotted against the nominal cut-off value (50% transmission) of the filters.

DNA is certainly one of the targets for damaging UV-B radiation (Peak *et al.*, 1985), and the most common damage is the formation of

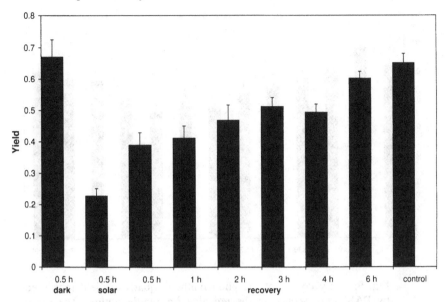

Fig. 8. Inhibition of the photosynthetic quantum yield induced by solar radiation at the surface, and recovery, in the green alga *Udotea petiolata* as measured by PAM fluorescence. The algae were harvested at 7 m depth in the shade of a rock and exposed at the surface. Recovery was in a rock pool (0 m), in darkness.

pyrimidine dimers, which has been found also in animal tissues (Yasuhira, Mitani & Shima, 1992). However, most organisms possess the enzyme photolyase which removes and replaces the thymine dimers; this enzyme is induced by UV-A and blue light (Yamamoto *et al.*, 1983; Hirosawa & Miyachi, 1983). Other UV-B effects do not depend on DNA damage, as indicated by the short response time and the lack of photoreactivation (Häder *et al.*, 1986). In these cases, photodynamic reactions can be the mechanism by which UV-B affects the cells (Ito, 1983). When a molecule absorbs a high-energy photon, the excess excitation energy can lead to the formation of singlet oxygen or free radicals. Both types of photodynamic response result in aggressive molecular species which destroy membranes and other cellular components. Photodynamic reactions have been found to play a role in some UV-B-induced types of damage, for example, in the ciliate *Stentor coeruleus* (Häder & Häder, 1991). In other systems the role of photodynamic responses could be excluded by the use of quenchers and scavengers (Häder *et al.*, 1986). In these cases, proteins were found to be the targets

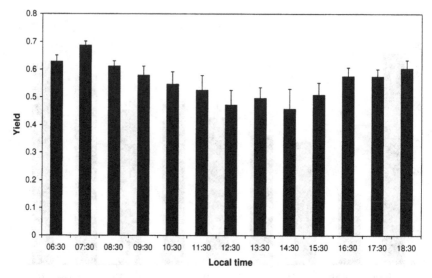

Fig. 9. Daily course of the photosynthetic quantum yield measured by PAM fluorescence in the red alga *Laurencia obtusa*. The algae were growing at a depth of 0 m, and measurements were made *in situ*.

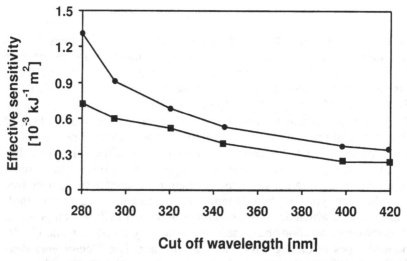

Fig. 10. Polychromatic action spectrum of the inhibition of motility in the flagellate *Euglena gracilis*. 50% motility, circles; 0% motility, squares.

of UV-B radiation. Aromatic amino acids strongly absorb short wavelength radiation. Biochemical analysis indicated the loss of proteins involved in photoperception and motility (Häder & Brodhun, 1991; Häberlein & Häder, 1992). Also, in the photosynthetic apparatus several targets of UV-B radiation have been identified including the D1/D2 protein complex associated with photosystem II (Renger *et al.*, 1989). Other targets are the water splitting site of the photosynthetic apparatus and the reaction centre of photosystem II (Bhattacharjee *et al.*, 1987; Bhattacharjee & David, 1987).

Protective strategies

Higher plants produce UV-B-absorbing substances which are synthesised and stored in the epidermal layer when induced by solar radiation (Beggs, Schneider-Ziebert & Wellmann, 1986; Tevini, Thoma & Iwanzik, 1983; Murali & Teramura, 1985). In phytoplankton too, a group of UV-absorbing substances has been found, which could be identified as mycosporine-like amino acids (Carretto *et al.*, 1990; Raven, 1991). These substances are even passed to the next trophic level in the food web by predation. The occurrence and induction by UV of screening pigments has also been recorded in several tropical red algae.

Cyanobacteria have been found to produce a UV-B-induced shielding substance, called scytonemine, which is incorporated into the slime sheaths of the organisms. The production of this substance very effectively prevents UV-B-dependent bleaching of chlorophyll (Garcia-Pichel & Castenholz, 1991). Other cyanobacteria have been found to produce shock proteins in response to UV irradiation (Shibata, Baba & Ochiai, 1991).

Consequences of UV-B damage in aquatic ecosystems

Loss of biomass productivity in aquatic ecosystems is one of the most important consequences of enhanced solar UV-B radiation. While aquatic ecosystems represent rather a small standing crop, their productivity equals that of terrestrial ecosystems (Houghton & Woodwell, 1989), and because of their enormous size even a small loss in biomass productivity has significant adverse effects on aquatic ecosystems (Häder *et al.*, 1995). Furthermore, any sizeable decrease in primary productivity is bound to result in a significant reduction in fisheries catch (Hardy & Gucinski, 1989).

Aquatic ecosystems are a major sink for atmospheric carbon dioxide; it is estimated that they incorporate about 10^{11} tonnes of carbon annually into organic material (Houghton & Woodwell, 1989; Siegenthaler & Sarmiento, 1993). This figure is again similar to that for terrestrial ecosystems. In both cases, most of this is released during the decay of the organic material. This natural carbon cycle is disturbed by the release of about 5 Gt of carbon from fossil fuel burning and another 2 Gt from (mostly tropical) deforestation. However, only 3 of the total of 7 Gt actually accumulate in the atmosphere. The remaining 4 Gt are believed to be removed from the cycle by the biological pump in the oceans: organic and inorganic carbon falls out of the upper layers of the water column in the form of oceanic snow which is deposited in the deep sea. Thus, a decrease in the phytoplankton populations will result in an increase in the atmospheric CO_2 concentration, augmenting the greenhouse effect and the resulting sea level rise (Schneider, 1989).

The sensitivity of phytoplankton to solar UV-B radiation varies among species; therefore changes in the species composition have been predicted as a consequence of ozone depletion (McLeod & McLachlan, 1959). This effect will continue through the subsequent trophic levels in the food web.

Due to intensive studies during the past few years there is ample evidence that aquatic habitats are under considerable UV-B stress even at ambient ultraviolet radiation levels. Any substantial depletion of the ozone layer will, therefore, have significant detrimental effects on the aquatic ecosystems.

Acknowledgements

This work was supported by financial aid from the Bundesminister für Forschung und Technologie (project KBF 57), the European Community (Environment programme, EV5V-CT91-0026; EV5V-CT94-0425; DG XII, Environmental Programme) and the State of Bavaria (BayForKlim).

References

Baker, K.S. & Smith, R.C. (1982). Spectral irradiance penetration in natural waters. In *Marine Ecosystems: The Role of Solar Ultraviolet Radiation* (Calkins, J., ed.), pp. 233–46. Plenum Press, New York.
Beggs, C.J., Schneider-Ziebert, U. & Wellmann, E. (1986). UV-B radiation and adaptive mechanisms in plants. In *Stratospheric Ozone Reduction: Solar Ultraviolet Radiation and Plant Life* (Worrest, R.C. & Caldwell, M. M., eds), Vol. G8, pp. 235–50. NATO ASI Series. Springer, Heidelberg.
Bhattacharjee, S.K. & David, K.A.V. (1987). UV-sensitivity of cyano-

bacterium *Anacystis nidulans*: Part II: a model involving photosystem (PSII) reaction centre as lethal target and herbicide binding high turnover B protein as regulator of dark repair. *Indian Journal of Experimental Biology,* **25**, 837–42.

Bhattacharjee, S.K., Mathur, M., Rane, S.S. & David, K.A.V. (1987). UV-sensitivity of cyanobacterium *Anacystis nidulans*: Part I: evidence for photosystem II (PSII) as a lethal target and constitutive nature of a dark-repair system against damage to PSII. *Indian Journal of Experimental Biology,* **25**, 832–6.

Blumthaler, M. & Ambach, W. (1990). Indication of increasing solar ultraviolet-B radiation flux in alpine regions. *Science,* **248**, 206–8.

Brown, P.C & Cochrane, K.L. (1991). Chlorophyll a distribution in the southern Benguela: possible effects of global warming on phytoplankton and its implication for pelagic fish. *Suid-Afrikaanse Tydskrif vir Wetenskap,* **87**, 233–42.

Büchel, C. & Wilhelm, C. (1993). *In vivo* analysis of slow chlorophyll fluorescence induction kinetics in algae: progress, problems and perspective. *Photochemistry and Photobiology,* **58**, 137–48.

Burns, N.M. & Rosa, F. (1980). *In situ* measurements of the settling velocity of organic carbon particles and ten species of phytoplankton. *Limnology and Oceanography,* **2**, 855–64.

Cabrera, S. & Montecino, V. (1987). Productividad primaria en ecosistemas limnicos. *Archives of Biology and Medicinal Experiment,* **20**, 105–16.

Carpenter, E.J. & Romans, K. (1991). Major role of the cyanobacterium *Trichodesmium* in nutrient cycling in the North Atlantic Ocean. *Science,* **254**, 1356–8.

Carreto, J.J., Carignana, M.O., Daleo, G. & de Marco, S.G. (1990). Occurrence of mycosporine-like amino acids in the red tide dinoflagellate *Alexandrium excavatum*: UV-photoprotective compounds. *Journal of Plankton Research,* **12**, 909–21.

Cullen, J.J. & Lesser, M.P. (1991). Inhibition of photosynthesis by ultraviolet radiation as a function of dose and dosage rate: results for a marine diatom. *Marine Biology,* **111**, 183–90.

Cullen, J.C., Neale, P.J. & Lesser, M.P. (1992). Biological weighting function for the inhibition of phytoplankton photosynthesis by ultraviolet radiation. *Science,* **258**, 646–50.

Donkor, V. & Häder, D-P. (1991). Effects of solar and ultraviolet radiation on motility, photomovement and pigmentation in filamentous, gliding cyanobacteria. *FEMS Microbiology and Ecology,* **86**, 159–68.

Donkor, V.A., Amewowor, D.H.A.K. & Häder, D-P. (1993a). Effects of tropical solar radiation on the motility of filamentous cyanobacteria. *FEMS Microbiology and Ecology,* **12**, 143–8.

Donkor, V.A., Amewowor, D.H.A.K. & Häder, D-P. (1993b). Effects of tropical solar radiation on the velocity and photophobic behavior of filamentous gliding cyanobacteria. *Acta Protozoologica*, **32**, 67–72.

Eggersdorfer, B. & Häder, D-P. (1991*a*). Phototaxis, gravitaxis and vertical migrations in the marine dinoflagellate, *Prorocentrum micans. European Journal of Biophysics*, **85**, 319–26.

Eggersdorfer, B. & Häder, D-P. (1991*b*). Phototaxis, gravitaxis and vertical migrations in the marine dinoflagellates, *Peridinium faeroense* and *Amphidinium caterii. Acta Protozoologica*, **30**, 63–71.

Galland, P., Keiner, P., Dörnemann, D., Senger, H., Brodhun, B. & Häder, D.-P. (1990). Pterin- and flavin-like fluorescence associated with isolated flagella of *Euglena gracilis. Photochemistry and Photobiology*, **51**, 675–80.

Garcia-Pichel, F. & Castenholz, R.W. (1991). Characterization and biological implications of scytonemin, a cyanobacterial sheath pigment. *Journal of Phycology*, **27**, 395–409.

Gerber, S., Biggs, A. & Häder, D-P. (1996). A polychromatic action spectrum for the inhibition of motility in the flagellate *Euglena gracilis. Acta Protozoologica*, **35**, 161–5.

Gerber, S. & Häder, D-P. (1992). UV effects on photosynthesis, proteins and pigmentation in the flagellate *Euglena gracilis*: biochemical and spectroscopic observations. *Biochemistry and Systematic Ecology*, **20**, 485–92.

Gerber, S. & Häder, D-P. (1993). Effects of solar irradiation on motility and pigmentation of three species of phytoplankton. *Environmental and Experimental Biology*, **33**, 515–21.

Gosink, J.J., Irgens, R.L. & Staley, J.T. (1993). Vertical distribution of bacteria in Arctic sea ice. *FEMS Microbiology Ecology*, **102**, 85–90.

Häberlein, A. & Häder, D-P. (1992). UV effects on photosynthetic oxygen production and chromoprotein composition in the freshwater flagellate *Cryptomonas S2. Acta Protozoologica*, **31**, 85–92.

Häder, D-P. (1987). Polarotaxis, gravitaxis and vertical phototaxis in the green flagellate, *Euglena gracilis. Archives of Microbiology*, **147**, 179–83.

Häder, D-P. (1991*a*). Effects of enhanced solar ultraviolet radiation on aquatic ecosystems. In *Biophysics of Photoreceptors and Photomovements in Microorganisms*. (Lenci, F., Ghetti, F., Colombetti, G., Häder, D-P. & Song, P-S. eds), pp. 157–72. Plenum Press, New York, London.

Häder, D-P. (1991*b*). Phototaxis and gravitaxis in *Euglena gracilis*. In *Biophysics of Photoreceptors and Photomovements in Microorganisms*. (Lenci, F., Ghetti, F., Colombetti, G., Häder, D-P. & Song, P-S., eds), pp. 203–21. Plenum Press, New York, London.

Häder, D-P. (1995*a*). Influence of ultraviolet radiation on phytoplankton ecosystems. In *Algae, Environment and Human Affairs*.

(Wiessner, W., Schnepf, E. & Starr, R.C., eds), pp. 41–55. Biopress Limited, Bristol.

Häder, D-P. (1995b). Novel method to determine vertical distributions of phytoplankton in marine water columns. *Environmental and Experimental Botany*, **35**, 547–55.

Häder, D-P. (1996). Effects of UV radiation on phytoplankton. *Advances in Microbial Ecology*, in press.

Häder, D-P. & Brodhun, B. (1991). Effects of ultraviolet radiation on the photoreceptor proteins and pigments in the paraflagellar body of the flagellate, *Euglena gracilis*. *Journal of Plant Physiology*, **137**, 641–6.

Häder, D-P. & Häder, M. (1990a). Effects of solar radiation on motility, photomovement and pigmentation in two strains of the cyanobacterium, *Phormidium uncinatum*. *Acta Protozoologica*, **29**, 291–303.

Häder, D-P. & Häder, M. (1990b). Effects of UV radiation on motility, photo-orientation and pigmentation in a freshwater *Cryptomonas*. *Journal of Photochemistry and Photobiology B: Biology*, **5**, 105–14.

Häder, D-P. & Häder, M.A. (1991). Effects of solar radiation on motility in *Stentor coeruleus*. *Photochemistry and Photobiology*, **54**, 423–28.

Häder, D-P. & Hoiczyk, E. (1992). Gliding motility. In *Algal Cell Motility*. (Melkonian, M. ed.), pp. 1–38. Chapman & Hall, New York, London.

Häder, D-P. & Worrest, R.C. (1991). Effects of enhanced solar ultraviolet radiation on aquatic ecosystems. *Photochemistry and Photobiology*, **53**, 717–25.

Häder, D-P. & Schäfer, J. (1994a). *In-situ* measurement of photosynthetic oxygen production in the water column. *Environmental Monitoring Assessment*, **32**, 259–68.

Häder, D-P., Colombetti, G., Lenci, F. & Quaglia, M. (1981). Phototaxis in the flagellates, *Euglena gracilis* and *Ochromonas danica*. *Archives of Microbiology*, **130**, 78–82.

Häder, D-P & Schäfer, J. (1994b). Photosynthetic oxygen production in macroalgae and phytoplankton under solar irradiation. *Journal of Plant Physiology*, **144**, 293–9.

Häder, D-P., Watanabe, M. & Furuya, M. (1986). Inhibition of motility in the cyanobacterium, *Phormidium uncinatum*, by solar and monochromatic UV irradiation. *Plant and Cell Physiology*, **27**, 887–94.

Häder, D-P., Rhiel, E. & Wehrmeyer, W. (1988). Ecological consequences of photomovement and photobleaching in the marine flagellate *Cryptomonas maculata*. *FEMS Microbiology and Ecology*, **53**, 9–18.

Häder, D-P., Worrest, R.C. & Kumar, H.D. (1989). Aquatic ecosystems. pp. 39–48. UNEP Environmental Effects Panel Report.

Häder, D-P., Häder, M., Liu, S.-M. & Ullrich, W. (1990a). Effects of solar radiation on photoorientation, motility and pigmentation in a freshwater *Peridinium*. *BioSystems*, **23**, 335–43.

Häder, D-P., Liu, S.-M., Häder, M. & Ullrich, W. (1990b). Photoorientation, motility and pigmentation in a freshwater *Peridinium* affected by ultraviolet radiation. *General Physiology and Biophysics*, **9**, 361–71.

Häder, D-P., Worrest, R.C. & Kumar, H.D. (1991). Aquatic ecosystems. pp. 33–40. UNEP Environmental Effects Panel Report.

Häder, D-P., Worrest, R.C., Kumar, H.D. & Smith, R.C. (1994). Effects of increased solar ultraviolet radiation on aquatic ecosystems. pp. 65–77. UNEP Environmental Effects Panel Report.

Häder, D-P., Worrest, R.C., Kumar, H.D. & Smith, R.C. (1995). Effects of increased solar ultraviolet radiation on aquatic ecosystems. *Ambio*, **24**, 174–80.

Häder, D-P., Porst, M., Herrmann, H., Schafer, J. & Santas, R. (1996). Photoinhibition in the mediterranean green alga *Halimeda tuna* measured *in situ*. *Photochemistry and Photobiology*, **64**, 428–34.

Hardy, J. & Gucinski, H. (1989). Stratospheric ozone depletion: implications for marine ecosystems. *Oceanographic Magazine*, **2**, 18–21.

Hirosawa, T. & Miyachi, S. (1983). Inactivation of Hill reaction by long-wavelength ultraviolet radiation (UV-A) and its photoreactivation by visible light in the cyanobacterium, *Anacystis nidulans*. *Archives of Microbiology*, **135**, 98–102.

Houghton, R.A. & Woodwell, G.M. (1989). Global climatic change. *Scientific American*, **260**, 18–26.

Huang, T.-C. & Chow, T.-J. (1988). Comparative studies of some nitrogen-fixing unicellular cyanobacteria isolated from rice fields. *Journal of General Microbiology* 134, 3089–97.

Ignatiades, L. (1990). Photosynthetic capacity of the surface microlayer during the mixing period. *Journal of Plankton Research*, **12**, 851–60.

Ito, T. (1983). Photodynamic agents as tools for cell biology. In *Photochemical and Photobiological Reviews*. (Smith, K.C., ed.), Vol. 7, pp. 141–86. Plenum Press, New York.

Jerlov, N.G. (1970). Light – general introduction. In *Marine Ecology*. (Kinne, O., ed.), Vol. 1, pp. 95–102.

Karentz, D. (1991). Ecological considerations of Antarctic ozone depletion. *Antarctic Science*, **3**, 3–11.

Krause, G.H. & Weis, E. (1991). Chlorophyll fluorescence and photosynthesis: the basics. *Annual Review of Plant Physiology and Plant Molecular Biology*, **42**, 313–49.

Kumar, A. & Kumar, H.D. (1988). Nitrogen fixation by blue-green algae. In *Plant Physiology Research.* (Sen, S.P., ed.), pp. 15–22. New Dehli: Society for Plant Physiology and Biochemistry, First International Congress of Plant Physiology.

Larkum, A.W.D. & Wood, W.F. (1993). The effect of UV-B radiation on photosynthesis and respiration of phytoplankton, benthic macroalgae and seagrasses. *Photosynthesis Research,* **36,** 17–23.

Lohrenz, S.E., Arnone, R.A., Wiesenburg, D.A. & DePalma, I.P. (1988). Satellite detection of transient enhanced primary production in the western Mediterranean Sea. *Nature,* **335,** 245–7.

Lüning, K. (1985). In *Meeresbotanik.* (Thieme, G., ed.), Stuttgart, New York.

Madronich, S., McKenzie, R.L., Caldwell, M.M. & Björn, L.O. (1994). Changes in ultraviolet radiation reaching the Earth's surface. pp. 1–13. Environmental Effects Panel Report, United Nations, Environmental Program.

Madronich, S., McKenzie, R.L., Caldwell, M.M. & Björn, L.O. (1995). Changes in ultraviolet radiation reaching the earth's surface. *Ambio,* **24,** 143–52.

McLeod, G.C. & McLachlan, J. (1959). The sensitivity of several algae to ultraviolet radiation of 2537 Å. *Physiologia Plantarum,* **12,** 306–9.

Maske, H. (1984). Daylight ultraviolet radiation and the photoinhibition of phytoplankton carbon uptake. *Journal of Plankton Research,* **6,** 351–57.

Murali, N.S. & Teramura, A.H. (1985). Effects of ultraviolet-B irradiance on soybean. VI. Influence of phosphorus nutrition on growth and flavonoid content. *Physiologia Plantarum,* **63,** 413–16.

Nultsch, W. & Agel, G. (1986). Fluence rate and wavelength dependence of photobleaching in the cyanobacterium *Anabaena variabilis. Archives of Microbiology,* **144,** 268–71.

Padhy, R.N. (1985). Cyanobacteria employed as fertilizers and waste disposers. *Nature,* **317,** 475–6.

Peak, J.G., Peak, M.J., Sikorski, R.S. & Jones, C.A. (1985). Induction of DNA-protein crosslinks in human cells by ultraviolet and visible radiations: action spectrum. *Photochemistry and Photobiology,* **41,** 295–302.

Piazena, H. & Häder, D.-P. (1994). Penetration of solar UV irradiation in coastal lagoons of the Southern Baltic Sea and its effect on phytoplankton communities. *Photochemistry and Photobiology,* **60,** 463–9.

Raven, J.A. (1991). Responses of aquatic photosynthetic organisms to increased solar UVB. *Journal of Photochemistry and Photobiology, B: Biology,* **9,** 239–44.

Renger, G., Völker, M., Eckert, H.J., Fromme, R., Hohm-Veit, S. & Gräber, P. (1989). On the mechanisms of photosystem II deterioration by UV-B irradiation. *Photochemistry and Photobiology*, **49**, 97–105.

Rowland, F.S. (1989). Chlorofluorocarbons and the depletion of stratospheric ozone. *American Scientist*, **77**, 36–46.

Schneider, S.H. (1989). The changing climate. *Scientific American*, **261**, 38–47.

Schreiber, U., Schliwa, U. & Bilger, W. (1986). Continuous recording of photochemical and non-photochemical chlorophyll fluorescence quenching with a new type of modulation fluorometer. *Photosynthesis Research*, **10**, 51–62.

Schreiber, U., Bilger, W. & Neubauer, C. (1994). Chlorophyll fluorescence as a nonintrusive indicator for rapid assessment of *in vivo* photosynthesis. In *Ecophysiology of Photosynthesis. Ecological Studies, Volume 100.* (Schulze, E.D. & Caldwell, M.M., eds), pp. 49–70. Springer, Berlin.

Schreiber, U., Endo, T., Mi, H. & Asada, K. (1995). Quenching analysis of chlorophyll fluorescence by the saturation pulse method: particular aspects relating to the study of eukaryotic algae and cyanobacteria. *Plant Cell Physiology*, **36**, 873–82.

Shibata, H., Baba, K. & Ochiai, H. (1991). Near-UV irradiation induces shock proteins in *Anacystis nidulans* R-2; possible role of active oxygen. *Plant Cell Physiology*, **32**, 771–6.

Siegenthaler, U. & Sarmiento, J.L. (1993). Atmospheric carbon dioxide and the ocean. *Nature*, **365**, 119–25.

Sinha R.P. & Kumar, A. (1992). Screening of blue-green algae for biofertilizer. In *Proceedings of the National Seminar on Organic Farming.* (Patil, P.L., ed.), pp. 95–7. Pune, India.

Sinha, R.P., Kumar, H.D., Kumar, A. & Häder, D.-P. (1995). Effects of UV-B irradiation on growth, survival, pigmentation and nitrogen metabolism enzymes in cyanobacteria. *Acta Protozoologica*, **34**, 187–92.

Smith, R. (1989). Ozone, middle ultraviolet radiation and the aquatic environment. *Photochemistry and Photobiology*, **50**, 459–68.

Smith, R.C., Prezelin, B.B., Baker, K.S., Bidigare, R.R., Boucher, N.P., Coley, T., Karentz, D., MacIntyre, S., Matlick, H.A., Menzies, D., Ondrusek, M., Wan, Z. & Waters, K.J. (1992). Ozone depletion: ultraviolet radiation and phytoplankton biology in Antarctic waters. *Science*, **255**, 952–9.

Storch, T.A., Saunders, G.W. & Ostrofsky, M.L. (1990). Diel nitrogen fixation by cyanobacterial surface blooms in Sanctuary Lake, Pennsylvania. *Applied Environmental Microbiology*, **56**, 466–71.

Tevini, M., Thoma, U. & Iwanzik, W. (1983). Effects of enhanced UV-B radiation on germination, seedling growth, leaf anatomy and pigments of some crop plants. *Zeitschrift für Pflanzenphysiologie*, **109**, 435–48.

Walsby, A.E., Kinsman, R. & George, K.I. (1992). The measurement of gas volume and buoyant density in planktonic bacteria. *Journal of Microbiological Methods*, **15**, 293–309.

Wei, D.-W. (1991). On the formation of the Antarctic ozone hole and its trend predictions. *Science in China B*, **34**, 95–103.

Weis, E. & Berry, J. (1987). Quantum efficiency of photosystem II in relation to energy-dependent quenching of chlorophyll fluorescence. *Biochimica et Biophysica Acta*, **894**, 198–208.

Yamamoto, K.M., Satake, M., Shinagawa, H. & Fujiwara, Y. (1983). Amelioration of the ultraviolet sensitivity of an *Escherichia coli* recA mutant in the dark by photoreactivating enzyme. *Molecular General Genetics*, **190**, 511–15.

Yasuhira, S., Mitani, H. & Shima, A. (1992). Enhancement of photo-repair of ultraviolet-induced pyrimidine dimers by preillumination with fluorescent light in the goldfish cell line. The relationship between survival and yield of pyrimidine dimers. *Photochemistry and Photobiology*, **55**, 97–101.

J.E. CORLETT, J. STEPHEN, H.G. JONES,
R. WOODFIN, R. MEPSTED and N.D. PAUL

Assessing the impact of UV-B radiation on the growth and yield of field crops

Introduction

After more than 20 years' research into the effect on plants of elevated levels of UV-B radiation, one might expect it to be a relatively easy matter to predict the impact of future changes in UV-B climate on the growth and yield of important field crops, but this is not the case. Predicting future changes in ground-level UV-B is itself proving very difficult because of uncertainty in trends for future emissions of ozone-destroying chemicals, incomplete knowledge of the atmospheric chemistry involved (see chapters by Webb and by Pyle, this volume) and possible interactions between UV-B and other climate change variables such as CO_2 and temperature. Even if future trends were known, recent reviews (for example, Caldwell & Flint, 1994; Fiscus & Booker, 1995) have highlighted the methodological limits of many studies of plant responses to UV-B which make problematic any direct extrapolation to predict likely yield losses at the field-scale. This review of the existing knowledge on field crop responses to UV-B focuses specifically on potential effects on yield. We consider controlled environment, glasshouse and field studies with two questions in mind: (a) will elevated UV-B irradiance alter yields or quality of field-grown crops, and (b) are there beneficial/adaptive effects of UV-B which might be exploited?

Unless otherwise stated, the UV-B quantities discussed below are all in terms of the Caldwell generalised plant action spectrum normalised to 300 nm (PAS300; Caldwell, 1971; Caldwell et al., 1986) as discussed by Holmes (this volume).

Controlled environment and glasshouse studies of field crops

Controlled environments

While the limits of controlled environment experiments for assessing crop yield are generally recognised, they are ideal for studying detailed processes underlying plant responses to UV-B where variability in environmental conditions needs to be minimised. However, although the 'climate' in controlled environments (CE) is by definition highly controlled both in space and time, careful experimental design is essential as plant-to-plant, cabinet-to-cabinet and within cabinet variability will still be important (Potvin & Tardif, 1988). Apparent differences have to be treated with caution where treatments have not been replicated in space or time to eliminate chamber effects. In the specific context of predicting changes in yield under field conditions, extrapolating from observed CE responses is also problematic because most CE experiments are of short duration, with plants in pots and unrealistic patterns of variables such as temperature and humidity.

In addition to these general constraints on CE studies, there are a number of additional elements, specific to UV-B investigations, which need to be considered when assessing existing research.

Light environment

If conclusions are to be drawn regarding likely plant responses in natural environments then the chamber conditions should be within the range likely to be experienced by the plants in the field. Caldwell & Flint (1994) found that in papers published between 1990 and 1993, where the light environment had been adequately reported, UV-B treatments tended to be high (ranging from 50 to 450% of mid-summer clear-sky values at 41° N) and irradiances (in terms of photosynthetically active wavelengths; PAR) were rarely as high as 50% of mid-summer clear-sky values (41° N). The importance of PAR in determining the magnitude of plant response to UV-B was reported as early as 1964 (Jagger, 1964) and has been demonstrated in controlled environment and glasshouse studies by several groups (for example, Teramura, 1980; Warner & Caldwell, 1983; Mirecki & Teramura, 1984; Cen & Bornman, 1990). Increased PAR generally decreases the impact of enhanced UV-B on variables such as efficiency and capacity of photosynthesis, leaf area, plant height and pigment concentrations. Despite these reports, recent CE studies have continued to use low levels of PAR (for example, 60 μmol m^{-2} s^{-1}; Takeuchi *et al.*, 1996) and, on occasions, very high UV-B doses in the absence of any PAR (for example, Taylor *et al.*, 1996).

Control treatments

A feature of many other CE experiments that has received little critical discussion is that control treatments often receive no UV-B. This raises the question as to whether a treatment with no UV-B is justifiable as a 'control' when assessing the sensitivity of field crops that will rarely, if ever, experience a daylight spectrum without UV-B. A zero UV-B treatment might sensibly be compared with a +UV-B treatment if the experimental aim is to study the importance to plant processes of UV-B *per se* or to assess responses of plants moved from −UV-B to +UV-B environments, but such experiments give few insights into the impact of increasing UV-B on plants grown in agricultural environments.

Reporting of treatments

The important variables to report are the integrated daily totals for both PAR and UV-B (Fiscus & Booker, 1995). The practice of reporting UV-B irradiance as a percentage of midday clear-sky values can be misleading since this substantially over-estimates the mean daily dose, even in the absence of cloud. Thus a UV-B irradiance of 360 mW m^{-2} reported by Taylor *et al.* (1996) as 200% of maximum ambient irradiance for a summer day in northern England actually equates to a daily total of 20.7 kJ m^{-2}, which is >500% of the mean daily total for July at 55° N. Ideally, because UV-A as well as PAR can have an ameliorating effect on plant responses to UV-B, integrated daily UV-A doses should also be reported (Caldwell, Flint & Searles, 1994; Fiscus & Booker, 1995). The practice of expressing dosage in terms of a specific level of ozone depletion is useful for placing treatments in an environmental context but may be misleading unless it is clear how calculations have been made.

Many responses to UV-B found in CE studies might be deleterious to crop growth and yield if they occur in the field. These include DNA damage (Taylor *et al.*, 1996), down-regulation of photosynthetic genes (Jordan *et al.*, 1992), reduced photosynthetic capacity (Warner & Caldwell, 1983), reduced photosystem II efficiency (ϕ_{PSII}) and reduced biomass accumulation (Tevini *et al.*, 1991). Potentially adaptive or protective responses include increases in leaf thickness and increased levels of UV-B absorbing pigments in leaves (Warner & Caldwell, 1983; He *et al.*, 1993).

Glasshouse experiments

Glasshouse experiments for assessing impacts on yield have the advantage over those in CE in terms of space and cost, allowing greater replication and longer growing periods. Ideally, as with CE experiments, climate variables including PAR, UV-A and UV-B should be similar to those likely to be experienced in the field. Climate variables may be well controlled but glasshouse structures and glass reduce PAR and UV-A levels below ambient while UV-B is excluded (Middleton & Teramura, 1993). The UV-B : PAR ratio is important, as for CE experiments, and it should be noted that increasing PAR levels in the absence of UV-B may induce plant responses similar to those resulting from enhancement of UV-B under low PAR (Cen & Bornman, 1990).

The responses of crop species to UV-B reported in glasshouse studies include reductions in: maximum photosynthetic rate (P_{max}), efficiency of CO_2 fixation (ϕ_{CO_2}), and efficiency of photosystem II (ϕ_{PSII}) (Nogués & Baker, 1995), internode lengths (Barnes, Flint & Caldwell, 1990), stomatal conductance (Mirecki & Teramura, 1984), dry matter production, leaf area and height (Barnes *et al.*, 1993). These responses suggest that yields might be reduced in the field. Increases in UV-B absorbing pigments and chlorophyll have also been observed (for example, Mirecki & Teramura, 1984) which may have adaptive significance (see below).

Case study: pea

Chambers supplying PAR levels of 800–950 μmol m^{-2} s^{-1} (SON-T HPS and HQI Metal halide lamps) were supplemented with UV-A (Philips TLD36W 'black lights') and UV-B (Philips TL-40). UV-B was given as a square-wave centred on the middle of the 16 h photoperiod. The daily dose was varied both by exposure time and dimming of the UV-B lamps. Air temperatures were held at 20 °C during the day and allowed to fall to 10 °C during the night.

In initial studies, comparing six lines of pea with contrasting leaf surface waxiness (R. Gonzales, unpublished observations) either zero UV-B or a PAS300 dose of 6.5 kJ m^{-2} d^{-1} were applied, starting pre-emergence. Wax on adaxial and abaxial leaf surfaces increased in all but one cultivar, but UV-B reflectance was not significantly affected. Increased accumulation of UV-absorbing pigments (measured as absorbance at 300 nm: A300) was observed in all lines together with reductions in height, leaf and stipule areas and dry weight. These reductions were greatest in lines showing low constitutive absorbance at 300 nm (which varied greatly between cultivars) but were not correlated to the UV-B-induced increases in A300. The most morphologically responsive cultivar tested (JI812) showed the largest change in A300 between control and treated plants. There were no detectable treatment effects on either gas exchange or chlorophyll fluorescence within the errors of the experiment.

In a subsequent study, two lines known from the initial screen to differ in their UV-B response, were treated with a range of four UV-B doses from 2.3 to 9.2 kJ m^{-2} d^{-1} (R. Gonzales, unpublished observations). The more sensitive line (JI1389) showed progressive reductions in most measures of growth (height, leaf area, dry weight) as UV-B dose increased, but branch dry weight increased between 6.9 and 9.2 kJ m^{-2} d^{-1}. In the less sensitive line (Scout), branch dry weight increased with increasing dose above 4.6 kJ m^{-2} d^{-1}, but for height, leaf area and total dry weight there was no progressive dose response. A300 increased progressively for both cultivars with UV-B dose. Internode length was reduced in JI1389 but not in Scout and this was due to fewer cells per internode rather than cell length. No effects on gas exchange or chlorophyll fluorescence could be detected in either line, even at the highest doses.

Mechanisms of growth reductions were investigated in more detail in a pea cultivar (Guido) found to respond to UV-B in the field (Mepsted *et al.*, 1996). Plants were exposed in CE to low (2.16 kJ m^{-2} d^{-1}) or high (9 kJ m^{-2} d^{-1}) UV-B with four treatments according to length and sequence of exposure (high-high, high-low, low-high and low-low) allowing separation of effects on cell expansion and division. The high UV-B treatment was found to reduce leaf area primarily by reducing cell division. When plants were transfered from high to low UV-B, compensatory growth allowed some degree of recovery from the earlier effects of high UV-B (R. Mepsted, unpublished observations).

Field studies of agricultural and horticultural crops

Field studies for assessment of growth and yield

Field experiments designed to assess the impact of future ozone depletion in terms of enhanced UV-B radiation on crop growth and yield would ideally simulate a likely future radiation climate (both in spatial and temporal distribution of energy). This is difficult in practice because:

1. The amount of future ozone depletion is uncertain, and any depletion will vary seasonally and geographically,
2. UV-B irradiances and spectral distribution vary continuously diurnally and seasonally due to solar angle, cloud cover and air pollution levels (see chapter by Webb, this volume),
3. The output of fixed-spectrum lamps must be weighted using a plant action spectrum – however, this presupposes prior knowledge of how different species and processes will

respond to UV-B (see chapters by Holmes and by Paul, this volume),

4. The broad-band UV-B sensors most often used in field-based systems do not have spectral responses appropriate to real plant action spectra.

Problems 3 and 4 may be partially overcome by regular calibration of sensors against an appropriate action spectrum in daylight and under lamps. Many field studies have also ignored seasonal and diurnal changes in UV-B irradiance by applying a constant biologically weighted UV-B supplement for a certain number of hours per day, usually centred on solar noon (square-wave). This supplement is generally calculated for clear sky conditions at the summer solstice from the predicted consequences of a fixed percentage ozone depletion, for example, Sullivan and Teramura (1990), Ziska *et al.* (1993), Ambasht and Agrawal (1995). In this context, as noted previously by Fiscus and Booker (1995), the model of Green, Cross and Smith (1980) may overestimate ground-level UV-B and hence overestimate UV-B supplements.

Other problems with the square-wave approach have been highlighted by Sullivan *et al.* (1994), who compared the response of soybean given a square-wave treatment (two-step and adjustment every 21 d for seasonal effects) with the response of the same cultivar given a proportional addition to ambient UV-B (modulated). Daily average total UV-B dose was up to 30% greater in the square-wave than the modulated treatments because only the modulated system made adjustments for cloud cover. Interestingly, seed yields were in the order modulated > control > square wave, where the control treatment was irradiated by lamps filtered by polyester/mylar film (absorbs below 318 nm).

The UV-B-emitting fluorescent tubes used in most growth chamber, greenhouse and field experiments emit some UV-C (200–290 nm) and UV-A (320–400 nm) radiation in addition to the required UV-B (290–320 nm). In order to determine the effects on plants of UV-B specifically, radiation from the tubes must be filtered. Celluose diacetate plastic film, which is opaque to wavelengths < 290 nm, is used to prevent UV-C radiation (not present in sunlight at ground level) from reaching experimental plants. Mylar (a Du Pont Co. tradename) is one of several types of polyester film opaque to wavelengths < 320 nm, and is often used in experiments to prevent both UV-C and UV-B from reaching experimental plants. In such experiments, the effect of UV-B radiation on plant response is assumed to be the difference between the UV-B + UV-A treatment (cellulose diacetate filter) and the UV-A treatment (Mylar or equivalent polyester filter).

In some field experiments, polyester filters have not been used. For example, Mepsted *et al.* (1996) compared what was technically an enhanced UV-B + UV-A treatment (cellulose diacetate filter) with an ambient radiation control treatment. It was calculated that, with a 25% PAS300 enhancement to summer daylight, the cellulose diacetate filtered lamps would enhance ambient UV-A by only 0.2–0.3%, and this was considered insignificant. However, in at least one field experiment where UV-B + UV-A (cellulose diacetate), UV-A (polyester) and ambient control treatments have all been present, significant plant response differences between UV-A and ambient control treatments have been detected (McLeod *et al.*, this volume). Whether these treatment differences were due to the very small quantity of additional UV-A or to some other feature of the irradiation system is not known.

It is not just the use of different types of irradiation system which can make comparisons between experiments difficult. Other features of experimental design, for example, plot size and replication, also vary considerably from experiment to experiment, and this adds to the difficulties of comparison and interpretation of results. Table 1 shows that many field experiments have used rather small plots and sometimes treatments are not replicated. Because of the variable environmental conditions in the field, large plot sizes and many replicate plots may be needed to enable the detection of small treatment effects. The power of an experiment, expressed as the smallest percentage difference between treatment means required in order for there to be a given chance of detecting that difference, is a useful but seldom published measure of experimental variability.

Table 1 summarises recent field experiments with field crops and shows that, of those employing modulated systems, only two reported both crop biomass and economic yields for plants grown to maturity in field plots rather than pots. Using pots in the field rather than direct planting may reduce variability and thus make treatment differences easier to detect, but extrapolation to likely UV-B impact on field grown crops is difficult both because of the unnatural root environment, and because in some cases plants have been raised in the glasshouse (no UV-B) prior to field exposure (for example, Ziska *et al.*, 1993). In addition, yield per plant may not account for the effects of varying plant population on yield per unit area which is the key variable for relating results to agricultural yield.

Case study: pea and winter barley

A field experiment at HRI-Wellesbourne (52° N) was designed to quantify the effect on pea of a realistic increase in surface UV-B radiation under UK conditions (Fig. 1). Supplementary UV-B was applied to

Table 1. *Summary of recent field studies using elevated UV-B radiation*

Author	Species	Cvs	Reps	Plot size (m²)	UV-B Modulated or Square Wave?	Modelled ozone loss (%)	Grown to Maturity?	Biomass or Economic yield?	Polyester or Unpowered control?
Flint (1985)	faba bean	1	3	2.2	M	6, 32	N, 40 d	no	P
Murali (1986)	soybean	1	1	11.25	SW	25	Y	B	P
Beyschlag (1988)	wheat/wild oat	1	?	?	M	20	N	no	P
Barnes (1988)	wheat/wild oat	1	var	0.72	M	16–40	82 d	no	P
Teramura (1990)	soybean	2	4	2 rows	SW 2 step	16,25	Y	E	P
Sullivan (1990)	soybean	1	1	?	SW 2 step	25	Y	B and E	P
Sinclair (1990)	soybean	6	1	1.44	SW 2 step	16	Y	B and E	P and U
Goyal (1991)	linseed	1	1	?	SW 1,4 h	?	Y	B and E	U
Ziska (1993)	cassava	1	2	12 pots	SW 2 step	15	N, 95 d	B	P
D'Surney (1993)	soybean	2	2	pots	M	?	N, 61 d	B	P
Sullivan (1994)	soybean	1	2–4	?	M and SW	25	Y	B and E	P
Caldwell (1994)	soybean	1	5	0.28	M	36	N, 45 d	B	n/a
Ambasht (1995)	maize	1	3	1.2	SW 3 h	20	Y	B	P
Kim (1996)	rice	3	4	pots	M	38	Y	B and E	P
Day (1996)	pea	1	?	16 pots	M	16,24	N, 25 d	B	U
Mepsted (1996)	pea	4	5	2.7	M	15	Y	B and E	U

Only first author of multi-author papers is shown.

Fig. 1 UV-B arrays in the field over a pea crop at HRI Wellesbourne, Warwickshire, UK.

treated plots as a proportional addition to the UV-B dose received under control plots. The effect of ozone depletion on UV-B was calculated using the modified radiation transfer model of Björn and Murphy (1985) and the PAS300 addition calculated from the difference in dose between ambient ozone (mean 1979–1991) and an average year-round depletion of 15%, taking account of the seasonal variation in ozone depletion (Hermann, McPeters & Larko, 1993). For full experimental details, see Mepsted *et al.* (1996).

Taking all cultivars together, enhanced UV-B resulted in small reductions in the number of stems and total stem length per plant and significant ($p < 0.05$) decreases in the dry weight of peas (10%, Table 2), pods (10%) and stems (8%) per plant. UV-B treatment had no effect on the number of peas per pod or average pea weight, but did significantly reduce (12%) the number of pods per plant. This decrease in pod number was partly due to enhanced abscission of pods during the final month of plant growth. There was no detectable treatment effect on ϕ_{PSII} as measured using chlorophyll fluorescence induction, or on CO_2 assimilation rate per unit leaf area. These results are consistent with controlled environment experiments (above), and suggest that reductions in yield may result primarily from effects of UV-B on growth and morphology rather than from decreased assimilation per unit leaf area.

Table 2. *Pea yield (t ha^{-1} ± standard error of mean) for control and UV-B-treated plots compared with UK mean yields 1991–1995*

Cultivar	Control	UV-B	UK mean
Guido	1.84 ± 0.12	1.43 ± 0.15	4.5
Montana	2.05 ± 0.15	1.92 ± 0.17	5.4
Orb	1.56 ± 0.15	1.23 ± 0.18	4.8
Princess	1.61 ± 0.18	1.62 ± 0.17	4.8
All cultivars	1.76 ± 0.15	1.55 ± 0.13	4.9

Table 2 shows that pea yields in the control plots were roughly 35% of the expected yields for these cultivars in the UK (Anon., 1996). This was due to technical problems delaying planting (June rather than March) and damage caused by pea aphid (*Acyrthosiphon pisi* Harris). The size of the standard errors in Table 2 emphasises the degree of variability that is very often a feature of field experiments, especially where plot sizes are small.

The consequences of small plots were also apparent in a subsequent field experiment, with the same modulated UV-B system, simulating a 15% reduction in the level of stratospheric ozone. The effects of elevated UV-B irradiance on growth and yields of four varieties of winter barley were investigated. Few significant differences were detected and there were no detectable differences in biomass or economic yield due to UV-B treatment ($p > 0.05$), but this was not particularly surprising given that the experiment only had the power to detect treatment effects greater than 10–25% ($p = 0.05$; J. Stephen *et al.*, unpublished observations).

Summary of results from field experiments

Field experiments have produced varying predictions as to the impact of UV-B on crop economic and biomass yields. This is perhaps only to be expected given the range of species tested and the different treatment regimes. Decreases in biomass yield have been reported for some cultivars of soybean (Sullivan & Teramura, 1990; D'Surney *et al.*, 1993) and for pea (Mepsted *et al.*, 1996), while UV-B is reported to enhance biomass production by maize (Ambasht & Agrawal, 1995). The majority of field experiments (using pots or field planting) have shown no effect of elevated UV-B on above-ground biomass (for example, Barnes *et al.*, 1988; Sinclair, N'Diaye & Biggs, 1990; Ziska *et al.*, 1993;

Kim *et al.*, 1996; Day, Howells & Ruthland, 1996). Economic yield (seed yield) has been reduced in the field by elevated UV-B in linseed (Goya, Jain & Ambrish, 1991) and pea (Mepsted *et al.*, 1996), increased or decreased or unaffected in soybean depending on climatic conditions (Teramura, Sullivan & Lyndon, 1990; Sullivan & Teramura, 1990; Sinclair, N'Diaye & Biggs, 1990; Sullivan *et al.*, 1994) and unaffected in rice (Kim *et al.*, 1996) or winter barley (J. Stephen *et al.*, unpublished observations). Teramura, Sullivan and Lyndon (1990) reported soybean yields over five full growing seasons and demonstrated both how cultivars of the same species may differ in their response to UV-B, and how UV-B effects interact strongly with weather variables such as rainfall (amount and distribution), irradiance and air temperature. UV-B supplements decreased yield of cv. Essex relative to the control in four out of five seasons but increased yield in the fifth year when there was drought during early growth. The yield of another cultivar (Williams) responded positively to UV-B in all but the dry year. Yields in this, as in many other papers, are reported in terms of g plant^{-1} and assuming identical plant populations in each season, grain yields varied in the control plots from 3.6 to 9.5 t ha^{-1}.

Many field experiments report increases in UV-absorbing pigments in response to elevated UV-B (Murali & Teramura, 1986; Flint, Jordan & Caldwell, 1985; D'Surney *et al.*, 1993; Caldwell *et al.*, 1994; Day *et al.*, 1996). Other significant responses include reductions in mid-day ϕ_{PSII} (Ziska *et al.*, 1993), decreased chlorophyll content during early growth (Kim *et al.*, 1996), increased DNA damage (D'Surney *et al.*, 1993), changes in competitive advantage (Barnes *et al.*, 1988) and increases or decreases in specific leaf weight (Murali & Teramura, 1986; Sullivan *et al.*, 1994).

Discussion

Will elevated UV-B irradiance alter the yield or quality of field crops?

It is not the purpose of this review to discuss the likely future changes in stratospheric ozone or in UV-B irradiance reaching the ground, but rather to ask whether existing data from field studies can help us predict the likely effects on crop yield and quality if ozone depletion occurs within the range simulated by these experiments, i.e. 15–40% (Table 1). Given a specific focus on economic yield, we may apply a range of criteria to assess the value of different studies in quantifying future changes. For example, we might argue that plants must be grown in soil rather than pots, using a modulated UV-B system and polyester controls, in which case only one experiment fulfils all the criteria (Sullivan *et al.*, 1994). Alternatively, if we argue that sensible agronomic conclusions can only be drawn from adequately replicated experiments repeated over several growing seasons then only Teramura *et al.* (1990) would fulfil the criteria. The need for experiments over several seasons is highlighted by the many field studies on soybean which point to the conclusion that yield response to UV-B is highly dependent on cultivar/climate interactions (Teramura *et al.*, 1990). Thus, while there is evidence that seed yields in linseed (Goya *et al.*, 1991) and pea (Mepsted *et al.*, 1996) may be reduced by moderate increases in UV-B irradiance, these results need to be confirmed by multi-season trials. Given that the most consistently seen response of crops to UV-B is an increase in UV-B absorbing pigments, farmers may avoid any commercial impact of rising UV-B levels if (a) strategies of UV-B tolerance have no yield penalty and (b) if the value of the product is not affected by pigment concentrations.

Given the high cost of multi-year field trials, with large plots and adequate replication, screening large numbers of cultivars of any one species appears to be practicable only via short-term experiments under controlled conditions. If economic yield cannot be assessed in this type of screening then what characteristics should be used? Our experience suggests that pea cultivars that have higher constitutive levels of UV-B absorbing pigments are less responsive to UV-B morphologically. Photosynthetic responses to UV-B have been undetectable in peas given realistic PAR, UV-A and UV-B treatments from emergence, so perhaps morphological responses such as changes in internode length and branching pattern are the more suitable traits for screening. It is often difficult to predict the effect morphological changes will have on economic yield, but it would be interesting to test whether non-responsivess both of pigment concentration and morphology in the CE or glasshouse was correlated with 'non-responsiveness' of yield in the field.

Are there beneficial/adaptive effects of UV-B which might be exploited?

Several authors have studied the interaction between elevated UV-B and water deficit in the field (Murali & Teramura, 1986; Sullivan & Teramura, 1990). In both studies, increasing UV-B levels caused decreases in biomass yield under well-watered conditions but had no effect on the yield of water-stressed plants. These authors suggested that the plant responses to water deficit may have contributed to tolerance of increased UV-B, for example through increased levels of UV-B absorbing compounds and increased specific leaf weight. It is worth asking whether the converse might also be true; that is, might the responses of plants to increasing UV-B increase tolerance of water deficits? Under well-watered conditions in CE experiments, plants may respond to UV-B by reducing elongation growth, and leaf area and by increasing leaf thickness and leaf-surface waxiness. Such changes seem likely to modify plant responses to drought, and it is arguable that they might be exploited under certain circumstances. For example, there might be advantages in supplying UV-B to plants growing in 'artificial' environments where UV-B is normally excluded (CE or glasshouse).

Many plant species fail to thrive or develop normally under CE or glasshouse conditions, and the absence of UV-B might be a contributory factor. PAR levels in controlled environments and glasshouses are often less than half of that in full sunlight, but enhancing UV-B under low PAR may induce similar plant responses to those resulting from high PAR. Moving plants from conditions of low PAR and no UV-B (CE or glasshouse) to high PAR and high UV-B (field) is likely to be particularly damaging. Latimer and Mitchell (1987) reported that supplemental PAR irradiation of eggplant (*Solanum melongena* L.) raised in the glasshouse increased subsequent growth on transplanting to the field relative to plants that had received no extra PAR. While stating that many of the observed plant responses to high PAR, such as increased specific leaf weight, reduced leaf area and increased pigment production are also likely to contribute to subsequent protection from UV-B, they did not test or speculate as to whether a low level of supplemental UV-B in the glasshouse might have a similar effect to supplemental PAR and so pre-condition plants to cope with the sudden change in light climate on transplanting. If responses to UV-B contribute towards tolerance of water deficits, then pre-treated transplants might also be more able to cope with transient drought during the establishment phase.

Conclusions

Despite the large crop responses and significant damage to plants reported from CE and glasshouse experiments, evidence from the field suggests that changes in UV-B radiation resulting from a 15–40% ozone depletion may have only limited impact on commercially grown crops. Subtle changes in crop morphology and pigment concentrations appear more likely than large changes in productivity. However, the impact of UV-B does vary significantly depending on crop species, variety and

climate variables other than UV-B. As field data have been gathered
for only a few species, varieties and locations, there is a need still for
multi-year and multi-variety field studies. This may be complemented
by extensive screening in controlled environments that closely simulate
the natural light environment.

References

Ambasht, N.K. & Agrawal, M. (1995). Physiological responses of
field grown *Zea mays* L. plants to enhanced UV-B radiation. *Bio-
tronics*, **24**, 15–23.

Anon. (1996). *Pulse Variety Handbook*. National Institute of Agricultural Botany, Cambridge.

Barnes, P.W., Jordan, P.W., Gold, W.G., Flint, S.D. & Caldwell,
M.M. (1988). Competition, morphology and canopy structure in
wheat (*Triticum aestivum* L.) and wild oat (*Avena fatua* L.) exposed
to enhanced ultraviolet-B radiation. *Functional Ecology*, **2**, 319–30.

Barnes, P.W., Flint, S.D. & Caldwell, M.M. (1990). Morphological
responses of crop and weed species of different growth forms to
ultraviolet-B radiation. *American Journal of Botany*, **77**, 1354–60.

Barnes, P.W., Maggard, S., Holman, S.R. & Vergara, B.S. (1993).
Intraspecific variation in sensitivity to UV-B radiation in rice. *Crop
Science*, **33**, 1041–6.

Björn, L.O. & Murphy, T.M. (1985). Computer calculation of solar
ultraviolet radiation. *Physiologia Vegetale*, **23**, 555–61.

Caldwell, M.M. (1971). Solar UV irradiation and the growth and
development of higher plants. In *Photophysiology*. (Giese, A.C.
ed.), Vol 6. pp. 131–77. Academic Press, New York.

Caldwell, M.M. & Flint, S.D. (1994). Stratospheric ozone reductions,
solar UV-B radiation and terrestrial ecosystems. *Climate Change*,
28, 375–94.

Caldwell, M.M., Camp, L.B., Warner, C.W. & Flint, S.D. (1986).
Action spectra and their role in assessing biological consequences
of solar radiation change. In *Stratospheric Ozone Reduction, Solar
Ultraviolet Radiation and Plant Life*. (Worrest, R.C. & Caldwell,
M.M., eds), pp. 87–111. Springer-Verlag, Berlin.

Caldwell, M.M., Flint, S.D. & Searles, P.S. (1994). Spectral balance
and UV-B sensitivity of soybean: a field experiment. *Plant Cell and
Environment*, **17**, 267–76.

Cen, Y.-P. & Bornman, J.F. (1990). The response of bean plants to
UV-B radiation under different irradiances of background visible
light. *Journal of Experimental Botany*, **41**, 1489–95.

Day, T.A., Howells, B.W. & Ruthland, C.T. (1996). Changes in
growth and pigment concentrations with leaf age in pea under
modulated UV-B radiation field treatments. *Plant Cell and Environment*, **19**, 101–8.

D'Surney, S.J., Tschaplinski, T.J., Edwards, N.T. & Shugart, L.R. (1993). Biological responses of two soybean cultivars exposed to enhanced UV-B radiation. *Environmental and Experimental Botany*, **33**, 347–56.

Fiscus, E.L. & Booker, F.L. (1995). Is increased UV-B a threat to crop photosynthesis and productivity? *Photosynthesis Research*, **43**, 81–92.

Flint, S.D., Jordan, P.W. & Caldwell, M.M. (1985). Plant protective response to enhanced UV-B radiation under field conditions: leaf optical properties and photosynthesis. *Photochemistry and Photobiology*, **41**, 95–9.

Goya, A.K., Jain, V.K. & Ambrish, K. (1991). Effect of supplementary ultraviolet-B radiation on the growth, productivity and chlorophyll of field grown linseed crop. *Indian Journal of Plant Physiology*, **34**, 374–7.

Green, A.E.S., Cross, K.R. & Smith, L.A. (1980). Improved analytical characterisation of ultraviolet skylight. *Photochemistry and Photobiology*, **31**, 59–65.

He, J., Huang, L.-K, Chow, W.S., Whitecross, M.I. & Anderson, J.M. (1993). Effects of supplementary ultraviolet-B radiation on rice and pea plants. *Australian Journal of Plant Physiology*, **20**, 129–42.

Hermann, J.R., McPeters, R. & Larko, D. (1993). Ozone depletion at northern and southern latitutes derived from January 1979 to December 1991 Total Ozone Mapping Spectrometer Data. *Journal of Geophysical Research*, **98**, 12 783–93.

Jagger, J. (1964). Photoreactivation and photoprotection. *Photochemistry and Photobiology*, **3**, 451–61.

Jordan, B.R., He, J., Chow, W.S. & Anderson, J.M. (1992). Changes in mRNA levels and polypeptide subunits of ribulose 1,5-bisphosphate carboxylase in response to supplementary ultraviolet-B radiation. *Plant Cell and Environment*, **15**, 91–8.

Kim, H.Y., Kobayashi, K., Nouchi, I. & Yoneyama, T. (1996). Enhanced UV-B radiation has little effect on growth, $\delta^{13}C$ values and pigments of pot-grown rice (*Oryza sativa*) in the field. *Physiologia Plantarum*, **96**, 1–5.

Latimer, J.G. & Mitchell, G.A. (1987). UV-B radiation and photosynthetic irradiance acclimate eggplant for outdoor exposure. *HortScience*, **22**, 426–9.

Mepsted, R., Paul, N., Stephen, J., Nogués, S., Corlett, J.E., Baker, N.R., Jones, H.G. & Ayres, P.G. (1996). Effects of enhanced UV-B radiation on pea (*Pisum sativum* L.) grown under field conditions. *Global Change Biology* (in press).

Middleton, E.M. & Teramura, A.H. (1993). The role of flavonol glycosides and carotenoids in protecting soybean from ultraviolet-B damage. *Plant Physiology*, **103**, 741–52.

Mirecki, R.M. & Teramura, A.H. (1984). Effects of ultraviolet-B irradiance on soybean. V. The dependence of plant sensitivity on the photosynthetic photon flux density during and after leaf expansion. *Plant Physiology*, **74**, 475–80.

Murali, N.S. & Teramura, A.H. (1986). Effectiveness of UV-B radiation on the growth and physiology of field-grown soybean modified by water stress. *Photochemistry and Photobiology*, **44**, 215–19.

Nogués, S. & Baker, N.R. (1995). Evaluation of the role of damage to photosystem II in the inhibition of CO_2 assimilation in pea leaves on exposure to UV-B radiation. *Plant Cell and Environment*, **18**, 781–7.

Potvin, C. & Tardif, S. (1988). Sources of variability and experimental designs in growth chambers. *Functional Ecology*, **2**, 123–30.

Sinclair, T.R., N'Diaye, O. & Biggs, R.H. (1990). Growth and yield of field-grown soybean in response to enhanced exposure to ultraviolet-B radiation. *Journal of Environmental Quality*, **19**, 478–81.

Sullivan, J.H. & Teramura, A.H. (1990). Field study of the interaction between solar ultraviolet-B radiation and drought on photosynthesis and growth in soybean. *Plant Physiology*, **92**, 141–6.

Sullivan, J.H., Teramura, A.H., Adamse, P., Kramer, G.F., Upadhyaya, A., Britz, S.J., Krizek, D.T. & Mirecki, R.M. (1994). Comparison of the responses of soybean to supplemental UV-B radiation supplied by either square-wave or modulated irradiation systems. In *Stratospheric Ozone Depletion/UV-B Radiation in the Biosphere*. (Biggs, R.H. & Joyner, M.E.B., eds), NATO ASI Series, Vol. I, 18. Springer-Verlag, Berlin.

Takeuchi, Y., Kubo, H., Kasahara, H. & Sakaki, T. (1996). Adaptive alterations in the activities of scavengers of active oxygen in cucumber cotyledons irradiated with UV-B. *Journal of Plant Physiology*, **147**, 589–92.

Taylor, R.M., Nikaido, O., Jordan, B.R., Rosamond, J., Bray, C.M. & Tobin, A.K. (1996). Ultraviolet-B-induced DNA lesions and their removal in wheat (*Triticum aestivum* L.) leaves. *Plant Cell and Environment*, **19**, 171–81.

Teramura, A.H. (1980). Effects of ultraviolet-B irradiances on soybean. *Physiologia Plantarum*, **48**, 333–9.

Teramura, A.H., Sullivan, J.H. & Lyndon, J. (1990). Effects of UV-B radiation on soybean yield and seed quality: a 6-year field study. *Physiologia Plantarum*, **80**, 5–11.

Tevini, M., Mark, U., Fieser, G. & Saile, M. (1991). Effects of enhanced solar UV-B radiation on growth and function of selected crop plant seedlings. In *Photobiology*. (Riklis, E., ed.), pp. 635–49. Plenum Press, New York.

Warner, C.W. & Caldwell, M.M. (1983). Influence of photon flux density in the 400–700 nm waveband on inhibition of photosynthesis by UV–B (280–320 nm) irradiation in soybean leaves: separation of indirect and immediate effects. *Photochemistry and Photobiology*, **38**, 341–6.

Ziska, L.H., Teramura, A.H., Sullivan, J.H. & McCoy, A. (1993). Influence of ultraviolet-B (UV-B) radiation on photosynthetic and growth characteristics in field-grown cassava (*Manihot esculentum* Crantz). *Plant Cell and Environment*, **16**, 73–9.

J. ROZEMA, J.W.M. VAN DE STAAIJ
and M. TOSSERAMS

Effects of UV-B radiation on plants from agro- and natural ecosystems

Introduction

Most reports on effects of enhanced UV-B radiation on terrestrial plants relate to agricultural crops cultivated under laboratory, climate room or greenhouse conditions (Caldwell & Flint, 1994a). Many early studies reported differential growth reduction and a decrease in yield of crops, mostly from temperate climate regions. In reviews by Caldwell, Terramura & Tevini (1989) and Caldwell and Flint (1994b), direct damage effects were stressed. Reduced plant growth under enhanced UV-B was expressed as a reduction of plant height, plant dry weight and leaf area. Also, photosynthetic activity was reported to be reduced under enhanced UV-B radiation, through direct effects on the photosynthetic process (photosystem II, in particular) or indirectly by effects of UV-B radiation on photosynthetic pigments or stomatal functioning. Sensitivity to enhanced UV-B was shown to vary among plant species and cultivars of a crop species.

Fewer studies have been made of UV-B effects on native plant species in their natural ecosystems. Where these have been carried out, negative effects of enhanced UV-B radiation, simulating realistic scenarios of stratospheric ozone depletion, tend to be less than predicted from greenhouse studies (see chapter by Corlett et al., this volume). Rather than direct effects of enhanced solar UV-B radiation, indirect UV-B effects may change structure and functioning of agro-ecosystems and natural ecosystems.

This chapter considers the effects of enhanced UV-B radiation on plants from agro-ecosystems and natural ecosystems. This topic has been reviewed in the past and more recently (Caldwell et al., 1989; Tevini & Teramura 1989; Caldwell et al., 1995). Therefore, only selected points are considered here in detail. The methodology of UV-B experimentation is considered, and effects of UV-B radiation on crop plants are discussed using the growth and physiological response of bean *(Phaseolus vulgaris)* as an example of an indoor type of study.

Finally, some recent developments in the study of direct and indirect effects of enhanced UV-B radiation on terrestrial plants are reviewed.

Methodology of UV-B experimentation

In assessing the effects of enhanced solar UV-B radiation on terrestrial plants, different experimental methods are followed. Until recently most experimental UV-B studies were done in climate rooms and glasshouses (Caldwell & Flint, 1994b); however, because of unfavourable (low) PAR : UV-B ratios and differences between PAR and the UV-lamp spectra and that of solar radiation, results of such indoor studies cannot easily be extrapolated to outdoor conditions. Results obtained by studying effects of different levels of natural solar UV-B radiation on terrestrial plants will be more realistic than those from indoor studies (Caldwell & Flint, 1994b). For outdoor studies there are, in practice, two ways of changing levels of UV-B radiation. Firstly, by the use of UV-lamps and appropriate filters, UV-B radiation can be enhanced. Based on actual outdoor UV-B radiation measurements or on model calculations (Green, Cross & Smith, 1980) and the use of a UV-B plant action spectrum (see chapter by Holmes, this volume), a weighted biologically effective UV-B dose (UV-B$_{BE}$) can be obtained and varied, simulating the increase of UV-B radiation at the earth's surface related to stratospheric ozone depletion. Secondly, filters can be used to reduce levels of ambient, natural solar UV-B radiation levels.

Outdoor UV-B supplementation systems

Some of the factors involved in designing an outdoor system are included in the following description of a system used in the Netherlands (Fig. 1). Philips TL 12/40 lamps were installed in metal racks, allowing the height of the UV-lamps above the plant to be varied, as well as the burning period of the lamps. For both the ambient and the enhanced solar UV-B treatment, UV-lamps were burned. For the ambient UV-B (about 5 kJ m^{-2} day^{-1} UV-B$_{BE}$ in June and July) UV-tubes were wrapped in mylar foil, which excludes radiation with a wavelength <315 nm. Mylar foil therefore excludes about 90% of the UV-B but transmits UV-A radiation. For the enhanced UV-B radiation (7.5 kJ m^{-2} day^{-1}, UV-B$_{BE}$ simulating about 15–20% depletion of stratospheric ozone for the Netherlands at June 21), tubes were wrapped in cellulose acetate foil, transmitting radiation >290 nm. The combination of cellulose acetate and Mylar polyester foils means that UV-A radiation emitted by the Philips TL 12/40 lamps was the same for the ambient and

Fig. 1. Outdoor UV-B supplementation system, using Philips TL 12/ 40 lamps installed in metal racks. For the ambient UV-B (about 5 kJ m^{-2} day $UV-B_{BE}$ in June and July) UV-tubes were wrapped in Mylar foil, and for the enhanced UV-B radiation (7.5 kJ m^{-2} day^{-1} $UV-B_{BE}$ simulating about 15–20% depletion of stratospheric ozone for the Netherlands at June 21), tubes were wrapped in cellulose acetate foil. This outdoor UV-B supplementation system has been used to assess the response of various crop and native plant species grown in pots to enhanced UV-B radiation, as well as the response of plants in monocultures and mixed cultures to elevated UV-B. (See text for further details.)

enhanced UV-B radiation treatments (see Preface at the beginning of this volume).

UV-B supplementation can be applied in a 'square wave' mode, switching on UV-B lamps for a fixed period around midday, when natural solar UV-B is greatest. A more gradual increase of supplemented

UV-B radiation can be achieved by a stepwise increase of the number of lamps burning in a rack. A more refined approach is to modulate the addition of UV-B; this involves a solar tracking UV-B supplementation system in which UV-B sensors track ambient solar UV-B radiation. UV-B sensors are placed beneath treatment (enhanced) and control (ambient) racks with UV-tubes, spaced such that one sensor will always be unshaded. The unshaded sensor is selected by the control system. Feedback control of output of UV-lamps is thereby not disturbed by shading effects. Lamp output is adjusted to give a constant multiple of UV-B radiation above ambient UV-B radiation. A detailed survey of outdoor UV-B supplementation systems is given by McLeod (1997).

UV-B filtration

Solid plastic filters of various types can be installed above plants in pots or above natural vegetation (Fig. 2). A variety of plastic filters can be used: acrylate filters transmit most solar UV-B and UV-A radiation; acrylate covered with Mylar polyester film excludes most solar UV-B and transmits UV-A; lexane filters absorb UV-B and UV-A radiation. The acrylate and lexane filters transmit about 95% of photosynthetically active radiation (PAR); transmission of PAR through acrylate filters with mylar polyester film is about 85% (Tosserams, Pais de Sa & Rozema, 1996). However, the acrylate and lexane filters need to be regularly cleaned carefully, and Mylar foil must be renewed frequently. Such a filter system allows a comparison of plant responses to ambient solar UV-B (about 5 kJ m^{-2} day^{-1} UV-B$_{BE}$) and below-ambient solar UV-B radiation (less than 1 kJ m^{-2} day^{-1} UV-B$_{BE}$) beneath the plastic filters. One big advantage of filtration of natural solar UV-B radiation is that the natural levels of PAR and UV-A wavebands can be maintained, although of course only ambient and below ambient UV-B radiation levels can be obtained.

Enhanced UV-B radiation and crop plants

To give some indication of the type of experimental approach used, the following section gives details of a typical 'indoor' type of UV-B experiment, carried out with bean plants, followed by a more general review of effects on a range of crop species.

Cultivation and UV-B treatment of bean plants

Seedlings of *Phaseolus vulgaris* were precultured for 6 days before the UV-B treatment started. Plants were grown in a glasshouse, temperature

Fig. 2. Outdoor UV-B filtration system, using solid plastic filters installed above plants in pots (see text for more detail): Acrylate (transmits most solar UV-B and UV-A radiation); Acrylate covered with Mylar polyester film (excludes most solar UV-B and transmits UV-A); Lexane filters (absorb UV-B and UV-A radiation). Natural precipitation was measured frequently and the area beneath the filters was watered to compensate for the rainfall intercepted by the filter surface. The plastic filters are 120×120 cm, and to avoid edge effects, only plants of the inner 60×60 cm were studied. In the summer months June, July, August 1994 no significant increase of temperature occurred beneath the filters in the dune grassland. In the autumn (October, November) condensation on the lower side of the filters may reduce transmittance of the filters, particularly in the morning hours.

varying between 23 °C (day) and 13 °C (night), relative humidity 80% (day) and 95% (night). UV-B tubes (Philips TL 12/40) were installed at both sides of 400 W HPI/T lamps, providing a PAR level of 300 μmol m^{-2} s^{-1} during a 16-hour photoperiod. Radiation from the UV-tubes was filtered through 0.1 mm thick cellulose acetate foil, absorbing wavelengths below 290 nm. Different levels of UV-B radiation were obtained by adjusting the distance between the UV-B tubes and the top of the bean plants. Weighted daily doses of UV-B_{BE} obtained were 6, 8 and 12 kJ m^{-2} day^{-1} simulating a 5%, 20% and 35% reduction of stratospheric ozone, respectively. This was based on model calculations (Green *et al.*, 1980) relating to a biologically effective UV-B dose of

5 kJ for clear sky conditions on June 21 in Amsterdam (52 °N latitude), and the generalised plant action spectrum, normalised at 300 nm (Caldwell, 1971). Control plants were also grown under the UV-B tubes, but with the addition of Mylar foil which effectively removed all the UV-B radiation (0 kJ m^{-2} day^{-1} at plant level). UV-B radiation levels were daily controlled with a UV-X sensor. The UV-X sensor was calibrated against an Optronics OL 752 Spectroradiometer. The spectroradiometer was calibrated with the OL-752-150 dual calibration and gain check source module. Cellulose acetate foil and Mylar foil were renewed twice a week to minimise the effects of foil degradation on output and spectral quality. There were 12 replications per UV-B radiation level.

Responses

Figure 3 shows the appearance of representative plants from the different treatments. Total plant dry weight of bean was reduced by 25% at

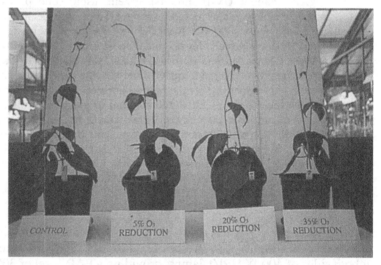

Fig. 3. Growth reduction of *Phaseolus vulgaris*, grown in a climatised greenhouse in response to increasing ultraviolet-B radiation, simulating a reduction of 5%, 20% and 35% in stratospheric ozone. Control plants received no UV-B radiation. Bean plants were grown in pots with commercial potting soil (Jongkind B.V., Aalsmeer). The four levels of UV-B radiation were obtained by adjusting the height of Philips 40W/12 tubes above the bean plants. UV-tubes were wrapped in cellulose acetate foil. (See also Tosserams & Rozema, 1995 and Rozema, Tosserams & Magendans, 1995.)

Table 1. *Effect of increasing levels of UV-B radiation on biomass parameters (g dry weight) of* Phaseolus vulgaris *after 18 days of growth*

UV-B$_{BE}$ (kJ m^{-2} d^{-1})	Root	Leaves	Stem	Total
0.0	0.72 ± 0.05	1.20 ± 0.06	0.67 ± 0.03	2.58 ± 0.13
6.0	0.60 ± 0.04	1.25 ± 0.06	0.55 ± 0.05	2.41 ± 0.13
8.0	0.65 ± 0.05	1.08 ± 0.06	0.43 ± 0.03	2.17 ± 0.11
12.0	0.63 ± 0.05	0.89 ± 0.04	0.42 ± 0.02	1.95 ± 0.06
Significance	*	*	*	*

Average values of 12 replications. Statistical significance of a one-way analysis of variance is indicated as * ($P < 0.05$); n.s., not significant.

the highest level of UV-B$_{BE}$ applied. This is related to biomass reduction of the aerial plant parts, leaves and stem, in particular (Table 1). Root dry weight was reduced by 14% at the highest UV-B radiation level. The relative growth rate of the bean plants was reduced by 23% with increased UV-B radiation. The growth reduction with enhanced UV-B might have been due to a decreased leaf area ratio (17.6% reduction at 12.0 kJ m^{-2} day^{-1}); the specific leaf area decreased by 18.1% indicating increased thickness of the bean leaves with enhanced UV-B radiation. However the leaf weight ratio was not significantly affected by enhanced UV-B radiation (Table 2), and there was no evidence that gas exchange of intact leaves was affected by increased UV-B radiation

Table 2. *Effect of increasing levels of UV-B irradiation on growth parameters of* Phaseolus vulgaris

UV-B$_{BE}$ (kJ m^2 day^{-1})	RGR (g kg^{-1} day^{-1})	LAR (m^2 kg^{-1})	SLA (m^2 kg^{-1})	LWR (kg kg^{-1})
0.0	133 ± 2.0	25.1 ± 0.81	54.9 ± 2.7	0.47 ± 0.02
6.0	125 ± 1.0	23.5 ± 0.45	45.4 ± 1.8	0.51 ± 0.03
8.0	114 ± 1.1	23.0 ± 1.40	46.5 ± 3.0	0.49 ± 0.01
12.0	102 ± 1.5	20.7 ± 2.56	45.0 ± 5.4	0.46 ± 0.02
Significance	*	*	*	ns

RGR, relative growth rate; LAR, leaf area ratio; SLA, specific leaf area; LWR, leaf weight ratio. Statistics as in Table 1.

Table 3. *Effect of increasing levels of UV-B irradiation on net photosynthesis, transpiration and water-use efficiency (WUE) of* Phaseolus vulgaris, *after 16 days of growth and exposure to UV-B irradiation*

UV-B$_{BE}$ (kJ m^{-2} day^{-1})	Net photosynthesis (μmol CO$_2$ m^{-2} s^{-1})	Transpiration (mmol H$_2$O m^{-2} s^{-1})	WUE (μmol CO$_2$. mmol H$_2$O^{-1})
0.0	24.2 ± 1.4	7.1 ± 0.9	3.7 ± 0.8
6.0	29.3 ± 1.5	9.2 ± 0.7	3.2 ± 0.2
8.0	25.1 ± 1.3	8.7 ± 0.7	2.9 ± 0.3
12.0	25.1 ± 1.0	8.3 ± 0.4	3.0 ± 0.1
Significance	ns	ns	ns

During the gas exchange measurement, leaves were exposed to a 1000 μmol m^{-2} s^{-1} PAR level at a temperature of 23 °C. Average values of four replications. Statistics as in Table 1.

(Table 3). Neither net leaf photosynthesis, nor leaf transpiration rates were reduced with enhanced UV-B radiation. Accordingly, there was no significant effect of increasing UV-B radiation on the water-use efficiency. UV-B absorbance of leaf extracts increased with enhanced UV-B up to 8 kJ m^{-2} day^{-1} ($P < 0.05$) (Fig. 4), while at the highest UV-B$_{BE}$ level (12 kJ m^{-2} day^{-1}) there was a decrease of UV-B absorbance.

The experimental results shown in Tables 1–3 and Fig. 4 indicate that the crop plant *Phaseolus vulgaris* is sensitive to enhanced solar UV-B radiation, a finding which has been confirmed by outdoor UV-B experiments with bean (Deckmyn, 1996; Antonelli *et al.*, 1997).

Sensitivity of species from crop systems and natural systems

The legumes *Pisum sativum* (Strid, Chow & Anderson, 1990) and *Vicia faba* are also UV-B sensitive (Rozema, Lenssen & van de Staaij, 1990; Rozema, Tosserams & Magendans, 1995), as are rye (*Secale cereale*) (Deckmyn, 1996) and cucumber (*Cucumis sativus*) (Adamse & Britz, 1992), but tomato (*Lycopersicon esculentum*), wheat (*Triticum aestivum*) and corn (*Zea mays*) are less sensitive to UV-B (Rozema *et al.*, 1992; Teramura & Sullivan, 1994; Caldwell & Flint, 1994*a*). It might be that sensitivity or resistance to enhanced UV-B radiation of a crop species relates to the level of UV-B radiation of the environment

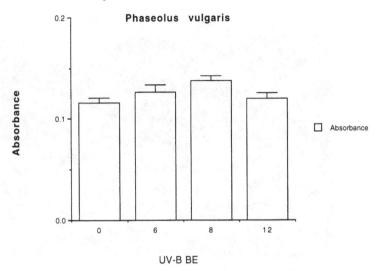

Fig. 4. The effect of increasing levels of UV-B$_{BE}$ on the absorbance of extracts of leaves of *Phaseolus vulgaris*, measured at 300 nm. Absorbance has been recalculated to a concentration of 1 mg fresh leaf per ml extract. Average values of four replications with s.e.m.

where the crop has its geographical origin, as has been shown for rice cultivars (Teramura, Sullivan & Sztein, 1991). It has also been suggested that UV-B sensitivity of crops may be linked with the process of breeding of crop plants from native varieties (Rozema, Lenssen & van de Staaij, 1990; van de Staaij *et al.*, 1995*b*; Teramura & Sullivan, 1994; Caldwell & Flint, 1994*b*). Native plant species tend to be less sensitive to enhanced solar UV-B radiation, although large differences between terrestrial native plant species do occur (Rozema *et al.*, 1995), for example, *Calamagrostis epigeios*, a grass species dominating coastal dune grasslands in the Netherlands (Fig. 5), appears to be highly UV-B resistant. With UV-B$_{BE}$ levels increasing to 14.9 kJ m^{-2} day^{-1}, simulating 44% stratospheric ozone depletion, no significant reduction of growth, photosynthesis or transpiration could be detected (Table 4). The content of flavonoids and lignin increased with enhanced UV-B radiation. A survey of UV-B sensitive and UV-B resistant plant species, mainly based on indoor UV-B studies in our group is presented in Table 5.

Possible causes of differences in sensitivity to enhanced UV-B between crop plants and native plant species are listed in Table 6, suggesting that accumulation of UV-B absorbing compounds is particularly

Fig. 5. UV-B supplementation (UV-tubes in racks) and UV-B filtration system installed in a dune grassland, Heemskerk, The Netherlands.

Table 4. *Summary of effects of enhanced solar UV-B on the
dune grassland species* Calamagrostis epigeios

Growth	0
Net photosynthesis	0
Transpiration	0
UV-B absorbing compounds (flavonoids)	+
Tannin	0/+
Lignin	+

0, no response; +, positive response; −, negative response
Based on Tosserams & Rozema (1995); Rozema, Tosserams & Magendans
(1995); Rozema *et al.* (1997c).

important in conferring resistance in native species. More general surveys of plant species ranked according to sensitivity or resistance to
UV-B will always be hindered by poor comparability of experimental
conditions (UV-B and PAR levels, spectral balance, different geographically latitudes relation to different ambient biologically effective
UV-B doses). This emphasises the need for careful recording of
irradiation protocols to allow for conversion of different experimental
methods.

Table 5. *Survey of UV-B-sensitive, -intermediate and UV-B-resistant plant species based on plant responses to enhanced UV-B radiation in indoor studies*

	Crop species	Native plant species	Source
UV-B-sensitive			
Pisum sativum	×		2
Phaseolus vulgaris	×		9
Vicia faba	×		14
Cucumis sativa	×		8
Secale cereale	×		12
Intermediate			
Lycopersicon esculentum	×		2
Triticum aestivum	×		3
Zea mays	×		3
Oryza sativa	×		4, 6
Plantago lanceolata		×	13
Elymus athericus		×	5, 7
Aster tripolium		×	1,2
Urtica dioica		×	13
Verbascum thapsus		×	13
UV-B-resistant			
Calamagrostis epigeios		×	10
Silene vulgaris		×	8, 11
Holcus lanatus		×	13
Epilobium hirsutum		×	13
Spartina anglica		×	1,8

Enhanced UV-B radiation simulated 20 (−30%) ozone depletion under clear sky.
1. van de Staaij, Rozema & Stroetenga, 1990.
2. Rozema, Lenssen & van de Staaij, 1990.
3. Rozema *et al.*, 1991.
4. Teramura, Sullivan & Ziska, 1990.
5. Rozema *et al.*, 1992.
6. Teramura, Sullivan & Sztein, 1991.
7. van de Staaij *et al.*, 1993.
8. Van de Staaij *et al.*, 1995*b*.
9. Deckmyn, Martens & Impens, 1994.
10. Tosserams & Rozema, 1995.
11. Van de Staaij *et al.*, 1995*a,b*.
12. Deckmyn, 1996.
13. Tosserams, Pais de Sa & Rozema, 1997.
14. Visser *et al.*, 1997.

Table 6. *Possible causes of differences in sensitivity to enhanced UV-B radiation between crop species and native plant species*

UV-B sensitivity of crop species
  I. UV-B resistance of crop species lost as a result of breeding procedure:
   • Breeding in UV-B-free environment. Glasshouses usually transmit no solar UV-B.
   • Selection for high relative growth rate, thin leaves.
   • Selection against morphological (leaf hairs, other epidermal structures) and morphogenetic characteristics giving resistance to UV-B radiation.
   • Selection against high content of tannins, flavonoids and other UV-B absorbing compounds, when these affect taste and quality.
  II. Area of origin of crop species may have been a low UV-B environment.

UV-B resistance of native plant species
  I. Environmental stresses, such as nutrient deficiency, may lead to increased levels of UV-B-absorbing compounds (flavonoids, polyphenolics). Similarly, plant–animal relationships such as herbivory may cause selection for plants high in tannins or flavonoids.
  II. Morphogenetic characteristics such as thick, erectophilous leaves and plant architecture may reduce UV-B damage to native plant species.

Direct and indirect effects of enhanced ultraviolet-B radiation on terrestrial plants

Enhanced UV-B radiation has, at least in the past, been considered an environmental stress factor, reducing biomass gain in many terrestrial plant species. Reduced plant growth has often been ascribed to disturbance of three primary targets of (enhanced) UV-B radiation. Firstly, UV-B may cause direct damage to DNA. Absorption of UV-B radiation by DNA may induce cyclobutane dimers, formation between nearby pyrimidines of the same DNA strand and formation of pyrimidine (6–4) pyrimidone. The latter are called the (6–4) photoproducts. DNA damage may be repaired by excision repair (relatively slow) or photo-repair by the photolyase enzyme system (see chapter by Taylor *et al.*, this volume). Secondly, UV-B radiation may damage photosystem II by disturbing sensitive sites (Caldwell *et al.*, 1989). Recently, however, Björn (1997) concluded that there is evidence that photosystem II is *not* the critical site for direct UV-B inhibition of light-saturated photosynthesis, a conclusion also reached by Baker *et al.* (this volume). It might be that enhanced UV-B radiation affects thylakoid membrane functioning, or enzymes of the Calvin cycle. Thirdly, UV-B radiation may directly disturb cell membrane functioning, by lipid peroxidation or damage to membrane proteins. Polyamines may be involved in ameliorating UV-B damage to membranes (Kramer, Krizek & Mirecki, 1992). According to a recent review, of these three possible primary targets for UV-B radiation in plants, DNA damage is probably more important than UV-B damage to photosynthesis and membrane lipids (Björn, 1997).

In contrast with these earlier views, there is growing evidence that enhanced UV-B does not have a significant direct effect on plant growth and primary production of natural ecosystems. Indirect effects of enhanced UV-B radiation, such as altered leaf angle, differential transmission and absorbance of UV-B radiation through stands of erectophilous or planophilous plants species, may have important consequences for the response of a stand to enhanced UV-B radiation (Rozema *et al.*, 1997*b*). For example, Deckmyn (1996) concluded that differences in sensitivity of *Phaseolus vulgaris* (planophilous plant species) and *Secale cereale* (erectophilous plant species) to UV-B related to differential attenuation of UV-B (and PAR) through stands of bean and rye. In addition, plant and ecosystem processes may change as a result of indirect effects of UV-B radiation (plant morphogenetic, phenological effects and changes in the chemical composition of plants under elevated UV-B).

Indirect effects of enhanced UV-B radiation cover a wide range of plant and ecosystem parameters (Table 7). Of the plant morphogenetic parameters, changes of leaf thickness, leaf angle and canopy architecture may have far reaching consequences for UV-B effects on ecosystems. Increased leaf thickness under enhanced UV-B (Johanson *et al.*, 1995; Rozema *et al.*, 1997*b*), often measured as a higher specific leaf weight (gram leaf weight per m^2 leaf area) or a lower specific leaf area (leaf area in m^2 per gram leaf weight), is likely to reduce UV-B damage to leaf cells. At the same time, reduced PAR levels in thicker leaves will reduce leaf photosynthesis. Deckmyn, Martens and Impens (1994) and Deckmyn (1996) indicated that UV-B/PAR ratios determined the sensitivity of bean and rye to UV-B radiation. In a model study Deckmyn (1996) predicted that a more planophilous leaf angle in an erectophilous stand of rye plants would reduce the UV-B/PAR ratio and therefore UV-B damage to rye plants. Of course, the above changes under enhanced UV-B will lead to changes in the competitive relations between plant species. Direct evidence for shifts in plant–plant relationships in ecosystems under elevated UV-B is limited (Caldwell *et al.*, 1995; Björn, 1997; Caldwell, this volume).

Table 7. *Direct and indirect effects of enhanced ultraviolet-B radiation on terrestrial plant growth and processes in terrestrial ecosystems*

I. Direct effects
  1. DNA-damage
    Cyclobutane dimer formation
    (6–4) photo-product formation
  2. Photosynthesis
    Disturbance of photosystem II
    Thylakoid membrane functioning
    Stomatal functioning
  3. Membrane functioning
    Peroxidation of unsaturated fatty acids
    Damage to membrane proteins
II. Indirect effects
  1. Plant morphogenetic effects
    Leaf thickness
    Leaf angle
    Plant architecture
    Biomass allocation
  2. Plant phenology
    Emergence
    Senescence
    Flowering
    Reproduction
  3. Chemical composition of plants
    Tannins
    Lignin
    Flavonoids

In addition to effects on plant growth, enhanced UV-B may have effects, both direct and indirect, on the decomposition of plant material. In a subarctic dwarf shrub ecosystem in Sweden and in a dune grassland ecosystem in The Netherlands decomposition of litter from plants grown under enhanced solar UV-B radiation was reduced compared with decomposition of control litter from plants grown under ambient solar UV-B radiation (Gehrke *et al.*, 1995; Rozema *et al.*, 1997c). In the dune grassland, plant material of *Calamagrostis epigeios* grown under elevated UV-B in the field had a significantly higher content of lignin than control plants, possibly due to a direct effect of UV-B radiation on the enzyme phenylalanine ammonia lyase (Beggs & Wellman,

1994). The increased plant lignin content would account for the reduced rate of litter decomposition (Rozema *et al.*, 1997c). Gehrke *et al.* (1995) found an increased content of tannins in leaf material of *Vaccinium* species, cultivated under enhanced UV-B radiation. Like lignin, an increased tannin content may slow down litter decomposition. However, litter exposed to enhanced solar UV-B decomposed, by photodegradation, more rapidly than under ambient solar UV-B (Rozema *et al.*, 1997c; Zepp, Callaghan & Erickson, 1995). Thus, in these studies, enhanced solar UV-B radiation caused opposing effects on litter decomposition; directly there was an increase, due to photodegradation, while indirectly there was a decrease, due to a change in the litter quality. A third effect of enhanced UV-B radiation on decomposition may be a direct effect on the decomposing fungi and other decomposer organisms (see chapter by Paul, this volume). Effects of UV-B radiation on litter decomposition are summarised in Table 8. At the moment there have been insufficient studies to predict what the long-term effect will be of enhanced UV-B on decomposition processes. When considering effects on decomposition, interactions with other climatic factors need to be considered, and particularly, in the case of litter, the increase in CO_2 (Rozema *et al.*, 1997a). Atmospheric CO_2 enrichment affects directly and indirectly the quantity and quality of decomposing plant litter. Under elevated CO_2, the nitrogen content of plants may decrease

Table 8. *Qualitative summary of direct and indirect effects of enhanced solar UV-B radiation on litter decomposition in natural ecosystems*

I. Direct effects of enhanced UV-B radiation on litter decomposition.
   1. Photodegradation. Plant litter exposed to enhanced UV-B decomposes more rapidly than under ambient UV-B. [a,b]
   2. Damage or disturbance to decomposer invertebrates, fungi and micro-organisms.[b]

II. Indirect effects of enhanced UV-B radiation on litter decomposition.
   1. Through an increased plant content of tannins or lignin, reducing microbial degradability of litter.[a,b] Increase of tannins and flavonoids may affect plant–animal relationships such as herbivory[c].

[a]Rozema *et al.* (1997c).
[b]Gehrke *et al.* (1995).
[c]Paul *et al.* (1997).

and the C/N ratio of resulting plant litter increase (Rozema, 1993). The result of the dilution of nitrogen in plants growing under elevated CO_2 may be a reduced rate of litter decomposition.

Finally, another consequence of changes in plant composition may be to affect feeding behaviour of insects; if there is a reduced nitrogen content, herbivorous insects may need more of such plant material to fulfil their nitrogen requirements. In addition, the tannin content of plant material may affect plant–animal relationships. In an ecosystem study Gwynn-Jones et al. (1997) demonstrated significant changes of insect herbivory in Vaccinium myrtillus and Vaccinium uliginosum following supplemental UV-B treatments (see also chapter by Moody et al., this volume, for a more detailed study of decomposition and herbivory).

Conclusions

Important physiological and biochemical responses to enhanced solar UV-B radiation on terrestrial plants have been determined from laboratory and climate room studies. Enhanced solar UV-B may be an environmental stress factor, disturbing membrane functioning and causing DNA damage, leading to reduced biomass and yield. However, in field studies on natural ecosystems, indirect effects of enhanced UV-B radiation on plant secondary metabolites and morphogenetic parameters (canopy architecture, competitive relationships between plant species, plant–animal relationships and litter decomposition) appear to be as important as direct effects of UV-B on plant growth and primary production. The detection of such, sometimes subtle, effects of enhanced solar UV-B on natural ecosystems requires careful and well-planned long-term experimentation.

Acknowledgements

We thank Mrs Nan Kasanpawiro for carefully word-processing the manuscript. We acknowledge Dr Peter Lumsden, the Society for Experimental Biology and Cambridge University Press for organising and supporting the workshop on UV-B radiation. The research of M. Tosserams was funded by the Dutch National Research Programme on Global Change (NRP, project number 850022). The authors are indebted to Lars Gerundt, Guido Kalis, Dagmar van Oijen, Jacqueline Smet and Femmie Smit, who performed UV-B experiments with bean plants. We also acknowledge the NV PWN Waterleidingbedrijf Noord-Holland for permission to work in the Dune Reserve.

References

Adamse, P. & Britz, S.J. (1992). Amelioration of UV-B damage under high irradiance. I. Role of photosynthesis, *Photochemistry and Photobiology*, **56**, 645–50.

Antonelli, F., Grifoni, D., Sabatini, F. & Zipoli, G. (1997). Morphological and physiological responses of bean plants to supplemental UV-radiation in a Mediterranean climate. *Plant Ecology*, **128**, 127–36.

Beggs, C.J. & Wellman, E. (1994). Photocontrol of flavonoid biosynthesis. In *Photomorphogenesis in Plants*. (Kendrick, R.E. & Kronenberg, G.H.M., eds.), 2nd edn, pp. 733–750. Kluwer Academic Publishers, Dordrecht.

Björn, L.O. (1997). Effects of ozone depletion and increased UV-B on terrestrial ecosystems. *International Journal of Environmental Studies*, **51**, 217–43.

Caldwell, M.M. (1971). Solar UV-B irradiation and the growth and development of higher plants. In *Photophysiology*. (Geise, A., ed.), pp. 131–70. Academic Press, New York.

Caldwell, M.M. & Flint, S.D. (1994a). Stratospheric ozone reduction, solar UV-B radiation and terrestrial ecosystems. *Climatic Change*, **28**, 375–94.

Caldwell, M.M. & Flint, S.D. (1994b). Solar ultraviolet radiation and ozone layer change: implications for crop plants. In *Physiology and Determination of Crop Yield*. Madison, American Society of Agronomy, 487–507.

Caldwell, M.M., Teramura, A.H. & Tevini, M. (1989). The changing ultraviolet climate and the ecological consequences for higher plants. *Trends in Ecology and Evolution*, **4**, 363–6.

Caldwell, M.M., Teramura, A.H., Tevini, M., Bornman, J.F., Björn, L.O. & Kulandaivelu, G. (1995). Effects of increased solar ultraviolet radiation on terrestrial plants. *Ambio*, **24**, 166–73.

Deckmyn, G. (1996). Effecten van verhoogde UV-B straling op planten. Doctorate Thesis Universitaire Instelling Antwerpen. pp. 74.

Deckmyn, G., Martens, C. & Impens, I. (1994). The importance of the ratio UV-B/photosynthetic active radiation (PAR) during leaf development as determining factor of plant sensitivity to increased UV-B irradiance: effects on growth, gas exchange and pigmentation of bean plants. (*Phaseolus vulgaris* cv label). *Plant Cell and Environment*, **18**, 1426–33.

Gehrke, C., Johanson, U., Callaghan, T.V., Chadwick, D. & Robinson, C.H. (1995). The impact of enhanced ultraviolet-B radiation on litter quality and decomposition processes in *Vaccinium* leaves from the Subarctic. *Oikos*, **72**, 213–22.

Green, A.E.S., Cross, K.R. & Smith, L.A. (1980). Improved analytic characterization of ultraviolet skylight. *Photochemistry and Photo-biology*, **31**, 59–65.

Gwynn-Jones, D., Lee, J.A., Callaghan, T.V. & Sonesson, M. (1997). Effects of enhanced UV-B radiation and elevated carbon dioxide concentrations on subarctic forest heath ecosystems. *Plant Ecology*, **129**, 242–9.

Johanson, U., Gehrke, C., Björn, L.O. & Callaghan, T.V. (1995). The effects of enhanced UV-B radiation on the growth of dwarf shrub in a subarctic heathland. *Functional Ecology*, **9**, 713–19.

Kramer, G.F., Krizek, D.T. & Mirecki, R.M. (1992). Influence of UV-B radiation on polyamines, lipid peroxidaton and membrane lipids in cucumber. *Phytochemistry*, **30**, 2101–8.

McLeod, A.R. (1997). Outdoor supplementation systems for studies of the effects of increased UV-B radiation. In *UV-B and Biosphere*. (Rozema, J., Gieskes, W.W.C., van de Geijn, S.C, Nolan, C. & de Boois, H., eds), Kluwer, Dordrecht.

Paul, N.G., Rasanayagam, M.S., Moody, S.A., Hatcher, P.E. & Ayres, P.G. (1997). The role of interactions between trophic levels in determining the effects of UV-B on terrestrial ecosystems. *Plant Ecology*, **128**, 296–308.

Rozema, J. (1993). Plant responses to atmospheric carbon dioxide enrichment: interactions with some soil and atmospheric conditions. *Vegetatio*, **104/105**, 173–90.

Rozema, J. (1995). Plants and high CO_2. In *Encyclopaedia of Environmental Biology*. Vol. 3, pp. 129–38. Academic Press, New York.

Rozema, J., Lenssen, G.M. & van de Staaij, J.W.M. (1990). The combined effect of increased atmospheric CO_2 and UV-B radiation on some agricultural and salt marsh species. In *The Greenhouse Effect and Primary Productivity in European Agro-ecosystems*. (Goudriaan, J., Van Keulen. H, & Van Laar, H.H., eds), pp. 68–71. Pudoc, Wageningen.

Rozema, J., Lenssen, G.M., Arp, W.J. & van de Staaij, J.W.M. (1991). Global change, the impact of the greenhouse effect and increased UV-B radiation on plants. In *Ecological Responses to Environmental Stresses*. (Rozema, J. & Verkleij, J.A.C., eds), pp. 220–31. Kluwer Academic, Dordrecht.

Rozema, J., van de Staaij, J.W.M., Costa, V., Torres Pereira, J.G.M., Broekman, R.A., Lenssen, G.M. & Stroetenga, M. (1992). A comparison of growth, photosynthesis and transpiration of wheat and maize in response to enhanced UV-B radiation. In *Global Climatic Changes on Photosynthesis and Plant Productivity*. (Abrol, Y., Wattal, P.N., Gnanam, A., Govindjee, Ort, D.R. & Teramura, A.H., eds), pp. 163–74. Asia Publishing House, Sittingbourne.

Rozema, J., Tosserams, M. & Magendans, E. (1995). Impact of enhanced solar UV-B radiation on plants from terrestrial ecosystems. In *Climate Change Research: Evaluation and Policy Impli-*

cations. (Zwerver, S., Van Rompaey, R.S.A.R, Kok, M.T.J. & Berk, M.M., eds), pp. 997–1004. Elsevier, Amsterdam.

Rozema, J. Lenssen, G.M., Tosserams, M. & Visser, A.J. (1997*a*). Effects of UV-B radiation on terrestrial plants. Interaction with CO_2 enrichment. *Plant Ecology*, **128**, 182–91.

Rozema, J., van de Staaij, J., Bjorn, L.O. & Caldwell, M.M. (1997*b*). UV-B as an evironmental factor in plant life: stress and regulation. *Trends in Ecology and Evolution*, **12**, 22–8.

Rozema, J., Tosserams, M., Nelissen, H.J.M., van Heerwaarden, I., Broekman, R.A. & Flierman, N. (1997*c*). Stratospheric ozone reduction and ecosystem processes: enhanced UV-B radiation affects chemical quality and decomposition of leaves of the dune grassland species *Calamagrostis epigeios*. *Plant Ecology*, **128**, 284–94.

Strid, A., Chow, W.S. & Anderson, J.M. (1990). Effects of supplementary UV-B radiation on photosynthesis of *Pisum sativum*. *Biochimica et Biophysica Acta,* **1020**, 260–8.

Teramura, A.H. & Sullivan, J.H. (1994). Effects of UV-B radiation on photosynthesis in *Pisum sativum*. *Biochimica et Biophysica Acta,* **1020**, 260–8.

Teramura, A.H., Sullivan, J.H. & Ziska, L.H. (1990). Interaction of elevated ultraviolet-B radiation and CO_2 on productivity and photosynthetic characteristics in wheat, rice and soybean. *Plant Physiology*, **94**, 470–5.

Teramura, A.H., Sullivan, J.H. & Sztein, A.E. (1991). Changes in growth and photosynthetic capacity of rice with increased UV-B radiation. *Physiologia Plantarum*, **83**, 373–80.

Tevini, M. & Teramura, A.H. (1989). UV-B effects on terrestrial plants. *Photochemistry and Photobiology,* **50**, 479–87.

Tosserams, M. & Rozema, J. (1995). Effects of ultraviolet-B radiation (UV-B) on growth and physiology of the dune grassland species *Calamagrostis epigeios*. *Environmental Pollution,* **89**, 209–14.

Tosserams, M., Pais de Sa, A. & Rozema, J. (1996). The effect of solar UV-B radiation on four plant species occurring in a coastal grassland vegetation in the Netherlands. *Physiologia Plantarum*, **97**, 731–9.

van de Staaij, J.W.M., Rozema, J. & Stroetenga, M. (1990). Expected changes in Dutch coastal vegetation resulting from enhanced levels of solar UV-B. In *Expected Effects of Climatic Change on Marine Coastal Ecosystems*. (Beukema, J.J., ed.), pp. 211–17. Kluwer Academic, Dordrecht.

van de Staaij, J.W.M., Lenssen, G.M., Stroetenga, M. & Rozema, J. (1993). The combined effects of elevated CO_2 levels and UV-B radiation on growth characteristics of *Elymus athericus*. *Vegetatio,* **104/105**, 433–9.

van de Staaij, J.W.M., Ernst, W.H.O., Hakvoort, H.W.J. & Rozema,

J. (1995a). Ultraviolet-B (280–320 nm) absorbing pigments in the leaves of *Silene vulgaris*: their role in UV-B tolerance. *Journal of Plant Physiology*, **147**, 75–80.

van de Staaij, J.W.M., Huijsmans, R., Ernst, W.H.O. & Rozema, J. (1995b). The effect of elevated UV-B (280–320 nm) radiation levels on *Silene vulgaris*: A comparison between a highland and a lowland population. *Environmental Pollution*, **90**, 357–62.

Visser, A.J., Tosserams, M., Groen, M.W., Magendans, G.H.W. & Rozema, J. (1997). The combined effects of CO_2 concentration and solar UV-B radiation on faba bean grown in open top chambers. *Plant Cell and Environment* (in press).

Zepp, R.G., Callaghan, T.V. & Erickson, D.J. (1995). Effects of increased solar ultra-violet radiation on biogeochemical cycles. *Ambio*, **24**, 181–7.

L.O. BJÖRN, T. CALLAGHAN, C. GEHRKE,
T. GUNNARSSON, B. HOLMGREN,
U. JOHANSON, S. SNOGERUP,
M. SONESSON, O. STERNER and S-G. YU

Effects on subarctic vegetation of enhanced UV-B radiation

Introduction

On-going depletion of the stratospheric ozone layer may lead to changes in the radiation climate that could seriously affect the biosphere. Therefore, many research groups around the world have conducted experiments to find out how organisms, in particular marine phytoplankton (see chapter by Häder, this volume) and terrestrial plants (see other chapters by Rozema *et al.*, Corlett *et al.*, Moody *et al.* and Caldwell), react to increased levels of UV-B radiation. Most experiments on terrestrial plants have been conducted with crop plants, and just a few with forest trees. Before our field experiments in Northern Scandinavia were started in 1991, no experiment had been conducted in which a natural ecosystem with plants, animals and microorganisms was exposed to increased UV-B.

UV radiation effects on growth, biomass production and survival of plants can be broadly subdivided into two classes, direct effects and indirect effects. Direct effects include radiation-induced changes in individual plants that affect their photosynthesis, cell division, or other life processes of direct importance to growth and development. Indirect effects include those that are mediated by radiation-induced changes in the plant environment, or changes in the plant which are of importance mainly for the plant's relation to other organisms. Much more effort has so far been expended trying to understand the direct effects, and they may be the most important ones in crops, growing as they do in a partly human-controlled environment, and mostly in monoculture. Indirect effects, however, may be more important for wild plants in a natural environment. Examples of indirect effects are effects on other plants which compete with the plant under consideration (see, for example, chapters by Caldwell and by Paul, this volume), effects on nutrient mobilisation, radiation effects on herbivores and microorganisms of importance to the plant, and effects on the plant which affect its palatability to herbivores and its resistance to microorganisms. These effects of radiation can therefore, be mediated through a complex chain of events involving several organisms.

Our choice of two ecosystems in Northern Scandinavia, at 68.3° N and within the Arctic Circle, for the first experiments may need some explanation, since even with substantial ozone depletion the radiation level at this high latitude would remain below what the biosphere near the equator is exposed to under undepleted conditions. There are several reasons for our choice, the three most important ones being:

1. The *relative* ozone depletion and the *relative* increase in UV-B anticipated in this part of the world are larger than at lower latitude (Madronich *et al.*, 1995), and this may be more important than *absolute* radiation levels. The organisms are presumably adapted only to low radiation levels in this area; for example, the epidermis of Arctic plants offers little protection to UV-B (Robberecht & Caldwell, 1978; Barnes, Flint & Caldwell, 1987).

2. Most plants in the area seldom reproduce sexually, and plant individuals become very old. There was *a priori* concern that radiation effects may accumulate in perennial plants over the years, as has been shown previously for trees.

3. Temperature remains low even during summer. Photochemical damage by UV-B radiation is rapid even at low temperature, while enzymatic repair processes take place more slowly at low temperature. Thus the relationship between damage and repair is unfavourable at low temperature. In fact, Takeuchi, Ikeda & Kasahara (1993) as well as Tevini & Mark (1993) found UV-B effects to be more pronounced at a lower than at a higher temperature, although the temperatures compared by them, 20 °C to 32 °C, were high for Arctic conditions.

Additional reasons for our choice were access to a well-equipped research station in the north, the Abisko Scientific Research Station of the Royal Swedish Academy of Sciences, and the fact that one of the ecosystems, a bog, is dominated by *Sphagnum fuscum*, the moss which is the most important peat former at this latitude. Even if the activity of this moss is modest in terms of annual biomass production per square metre, it is very important in the long run as a carbon sink, since such a large part of the biomass is deposited as peat.

This chapter gives some information from the arctic/sub-arctic region relating to changes in ozone; direct and indirect effects of elevated

UV-B on the ecosystem; and possible consequences of low temperature for repair processes.

Present and past variations in the ozone layer

Details of ozone levels from satellite and ground measurements giving global coverage are available in international reports. In addition, we have been able to make two kinds of more local comparisons of past and present ozone levels. First, we have compared values for Abisko interpolated from measurements at Tromsø to the north and Vindeln to the south during the summer of 1994 to a model (Björn, 1989) which is based on worldwide satellite measurements during 1970–1977 (Hilsenrath & Schlesinger, 1981). This comparison (Fig. 1) indicates that no depletion has taken place during the summer months. Secondly,

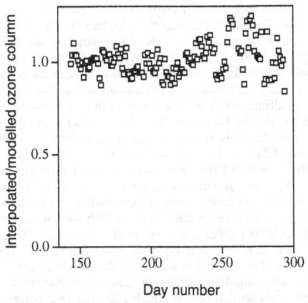

Day number

Fig. 1. Comparison of ozone levels during the summer 1994 with those a quarter of a century earlier. Values for 1994 were obtained by interpolation between measurements at Tromsø to the north of Abisko and at Vindeln to the south (see Björn & Holmgren, 1996 for details). Values for the 1970s were obtained from a model (Björn, 1989) based on satellite measurements (Nimbus-4) during the years 1970–1977 (Hilsenrath & Schlesinger, 1981). The graph shows the ratio of recent to past ozone averaged over the year. There is no systematic deviation from unity, i.e. no permanent ozone depletion.

we have the results from long-term ground-based measurements further south in Sweden recently made available by the Swedish Hydrological and Meteorological Institute (SMHI) (Josefsson, 1996). The latter indicate that the present rate of ozone depletion in Sweden, averaged over the year, is $0.8 \pm 0.3\%$ per year, based on the period 1979–1993. No trend analysis for the same period is available for the different seasons, but rough estimates for the winter half of the year for the period 1989–1994 amount to 1.0% per year depletion, and for the summer half of the years 1988–1993 to 0.4% per year depletion. Thus, the general trend is a depletion, although it is small in the summer, and it seems that, in summer 1994, the values at Abisko were not significantly different from those of 25 years previously. The tentative conclusion, though, is that there is likely to have been an increase, albeit small, in the level of UV-B at these latitudes.

We have also tried to get some idea of fluctuations of the UV-B irradiance during earlier times by analysing the flavonoid content of an ericaceous plant, *Cassiope tetragona*, from herbarium material collected in the Abisko area from 1820 onwards. This plant contains flavonoids with two types of flavonoid aglycones, myricetin and quercetin, and the content increases with the irradiance under which the plant grows (Fig. 2). Trying to reconstruct past radiation climate using this type of data involves many difficulties. The plants were collected at different dates, and the date has usually been recorded. On the other hand, the time of day is not known, and this may also be important. Finally, the material is derived from different clones growing in different locations, and the individual shoots which have been analysed faced different compass directions during their growth. Although we hope that, by analysing sufficient amounts of material, many of these difficulties could be overcome by statistics, there are sources of error which are not surmountable by statistics, including other meteorological factors that could affect flavonoid content, in particular temperature and precipitation. Temperature data for the Abisko region are available for about the last 130 years, and we have found no correlation between the flavonoid content of the current shoot section or that which grew the year before collection, and the temperature of the year of collection or that of the previous year or the year before that. This holds both for summer and annual mean temperatures. We have not looked at precipitation data yet. The flavonoid contents and some ratios in flavonoid content as a function of the year of collection were calculated. The only clear trend was in the ratio of myricetin to quercetin in shoots that had grown during the year before collection (Fig. 3). It is likely that long-term changes would

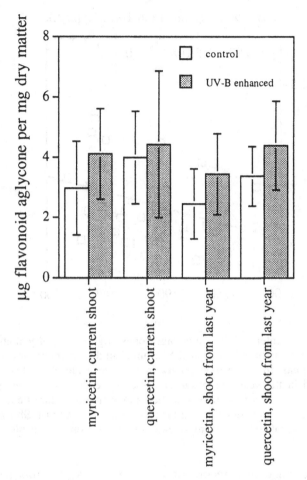

Fig. 2. Effect of UV-B enhancement on flavonoid aglycones (myricetin and quercetin) in *Cassiope tetragona* shoots. Plants growing in the wild near Abisko were transplanted to our experimental area and exposed during the vegetation season to radiation under lamp frames of the same type as used for the experiments with the heath ecosystem. UV-B enhancement corresponded to 15% ozone depletion under clear skies. Shoots from the current year of growth were used for analysis using HPLC.

238   L.O. BJÖRN *et al.*

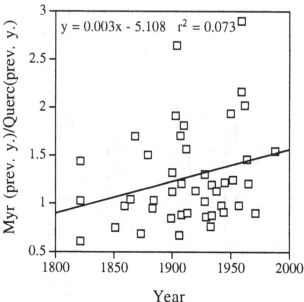

Fig. 3. Content of flavonoid aglycones (μg per mg dry matter) in herbarium material (Botanical Museum, Lund University) of *Cassiope tetragona* from Torne lappmark (mainly in or close to Abisko) collected in the years indicated. Analyses were done on shoots grown out during the year before collection and during the year of collection. Collections were done during the months June to August. Shown here is the calculated ratio of myricetin : quercetin from 1-year-old shoots.

have been revealed in a record of this sort, but so far this has not indicated any past fluctuations in UV-B level.

Effects of UV-B on microbial interactions and decomposition of organic material

Klironomos and Allen (1995) have clearly shown, in an experiment extended over 16 months, how radiation intercepted by plant shoots can have an enormous effect on the biosphere underground even if the plants themselves are not visibly affected by the different radiation treatments, and where both shoot and root biomass are only marginally decreased by increased radiation. Thus mycorrhizal vesicles were more than four times as frequent and non-mycorrhizal hyphae 2.5 times as frequent with UV-B, and spore production by mycorrhizal fungi was four times as high with UV-B as without. Growth of extra-radical fungal hyphae was three times as high with UV-B. Bacterial and actinomycete densities in the rhizosphere were approximately doubled under UV-B, and the composition of the rhizosphere flora was completely changed. The density of collembola was doubled by UV-B. These authors used a low irradiance of visible light (100 μmol m^{-2} s^{-1} for 16 h daily), which may be appropriate for tree seedlings adapted to the forest floor. However, the UV-B exposure of 6 μmol m^{-2} s^{-1} for 16 h daily, corresponding to an unweighted dose of about 130 kJ m^{-2} day^{-1}, may be artificially high for such an environment. Nevertheless, if such changes in the root environment and root do occur, there must be repercussions on the whole plant. Clearly, the consequences of changes in UV-B may be far less predictable than previously thought.

Our group (Gehrke *et al.*, 1995) has looked at microbial interactions primarily from the perspective of litter decomposition and remobilisation of nutrients, since nutrients are limiting for plant growth in the subarctic. *Vaccinium uliginosum* and *V. myrtillus* were treated in their natural mountain heath site with elevated UV-B (see below for details of irradiation). The leaf litter was collected, and its composition and decomposition were studied. It was found that the proportion of α-cellulose was decreased and that of tannins increased by UV-B enhancement, and, probably as a consequence of that, microbial activity was decreased. It was also found that UV-B irradiation during decomposition decreased colonisation by fungal decomposers. Thus increased UV-B may lead to a slowing down of nutrient mobilisation, although this has not yet been shown directly.

Direct effects of UV-B on plant growth and other plant processes

UV irradiation system

For irradiation of natural vegetation, six parallel, horizontal 40 W UV-B emitting fluorescent lamps have been used, supported 1.3 m above the ground on frames, each 1.5 m × 2.7 m. Initially Philips TL12/40W lamps were used, but these have now been replaced by Q-panel UV-B 313 lamps. The lamps and the lamp armatures are enclosed in plastic boxes, the bottoms of which consist either of UV-B transparent Plexiglass covered by one sheet of preirradiatied 0.13 mm thick cellulose diacetate (treatments) or 3 mm window glass (controls). The window glass absorbs all UV-B and some UV-A radiation. The lamps are switched on and off by timer-controlled switches, but not all at the same time. First, every second lamp is switched on, and after time t, the remaining three lamps in each frame are switched on. After a further time equal to 3t, the first three lamps are switched off, and after one more period of time t, the remaining three. In this way a stepwise change in irradiance rather than a square wave is obtained. The irradiation is centred at solar noon, and the time t determined from a computer programme to correspond to the required ozone depletion. The switch timers are adjusted every two weeks. Before use the lamps are pre-burnt to achieve a more stable output than new lamps. The central 70 cm of the two middle lamps in each frame are blocked off to achieve a more even irradiance at plant level. For irradiation of weighed amounts of leaf litter under the frames, the litter is enclosed in containers made of nylon netting and plexiglas.

Four replicate frames are used for the treatments, and four for the controls. The lamps are operated from mid-May to mid-September, which corresponds approximately to the snow-free period. The irradiation corresponds to 15% ozone depletion under cloud-free skies, or 19% under the cloud conditions of summer 1994.

For experiments with *Sphagnum* moss on a bog, smaller lamp systems were used, i.e. four Philips TL12/40W mounted in a square. This allowed areas of 10 cm × 10 cm to be homogeneously irradiated. In this case all four lamps in a frame were switched on and off simultaneously.

The results described below were obtained at Abisko in northern Sweden (68.35° N, 18.82° E, 360 m above sea level). Two similar experiments, with different vegetation types have also been started in Adventdalen, Spitzbergen (78.2° N, 15.8° E, and about 50 m above sea level).

UV-B enhancement slowed the growth of most of the mountain heath and bog plants investigated (Johanson *et al.*, 1995*a,b*; Gehrke *et al.*, 1996; C. Gehrke, personal communication). After two years, growth of evergreen dwarf shrubs (*Vaccinium vitis-idaea* and *Empetrum hermaphroditum*) in the mountain heath was repressed more than that of co-occurring deciduous dwarf shrubs (*Vaccinium myrtillus* and *V. uliginosum*, Table 1) indicating that over a long time period changes in their relative occurrence might result. However, there were no significant changes in species frequency over an observation period of four years, a not unexpected finding given the slow growth and clonal growth habit. This does not, however, preclude the possibility that large changes may take place over a longer period of time.

It can be seen in Table 1 that the stem growth of the deciduous ericaceans was repressed by about the same amount from year to year, but that the growth of the evergreen species was more repressed in the

Table 1. *Stem growth under enhanced UV-B in relation to control, untreated plants at the start of treatments*

Species	First year of treatment	Second year of treatment
Vaccinium myrtillus	81.8 ± 8.8%	88.7 ± 13.0%
Vaccinium uliginosum	90.8 ± 7.3%	90.4 ± 4.0%
Vaccinium vitis-idaea	104.7 ± 8.4%	72.6 ± 7.5%
Empetrum hermaphroditum	85.7 ± 4.2%	66.7 ± 4.1%

second treatment year than in the first. This indicates that radiation damage in the latter may accumulate from year to year. The same effect can be seen in the moss *Polytrichum commune*.

Leaf thickness was influenced in opposite directions in the evergreen *V. vitis-idaea* (thickening by UV-B enhancement) and in the deciduous *Vaccinium* species (Table 2). Also leaf dry weight and leaf area were affected in different ways in the various species.

There was one notable exception to the rule that growth was decreased by increased UV-B. When the moss *Hylocomium splendens* received additional water, growth was stimulated by increased UV-B (C. Gehrke, personal communication). Similar observations have been made by other groups with other plants (Musil & Wand, 1994; Tosserams & Rozema, 1995). In experiments with lichens, photosystem II quantum yield, as measured with a fluorescence technique, was increased after growth under UV-B (Sonesson, Callaghan & Björn, 1995). We are now starting a special project trying to understand such stimulations. Possibly they are associated in some way with ultraviolet-B induced changes in protein phosphorylation (Yu & Björn, 1996), which in turn affect enzyme and gene activities. The phosphorylation of most thylakoid proteins is decreased by UV-B, but we have also found polypeptides for which phosphorylation in the dark is very much increased by UV-B.

We conclude from these studies that regulating effects of UV-B radiation may be at least as important as damaging effects.

Temperature dependence of photodamage and photorepair

The temperature dependence of photorepair (see chapter by Taylor *et al.*, for detail) may be critical for the radiation effects on plants in cold climates. Formation of cyclobutane dimers can take place even in frozen

Table 2. Leaf properties under elevated UV-B. Values expressed as a percentage of controls

Species	Thickness (1992)	Thickness (1993)	Dry weight (1993)	Area (1993)	
				measured	calculated
Vaccinium myrtillus	95 ± 8*	91 ± 8***	86 ± 15 ns	86 ± 16 ns	100 ± 25
Vaccinium uliginosum	94 ± 6**	90 ± 5***	129 ± 17*	129 ± 20**	143 ± 59
Vaccinium vitis-idaea	109 ± 3***	104 ± 7***	116 ± 3 ns	114 ± 8 ns	112 ± 9
Empetrum hermaphroditum	–	–	107 ± 9 ns	–	–

The calculated relative area was obtained by division of relative dry weight by relative thickness. Significance levels: n.s. $P > 0.05$, * $0.05 > P\ 0.01$, ** $0.01 > P\ 0.001$, *** $P < 0.001$.

solution. Literature on the temperature dependence of enzymatic photo-reactivation in eukaryotic organisms is scarce and there is a need for further experiments; for example, there is controversy over which UV radiation-induced alteration in DNA is the most important one from a physiological viewpoint. Although formation of cyclobutane dimers is the most common change, it may not be the most important one (Zdzienicka *et al.*, 1992), since it occurs all over the DNA molecule, while formation of (6–4) photoproducts (Britt *et al.*, 1993), which are also photoreactive in plants, takes place in actively transcribed genes only (Mitchell *et al.*, 1993). Furthermore, it was recently shown by Taylor *et al.* (1996) that in plants the (6–4) photoproducts are not repaired as rapidly as the cyclobutane dimers. To circumvent the uncertainty, we have tried to investigate physiologically evident damage rather than molecular changes. For the first experiments we have used banana as a convenient model plant. Cells in banana peel turn brown after being damaged by UV radiation, and this browning can be prevented by visible light given after the UV irradiation (Fig. 4). In fact,

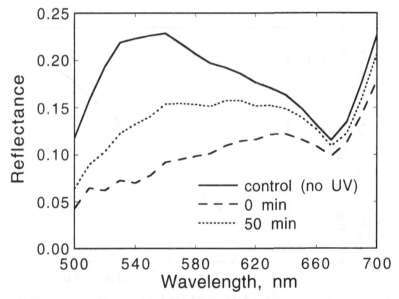

Fig. 4. Wavelength dependence of photoreactivation in banana peel. The banana was irradiated at 20 °C for 2 minutes with 4 J m^{-2} ultra-violet radiation of 253.7 nm wavelength delivered in 2 minutes, and then photoreactivated with white light for the number of minutes indicated at an irradiance of about 400 mol m^{-2} s^{-1}.

this was the way photoreactivation was discovered by Hausser and v. Oehmcke (1933). The degree of reactivation can be quantified using reflectance measurements (Fig. 5). Using this assay we have established that even in this tropical plant photoreactivation takes place down to 10 °C, at about half the rate for 20 °C, but cannot be detected at 6 °C. In our initial experiments unripe bananas were irradiated with 4 J m^{-2} of 253.7 nm UV-C for 2 min, after which the banana was brought to various temperatures in the dark for 10 min, and then photoreactivated at the chosen temperature for up to 45 min. After that the banana was kept at about 20 °C for a week, and the effect of the irradiations evaluated by reflectance spectroscopy. Figure 5 shows that photoreactivation occurs with time. Despite the low radiation levels in subarctic and arctic regions, it is possible that any damage caused will not be repaired so effectively, because of the possible inhibition of DNA repair by low temperature. It could, of course, be argued that subarctic plants will have photolyases which work well at lower temperatures, and so for experiments with subarctic plants the intention is to use a root growth assay (Burström & Gabrielson, 1964).

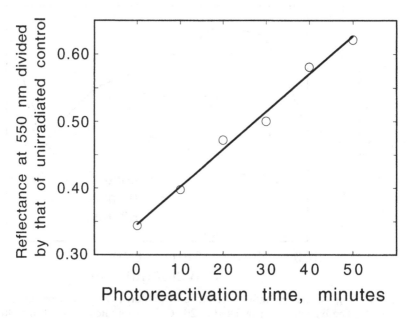

Fig. 5. Kinetics of photoreactivation at 20 °C. The reflectance at 550 nm was recorded and divided by the reflectance of non-irradiated banana peel.

Acknowledgements

This project was financed mainly by the Swedish Environmental Protection Agency (SNV) as part of a project of the Commission of the European Communities (Contract No. EV5V-CT91-0032), and we are also grateful to Astra Draco AB for support.

References

Barnes, P.W., Flint, S.D. & Caldwell, M.M. (1987). Photosynthesis damage and protective pigments in plants from a latitudinal arctic/ alpine gradient exposed to supplemental UV-B radiation in the field. *Arctic and Alpine Research,* **19**, 21–7.

Björn, L.O. (1989). Computer programs for estimating ultraviolet radiation in daylight. In *Radiation Measurement in Photobiology.* (Diffey, B.L., ed.), pp. 161–189. Academic Press, New York.

Björn, L.O. & Holmgren, B. (1996). Monitoring and modelling of the radiation climate at Abisko. *Ecological Bulletins,* **45**, 204–9.

Britt, A.B., Chen, J.-J., Wykoff, D. & Mitchell, D.L. (1993). A UV-sensitive mutant of *Arabidopsis* defective in the repair of pyrimidine-pyrimidone (6–4) dimers. *Science,* **261**, 1571–4.

Burström, H.G. & Gabrielsen, B.E. (1964). Localization of reactivation after UV inhibition of root growth. *Physiologia Plantarum,* **17**, 964–74.

Gehrke, C., Johanson, U., Callaghan, T.V., Chadwick, D. & Robinson, C.H. (1995). The impact of enhanced ultraviolet-B radiation on litter quality and decomposition processes in *Vaccinium* leaves from the Subarctic. *Oikos,* **72**, 213–22.

Gehrke, C., Johanson, U., Gwynn-Jones, D., Björn, L.O., Callaghan, T.V. & Lee, J.A. (1996). Effects of enhanced utraviolet-B radiation on terrestrial subarctic ecosystems and implications for interactions with increased atmospheric CO_2. *Ecological Bulletins,* **45**, 192–203.

Hausser, K.W. & v. Oehmcke, H. (1933). Lichtbräunung an Fruchtschalen. *Strahlentherapie,* **48**, 223–9.

Hilsenrath, E. & Schlesinger, B.M. (1981). Total ozone seasonal and interannual variation derived from the 7 year Nimbus-4 data set. *Journal of Geophysical Research,* **86**, 12087–96.

Johanson, U., Gehrke, C., Björn, L.O. & Callaghan, T.V. (1995*a*). The effects of enhanced UV-B radiation on the growth of dwarf shrubs in a subarctic heathland. *Functional Ecology,* **9**, 713–19.

Johanson, U., Gehrke, C., Björn, L.O., Callaghan, T.V. & Sonesson, M. (1995*b*). The effects of enhanced UV-B radiation on a subarctic heath ecosystem. *Ambio,* **24**, 108–13.

Josefsson, W. (1996). *Measurements of Total Ozone: National Environmental Monitoring 1993/1994.* SMHI/Swedish Environmental Protection Agency.

Klironomos, J.J. & Allen, M.F. (1995). UV-B-mediated changes on below-ground communities associated with the roots of *Acer saccharum*. *Functional Ecology*, **9**, 923–30.

Madronich, S., McKenzie, R., Caldwell, M.M. & Björn, L.O. (1995). Changes in ultraviolet radiation reaching the earth's surface. *Ambio*, **24**, 143–52.

Mitchell, D.L., Pfeifer, G.P., Taylor, J.-S.,0 Zdzienicka, M.Z. & Nikaido, O. (1993). Biological role of (6–4) photoproducts and cyclobutane pyrimidine dimers. In *Frontiers of Photobiology*. Proceedings of the 11th International Congress in Photobiology. (Shima, A., Ichahashi, M., Fujiwara, Y. & Takebe, H., eds), pp. 337–44. Elsevier, Amsterdam.

Musil, C.F. & Wand, S.J.F. (1994). Differential stimulation of an arid-environment winter ephemeral *Dimorphotheca pluvialis* (L.) Moench by ultraviolet-B radiation under nutrient limitation. *Plant Cell and Environment*, **17**, 245–55.

Robberecht, R. & Caldwell, M.M. (1978). Leaf epidermal transmittance of utraviolet radiation and its implications for plant sensitivity to ultraviolet-radiation induced injury. *Oecologia*, **32**, 277–87.

Sonesson, M., Callaghan, T.V. & Björn, L.O. (1995). Short-term effects of enhanced UV-B and CO_2 on lichens at different latitudes. *Lichenologist*, **27**, 547–57.

Takeuchi, Y., Ikeda, S. & Kasahara, H. (1993). Dependence on wavelength and temperature of growth inhibition induced by UV-B irradiation. *Plant and Cell Physiology*, **34**, 913–17.

Taylor, R.M., Nikaido, O., Jordan, B.R., Rosamond, J., Bray, C.M. & Tobin, A.K. (1996). Ultraviolet-B-induced DNA lesions and their removal in wheat (*Triticum aestivum* L.) leaves. *Plant Cell and Environment*, **19**, 171–81.

Tevini, M. & Mark, U. (1993). Effects of elevated ultraviolet-B radiation, temperature and CO_2 on growth and function of sunflower and corn seedlings. In *Frontiers of Photobiology*. Proceedings of the 11th International Congress Photobiology. (Shima, A., Ichahashi, M., Fujiwara, Y. & Takebe, H., eds), pp. 541–6. Elsevier, Amsterdam.

Tosserams, M. & Rozema, J. (1995). Effects of ultraviolet-B radiation (UV-B) on growth and physiology of the dune grassland species *Calamagrostis epigeios*. *Environmental Pollution*, **89**, 209–14.

Yu, S.-G. & Björn, L.O. (1996). Effects of ultraviolet-B radiation on the phosphorylation of thylakoid proteins. *Journal of Photochemistry and Photobiology B: Biology* (in press).

Zdzienicka, M.Z., Venema, J., Mitchell, D.L., van Hoffen, A., van Zeeland, A.A., Vrieling, H., Mullenders, L.H.F., Lohman, P.H.M. & Simons, J.W.I.M. (1992). (6–4) photoproducts and non-cyclobutane pyrimidine dimers are the main UV-induced mutagenic lesions in Chinese hamster cells. *Mutation Research*, **273**, 73–84.

A.R. McLEOD and K.K. NEWSHAM

Impacts of elevated UV-B on forest ecosystems

Introduction

Increases in solar UV-B (ultraviolet-B radiation, here taken as 280–315 nm) are expected to result from stratospheric ozone depletion (Blumthaler & Ambach, 1990; Kerr & McElroy, 1993; see also chapter by Webb, this volume). Consequently, a wide range of experimental studies have been undertaken to assess the impact of increases in UV-B on vegetation (Caldwell *et al.*, 1995). Many early studies were conducted in controlled environment chambers and greenhouses in which plants were exposed to supplemental levels of UV radiation from fluorescent lamps. More recently, field studies with supplemental UV-B radiation have also been undertaken. Initially these simply turned lamps on and off at a constant irradiance to give so-called 'square-wave' treatments (Sullivan *et al.*, 1994*a*), but increasingly, lamp irradiances have been modulated to maintain a proportional elevation above ambient UV-B (McLeod, 1997). The majority of studies have examined crop species, but trees, perhaps because of their size and vertical distribution of foliage, have been the subject of few investigations, even though they play a central role in the global carbon cycle. Of the 550 Gt carbon in the land biota, 65% is in forests and of the 1550 Gt carbon in soils and detritus, 50% is in forest soils (Cannell, 1995). Studies of the many interactions that occur between different trophic levels in an ecosystem and their role in decomposition and nutrient cycling have only been investigated in dwarf shrubs of the subarctic (Gehrke *et al.*, 1995), and studies of forest ecosystems have only recently begun (Newsham *et al.*, 1996). Many components of a forest ecosystem are exposed to a spatially and temporally variable environment of PAR (photosynthetically active radiation) and UV-B radiation. In this chapter we firstly review UV-B distribution in forests in terms of reflectance and transmission of both leaves and canopies. The early experimental investigations of UV-B effects on trees have been considered by Krupa and Kickert (1989), Teramura and Sullivan (1991), Bornman and Teramura (1993) and Sullivan (1994). We summarise these and review later studies which have assessed effects of UV-B on tree growth and physiology and effects on other components of the forest ecosystem (fungal pathogens, leaf surface micro-organisms, mycorrhizal fungi and invertebrates).

UV-B distribution in forest canopies

Experimental exposures of plants have usually measured UV-B at the top of the canopy in terms of irradiance (the UV-B per unit time and area falling on a horizontal plane) even though fluence rate (the UV-B passing through a sphere of unit surface area) may be a more appropriate measure of UV-B exposure for three-dimensional objects and the many orientations of leaves in a tree canopy (Björn & Teramura, 1993). Plant orientation relative to the angle of incidence of UV-B is important in determining plant response, and plants with horizontal foliage (such as low shrubs) are easier to irradiate experimentally in a controlled manner than plants with vertical foliage (see chapter by Holmes, this volume). This is particularly apparent when considering the vertical structure of a mature tree canopy. Whilst the upper foliage of a tree may be fully exposed to solar UV-B at a variety of angles of incidence, the lower leaves and understorey plants are subject to shading by branches, twigs and leaves. The presence of sunflecks and gaps creates a complex and highly variable radiation environment. Not only does the UV-B environment vary throughout the canopy but the spectral distribution of other wavelengths in the visible region is also modified (Grace, 1983).

Leaf reflectance and transmittance

At an individual leaf level, the surface reflectance and absorption of UV by pigments in the epidermis influences the sensitivity of plants to UV-B radiation. Studies of UV-B and the optical properties of leaves and canopies are summarised in Table 1. Leaf surface reflectance of UV-B in trees is generally less than 10% and may be affected by changes in leaf surface waxes and leaf hairiness. Karabourniotis *et al.* (1992) demonstrated that phenolic UV-screening pigments (including flavonoids) existed in the leaf hairs of *Olea* spp., *Populus* spp., *Quercus ilex* and *Eriobotrya japonica*, which they suggested were a shield against UV-B radiation. Subsequently, Karabourniotis, Kyparissis and Manetas (1993) showed that UV-B exposure of *Olea europaea* caused a decline in chlorophyll fluorescence only in de-haired leaves, while Grammatikopoulos *et al.* (1994) similarly demonstrated that UV-B effects on photosynthetic rate and stomatal conductance were much

Table 1. Studies of UV-B and the optical properties of the leaves and canopies of forest trees

Common name	Species	Methods	Reference	Observations
Western larch	Larix occidentalis	UV-B transmission through leaves measured with fibre-optic microprobe.	Day et al. (1992)	Epidermis of conifers attenuates UV-B. Epicuticular wax removal did not alter epidermal screening.
Subalpine fir	Abies lasiocarpa			
Limber pine	Pinus flexilis			
Lodgepole pine	Pinus contorta			
Engelmann spruce	Picea engelmannii			
Ponderosa pine	Pinus ponderosa			
Douglas fir	Pseudotsuga menziesii			
Common juniper	Juniperus communis			
Colorado spruce	Picea pungens			
Balsam fir	Abies balsamea	UV absorption and epidermal transmission spectra. Concentrations of UV-absorbing compounds.	Day (1993) Day et al. (1994)	Amounts of UV reaching the mesophyll were relatively low in evergreen species but higher in deciduous species.
Red spruce	Picea rubens			
Colorado spruce	Picea pungens			
Eastern hemlock	Tsuga canadensis	Fluorescing film and fibre-optic microprobe.	Day et al. (1993)	No epidermal transmitance in P. pungens.
White pine	Pinus strobus			
Red pine	Pinus resinosa			
Scots pine	Pinus sylvestris			
Black cherry	Prunus serotina			
White dogwood	Cornus florida			
Sassafras	Sasafras albidum			
Ailanthus	Ailanthus altissima			

Table 1. *contd.*

Common name	Species	Methods	Reference	Observations
Hazel alder Black locust Black oak	*Alnus serrulata* *Robinia pseudoacacia* *Quercus velutina*			
Red oak Black oak White oak Sugar maple Norway maple Hickory Sweetgum	*Quercus rubra* *Quercus velutina* *Quercus alba* *Acer saccharum* *Acer platanoides* *Carya tomentosa* *Liquidambar* *styraciflua*	Laboratory solar simulator. Spectral reflectance and transmission measurements.	Yang *et al.* (1995)	Transmittance of UV-B < 0.1%.
Douglas fir Sitka spruce Colorado spruce Blue spruce	*Pseudotsuga* *menziesii* *Picea sitchensis* *Picea pungens* *Picea pungens* var. *hoopsii*	Spectral reflectance measurements.	Clark & Lister (1975)	Surface wax contributed to ultraviolet reflectance.
Forests in USA, Chile, Panama, and Mexico		Field surveys of PAR and UV-B irradiance.	Brown *et al.* (1994)	About 1–2% of UV-B transmitted to forest floor in 4 forests with closed canopies. Considerable variation in the UV-B:PAR transmittance ratio.

Scots pine	*Pinus sylvestris*	Laboratory sun simulator.	Schnitzler *et al.* (1996*a*)	Identification and distribution of UV-B screening pigments.
Olive	*Olea europaea*	Removal of leaf	Karabourniotis *et al.*	Existence of UV-B
Golden olive Loquat Evergreen oak Poplar spp.	*Olea chrysophylla Eriobotrya japonica Quercus ilex Populus* spp.	peltate layers for spectroscopic examination and pigment extraction.	(1992)	absorbing pigments demonstrated in leaf hairs.
Olive	*Olea europaea*	Removal of peltate layer. Controlled environment irradiation. Measurement of chlorophyll fluorescence and reflectance spectra.	Karabourniotis *et al.* (1993)	UV-B exposure of de-haired leaves caused decline in chlorophyll fluorescence but not in intact leaves.
Olive	*Olea europaea*	Measurement of photosynthetic rates, leaf resistance to water vapour and chlorophyll fluorescence.	Grammatikopoulos *et al.* (1994)	UV-B exposure of de-haired leaves caused decrease in photosynthetic rate and leaf conductance to water vapour but not changes in photosynthetic capacity.

Table 1. *contd.*

Common name	Species	Methods	Reference	Observations
Evergreen oak	*Quercus ilex*	Removal of peltate layer. UV-B irradiation and determination of optical properties, and chlorophyll fluorescence.	Skaltsa *et al.* (1994)	Leaf hairs contain UV-absorbing pigments. PSII photochemical efficiency reduced by UV-B only when hairs removed.
Sweetgum Loblolly pine	*Liquidambar styraciflua* *Pinus taeda*	Comparison of UV-B treated and untreated foliage. Epidermal transmittance measured with fibre-optic microprobe. Concentrations of UV-B absorbing compounds determined.	Sullivan *et al.* (1996)	Epidermal transmittance about 1% in pine and 20% in sweetgum. Transmittance reduced to 15% in sweetgum with prior UV-B treatment. Concentration of UV-B absorbing compounds unaffected.

greater in de-haired leaves. Changes in photosynthetic capacity were not detected and this led to the suggestion that leaf hairs may prevent stomatal closure caused by UV-B radiation. Skaltsa *et al.* (1994) also reported that non-glandular leaf-hairs from *Quercus ilex* were more dense on the abaxial surface and contained UV-B absorbing flavonoids that functioned as effective UV-B filters. They demonstrated that experimental exposure to UV-B caused a reduction of photosystem II photochemical efficiency only when leaf hairs were removed.

The absorption of UV-B by the leaf epidermis was investigated in a number of plants, including nine coniferous species, by Day, Vogelmann and DeLucia (1992). In contrast to epidermal transmittances of 18–46% in a range of herbaceous dicotyledonous plants, which allowed UV-B penetration to the photosynthetic mesophyll, the epidermis of all the 1-year-old conifer needles studied attenuated all incident UV-B (Table 1). Removal of the epicuticular waxes did not significantly reduce the screening effect. In contrast, Clark and Lister (1975) found that the removal of surface waxes from needles of *Picea pungens* significantly reduced ultraviolet reflection. In a later study examining 42 plant species, including seven evergreen coniferous and seven deciduous trees, Day (1993) and Day, Howells and Rice (1994) measured UV absorption and epidermal transmittance spectra. Foliage of deciduous species generally contained lower concentrations of UV-B absorbing compounds (Day, 1993). Using different action spectra to calculate penetration of biologically effective UV-B, Day *et al.* (1994) noted that the amounts of UV-B radiation reaching the photosynthetic mesophyll were relatively low in evergreen species and higher in deciduous species. Day (1993) suggested that foliage of deciduous species may therefore be more susceptible to increased UV-B than evergreen species. However, Bornman & Teramura (1993) noted that conifer needles, in which UV-B attenuation is higher, may be particularly vulnerable when immature, shortly after needle elongation and emergence out of the protective bud scales.

The difference in epidermal transmittance between deciduous and coniferous species was further investigated by Sullivan *et al.* (1996) who compared UV-B irradiated and non-irradiated foliage of *Liquidambar styraciflua* and *Pinus taeda.* The epidermis of loblolly pine absorbed more than 99% of incident UV-B in both UV-B irradiated and non-irradiated foliage. However, the epidermis of *L. styraciflua* transmitted 20% of incident UV-B in non-irradiated foliage which was reduced to 15% with experimental UV-B irradiation. The concentration of UV-B absorbing compounds was unaffected by UV-B treatment and factors such as leaf anatomy, pigment localisation and qualitative differences in pigments were suggested as mechanisms influencing epidermal transmission of UV-B.

Recently, Schnitzler *et al.* (1996*a*) reported a UV-B induction of flavonoid biosynthesis in *Pinus sylvestris* seedlings in which the pigments were located in the epidermal cell layer of the needles. Up to 90% of the needle content of a group of diacylated flavonol glycosides were found in the epidermal layer (Schnitzler *et al.*, 1996*b*). However, the vast majority of phenolic compounds that were non-inducible, such as flavans and non-acylated flavonol glycosides, were located in the mesophyll. Model calculations suggested an almost complete shielding of mesophyll tissue from UV-B radiation, supporting the suggestion that, in contrast to many deciduous species, the foliage of Scots pine and other conifers is well protected against increases in UV-B.

Canopy reflectance and transmittance

The transmittance of UV-B through leaves of seven hardwood species was examined by Yang *et al.* (1995) and found to be <0.1% of that incident on foliage. Consequently, the transmission of UV-B through a dense forest canopy is likely to be extremely low. There have been only a limited number of studies of the UV-B environment in forest canopies. DeLucia, Day and Vogelmann (1991) observed that about 10% of diffuse UV-B reached the oldest leaves of *Picea engelmannii* in a clearcut forest. In a mixed forest, consisting predominantly of oaks (*Quercus alba, Q. velutina, Q. rubra, Q. prunus*), which had been previously defoliated by gypsy moth (*Lymantria dispar*), Yang *et al.* (1993) found that on average 25% of biologically weighted UV-B radiation reached the forest floor. A theoretical treatment of UV-B in different forest structures has been described by Allen, Gausman and Allen (1975).

The distribution of UV-B radiation in forests has been most recently examined by Brown, Parker and Posner (1994) who surveyed the spatial and temporal variation of UV-B in a number of forests. They found that as little as 1–2% of incident UV-B was transmitted to the lower levels of four widely varying forests with closed canopies. Canopy reflectance of UV-B was very low at about 1–2% of incident values. The vertical extinction of UV-B was very rapid with 40–70% of incident UV-B lost by absorption in the first 25% of the canopy. However, during the leafless season of a mixed deciduous forest, the geometric mean UV-B transmittance to the forest floor was 30%. Under closed canopies the transmittance of both UV-B and PAR was higher in sunflecks and gaps. However, the spatial variation of UV-B was less than PAR because of the greater diffuse component of incident solar UV-B.

Consequently, there was considerable variation in the UV-B:PAR transmittance ratio which was low in sunflecks (0.85), higher in the dark understorey (1.15) and highest at the shady edges of gaps (1.65). There have been no reports from experimental UV-B exposures of how UV-B:PAR transmittance ratio changes in the canopies of young saplings.

Brown *et al.* (1994) combined their measurements with a model of UV-B transmission through the atmosphere and estimated that a decline in stratospheric ozone would result in greatest increases in UV-B exposure in summer in the upper canopy and in spring in the lower canopy and understorey. These predictions, and the measured transmission during the leafless season of a deciduous forest, indicate potential for increased UV-B to influence understorey species and litter decomposition between leaf abscission in autumn and canopy development in spring.

Tree growth and physiology

There have been a wide range of exposure methods and different doses applied in studies of UV-B impacts on forest vegetation. Most of the early studies on tree growth examined conifers and only a few considered deciduous species (Table 2). The earlier studies have been reviewed by Krupa & Kickert (1989), Bornman and Teramura (1993) and by Sullivan (1994). More recent studies on forest tree species, some including outdoor exposures using modulated lamp supplementation, are shown in Table 2.

Studies of growth and photosynthesis

Filtration of ambient UV-B using Plexiglas (Bogenrieder & Klein, 1982) led to an increase in dry matter accumulation in four of five hardwood species tested, with one species showing no effect (Table 2). Different sensitivities to UV-B were demonstrated in two ecotypes of *Acer pseudoplatanus* from low and high elevations, with the dry matter production of the lowland ecotype reduced by UV-B but increased in the high elevation ecotype. Kaufmann (1978) undertook field supplementation of seedlings of two high elevation conifers (Table 2) but found no morphological changes after one growing season. Sullivan and Teramura (1988) suggested that this lack of deleterious effect was due to the low doses of UV-B applied (approximating 10–15% ozone depletion) or that high elevation species may be inherently tolerant of UV-B radiation. However, visible symptoms and increased mortality in

Table 2. Studies of UV-B effects on growth and physiology of forest trees

Common name	Species	Exposure method	Reference	Observations
Common ash Hornbeam Beech Norway maple Sycamore	Fraxinus excelsior Carpinus betulus Fagus sylvatica Acer platanoides Acer pseudoplatanus	Field exclusion with plexiglas. Species grown in competitive pairs.	Bogenrieder & Klein (1982) Gold & Caldwell (1983)	Increased growth of the first four species with UV-B exclusion. Changes in relative performance in competitive pairs. No effect on A. pseudoplatanus.
Engelmann spruce Lodgepole pine	Picea engelmannii Pinus contorta	Field supplementation and field exclusion.	Kaufmann (1978)	No effects on growth at the end of one season experiment with exclusion or supplementation.
Balsam fir Chinese chesnut Colorado blue spruce Norway spruce Red scarlet maple White dogwood	Abies balsamea Castanea mollissima Picea pungens Picea abies Acer rubrum Cornus florida	Glasshouse irradiation.	Semenuik (1978)	No effects after 12-week experiment.

Species (common)	Species (scientific)	Conditions	Reference	Effects
Lodgepole pine Loblolly pine Noble fir Ponderosa pine Slash pine Douglas fir White fir	*Pinus contorta* *Pinus taeda* *Abies procera* *Pinus ponderosa* *Pinus elliotti* *Pseudotsuga menziesii* *Abies concolor*	Controlled environment irradiation (Phytotron).	Kossuth & Biggs (1981)	Some deleterious effects on growth including reduced biomass, root-to-shoot ratio, height, leaf area in some species.
Lodgepole pine Red pine Loblolly pine White pine Pinyan pine Black pine Scots pine Engelmann spruce Fraser fir White spruce	*Pinus contorta* *Pinus resinosa* *Pinus taeda* *Pinus strobus* *Pinus edulus* *Pinus nigra* *Pinus sylvestris* *Picea engelmannii* *Abies fraseri* *Picea glauca*	Glasshouse irradiation	Sullivan & Teramura (1988)	Deleterious effects on growth and physiology.
Loblolly pine	*Pinus taeda*	Glasshouse irradiation: zero UV-B plus 3 levels.	Sullivan & Teramura (1989)	Deleterious effects on growth, photosynthesis and increases in UV absorbing pigments. Variable time responses of these parameters to different UV levels.

Table 2. *contd.*

Common name	Species	Methods	Reference	Observations
Scots pine	*Pinus sylvestris*	Controlled environment irradiation.	Fernbach & Mohr (1992)	UV-B and UV-A at high doses inhibited hypocotyl growth.
Norway spruce	*Picea abies*	Glasshouse irradiation with UV-B (1 level) ± cadmium in nutrient solution.	Dubé & Bornman (1992)	UV-B and cadmium combined caused changes in photosynthetic processes not seen with each stress alone.
Loblolly pine	*Pinus taeda*	Square-wave field supplementation. Seed sources from seven different latitudes.	Sullivan & Teramura (1992)	Variability in biomass response (positive, negative and no response) after 1 year. Reduced biomass in three seed sources tested after 3 years.
Jack pine	*Pinus banksiana*	Glasshouse irradiation with UV-B (2 levels) and CO_2 (2 levels).	Stewart & Hoddinott (1993)	Highest UV level decreased dry weight and shifted biomass partitioning towards leaf production.

Loblolly pine	*Pinus taeda*	Controlled environment irradiation (Phytotron). UV-B (three levels) and CO_2 (two levels). Chlorophyll fluorescence, O_2 evolution and leaf pigments measured.	Sullivan & Teramura (1994)	Growth response to CO_2 increase but photosynthetic capacity and quantum efficiency not altered. Highest UV-B level reduced biomass by 12% both with and without increased CO_2. Biomass preferentially allocated with UV-B to shoot at 350 ppm and to root at 650 ppm CO_2.
Sweetgum	*Liquidambar styraciflua*	Square-wave field supplementation	Sullivan et al. (1994b) Naidu et al. (1993) Dillenburg et al. (1995) Sullivan (1994)	Development of leaves, photosynthetic capacity and pigments modified.
Mahogony — — —	*Swietenia macrophylla* *Cecropia obtusifolia* *Tetragastris panamensis* *Calophyllum longifolium*	Field exclusion with plastic film at Barro Colorado Island, Panama.	Searles et al. (1995)	Decrease in UV-B absorbing compounds and specific leaf weight, and increase in plant height and leaf length with UV-B exclusion.

Table 2. *contd.*

Common name	Species	Methods	Reference	Observations
Stone pine Aleppo pine	*Pinus pinea* *Pinus halepensis*	Square-wave field supplementation.	Petropolou *et al.* (1995)	Reduced needle drop and PSII inactivation plus increased photosynthetic capacity with UV-B exposure.
Stone pine	*Pinus pinea*	Square-wave field supplementation.	Manetas *et al.* (1997)	UV-B alleviated summmer drought effects which reduced PSII photochemical efficiency and biomass accumulation. No UV-B effects on well-watered trees.
Pedunculate oak	*Quercus robur*	Modulated field supplementation. Comparison of effects beneath cellulose acetate (CA) and polyester (PY) filtered lamps with unenergised lamps.	Newsham *et al.* (1996)	Increased leaf thickness (with CA filtered lamps) and increased insect herbivory and lammas shoot elongation with both CA and PY filtered lamps.

the following season were suggested by Sullivan (1994) to indicate the possibility of cumulative effects on evergreen trees.

Basiouny and Biggs (1975) reported UV-B effects on peach seedlings grown inside a glasshouse in quartz sand and nutrient solutions with or without sufficient zinc. Zinc-deficient plants grown under UV-B were slightly stunted and had smaller leaves with more distinct zinc-deficiency symptoms than those grown without UV-B. However, Semeniuk (1978) reported no visual symptoms on seedlings of six orna-mental tree species in other glasshouse studies.

Kossuth and Biggs (1981) used UV-B lamps in a phytotron to expose seedlings of seven tree species to UV-B (Table 2). Biomass production of five species was significantly reduced (5–25%) at the highest treat-ment level of 9.56 W m^{-2} UV-B$_{DNA}$ (biologically effective UV-B weighted with the DNA damage action spectrum of Setlow (1974) and normalised at 300 nm) for 6 h each day. Low levels of PAR, as used in this study, are now known to enhance the detrimental effects of UV-B (Mirecki & Teramura, 1984). Nevertheless, the results of Kos-suth and Biggs (1981) suggested that species from higher elevations may be more resistant to UV-B damage. Many of the early studies of UV-B effects were often conducted inside chambers with low levels of PAR, for short durations (less than one season), with unrealistically high doses of UV-B, or inside glasshouses with treatments compared to controls that had no UV-B exposure (see also discussion in Corlett *et al.*, this volume). Thus it is difficult to extrapolate results to quantify and predict effects of ozone depletion on forest trees. In more recent studies, effects have been examined both on young tree seedlings in the glasshouse (Sullivan & Teramura, 1988,1989) followed by field vali-dation studies using outdoor lamp supplementation (Sullivan & Tera-mura, 1992).

Seedlings of ten conifer species were exposed in a glasshouse to zero and two elevated levels of UV-B for 22 weeks (Sullivan & Teramura, 1988)(Table 2). The height of three species and total biomass of six species were reduced by UV-B exposure, biomass increased in one species and the others were unaffected. They then exposed seedlings of the most UV-B sensitive species, *Pinus taeda*, in a glasshouse to UV-B doses corresponding to 0, 16, 25 and 40% ozone reductions for seven months (Sullivan & Teramura, 1989). They found effects on growth, photosynthesis and on foliar concentrations of pigments. The growth and photosynthetic capacity of seedlings exposed to a simulated 16% ozone reduction was reduced after only one month and effects persisted to result in lower biomass production. They reported no increase in UV-B absorbing compounds with this level of ozone depletion, but the simulated 25% ozone reduction produced an increase in UV-B absorbing pigments after six months, which paralleled an increase in photosynthetic rate and growth following an initial decline. The highest treatment, simulating a 40% ozone reduction, caused a more rapid increase in UV-B absorbing pigments and no effect on growth and photosynthesis until after four months. The authors concluded that photosynthesis was affected by direct effects on light-dependent processes and not via changes in needle chlorophyll or conductance. As it was impossible to predict whether such effects persisted, Sullivan and Teramura (1992) conducted a longer field validation study in which seedlings were exposed to a square-wave treatment using fluorescent lamps, corresponding to 16 and 25% ozone depletion. Seedlings from seven different latitudes (31 to 39° N) were exposed for one season and seedlings from four of these locations were exposed for a further two seasons. There was substantial variability in UV-B sensitivity between plants from different seed locations. After one year, seedlings from two sources were smaller and from one source were larger with UV-B exposure. After three years, the biomass was significantly reduced in plants from three of the four seed sources by 15–20% at the highest UV-B level. Root-to-shoot ratios were not altered and growth reduction was attributed to a lower carbon accumulation. The authors noted that long-term experiments are probably necessary to reveal cumulative effects of increased UV-B on growth.

Effects of UV-B on the photosynthesis of *P. taeda* seedlings were reported by Naidu *et al.* (1993). They found a 6% decrease in the chlorophyll fluorescence ratio (F_v/F_m) only in needles of the youngest age class but no effect on maximum photosynthetic rate, quantum yield and dark respiration in any of three age classes. Foliar nitrogen was also unaffected. However, the $^{13}C/^{12}C$ carbon isotope ratios in two age classes were more negative following UV-B treatment, suggesting that the long-term intracellular CO_2 concentration of the needles had been higher. This could have resulted from increases in stomatal conductance or from decreased photosynthetic rate. Naidu *et al.* (1993) concluded that the change in isotopic ratio was indicative of small chronic reductions in photosynthetic rate. In the same study, biomass was significantly reduced by 24, 16 and 18% in the roots, stems and needles, respectively, plants exposed to a simulated 25% ozone reduction compared to ambient. Root-to-shoot ratio, plant height and numbers of lateral branches were not affected, although average needle length was reduced in all four needle age classes. Reduced needle area was thought to have contributed more strongly to the overall growth reduction than small decreases in photosynthetic rate per unit of leaf area.

In one of the few studies into the effects of UV-B radiation on deciduous trees, Sullivan, Teramura and Dillenburg (1994*b*) exposed sweetgum (*Liquidambar styraciflua*) in a two-year field study to ambient or supplemental UV-B radiation using fluorescent lamps. The treatments provided a maximum daily supplementation of 3.1 and 5 kJ m^{-2} day^{-1} UV-B$_{PAS}$ (biologically effective UV-B weighted with the generalised plant action spectrum of Caldwell (1971) fitted with the mathematical function of Green, Sawada and Shettle (1974) and normalised at 300 nm), and corresponded to 16 or 25% ozone depletions over Beltsville, Maryland. Over two years, sweetgum was moderately tolerant to increased UV-B and showed little detrimental response in total biomass or photosynthetic capacity. The only growth reduction observed was a reduced leaf area and leaf biomass after a second season of exposure (Sullivan, 1994). However, there were subtle changes in leaf physiology, carbon allocation and growth of the species. Alteration in carbon allocation included an increase in branch number and root-to-shoot ratio (Sullivan *et al.*, 1994). Although final leaf size was unaffected, the rate of leaf elongation and accumulation of leaf area was slower in plants exposed to the lower UV-B treatment (Dillenburg, Sullivan & Teramura, 1995). Chlorophyll accumulation and development of photosynthetic capacity was more rapid in the higher than the lower UV-B treatment (Dillenburg *et al.*, 1995), chlorophyll *a* concentration and photosynthetic capacity being 20 and 15% higher, respectively, after the first month of leaf development (Sullivan, 1994). However, after 10 weeks the mature leaves showed a 16% reduction in these parameters. The accumulation of anthocyanins and flavonoids in foliage was not affected. Overall, the authors concluded that the change in the development of photosynthetic pigments and leaf development may have resulted in early senescence and reduced leaf area duration (Sullivan, 1994) and that in the absence of gross effects on biomass and productivity, changes in leaf ontogeny, photopigments and photosynthetic capacity could have significant consequences for sweetgum growing in elevated UV-B environments.

Investigation of UV-B effects on the many interactions between trophic levels in a forest ecosystem are difficult because of the physical size of forest trees. However, studies into the impacts of UV-B on leaf decomposition, micro-organisms, insects and nutrient cycling in the deciduous hardwood *Quercus robur* have begun (Newsham *et al.*, 1996). These studies use outdoor supplementation with lamp modulation to expose young oak saplings beneath ambient arrays of unenergised lamps, energised control lamps filtered with polyester (which transmits only UV-A) and energised treatment lamps filtered with cellulose diacetate (which transmits both UV-B and UV-A radiation). The treatment level is a 30% elevation above ambient UV-B$_{CIE}$ (biologically effective UV-B weighted with the erythemal action spectrum of McKinlay & Diffey (1987)). After one year of operation, this experimental design has revealed a typical effect of UV-B – an increase in leaf thickness, but also apparent effects of the small amount of supplemental UV-A radiation applied beneath control and treatment arrays on lammas shoot length (Fig. 1). These data raise concerns about the interpretation of data from other outdoor supplementation studies that have not used UV-A controls.

There has only been one assessment of UV-B impacts on tropical forest trees. Searles, Caldwell and Winter (1995) used plastic film to exclude UV-B from three native rain forest species and one economically important species at Barro Colorado Island, Panama. They found that ambient UV-B apparently increased UV-B absorbing compounds, reduced plant height, increased specific leaf weight and reduced leaf length. Total biomass and PSII (photosystem II) function were generally unaffected by UV-B exclusion. This study suggests that although ozone depletion in the tropics may be small, changes in the very intense levels of UV-B experienced at these latitudes may have considerable biological consequences.

Interactions of UV-B with other environmental factors

Recent studies on *Pinus pinea* using square-wave UV-B supplementation in the field, simulating a 15% ozone depletion over Patras, Greece, have revealed a possible mechanism by which increased UV-B may interact with summer drought. Petropolou *et al.* (1995) first demonstrated that, under summer conditions, irradiation of *P. pinea* and *P. halepensis* with UV-B led to reduced needle drop and PSII inactivation, while the photon yield of oxygen evolution and photosynthetic capacity were higher than in plants exposed only to ambient UV-B. Subsequently, Manetas *et al.* (1997) showed in one-year old seedlings of *P. pinea* that UV-B treatment had no effect on PSII photochemical efficiency and biomass accumulation in well-watered plants. Trees that did not receive additional irrigation suffered needle loss and reduced PSII efficiency during the dry summer but these symptoms were alleviated by UV-B treatments. This was attributed to an improved water economy of the unirrigated plants under elevated UV-B. Such an effect reveals the seasonal nature of UV-B impacts and the interaction of UV-B with other climatic variables. The importance of such an alleviation of drought effects will depend on evaporative demand in the

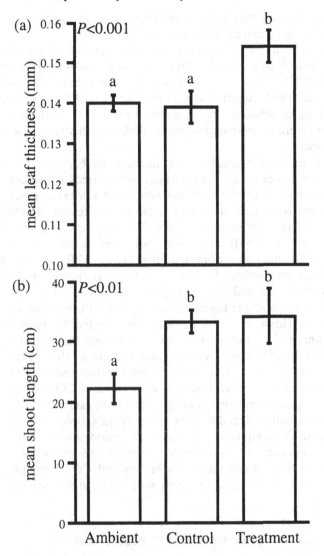

Fig. 1 Effects of ambient, polyester-filtered control (UV-A) and cellulose diacetate-filtered treatment (UV-A and UV-B) levels of radiation
from fluorescent lamps on (a) leaf thickness and (b) length of lammas
(summer) shoots of *Quercus robur* saplings. Treatment levels were
maintained outdoors at 30% above the ambient level of erythemally
weighted UV-B radiation. Values are means of four replicates ± standard deviation. Letters above each column indicate differences at the
P value indicated on the Figure following a two-way ANOVA and a
Tukey's test. (Modified from Newsham *et al.*, 1996.)

summer months and may have wider implications for tree response to UV-B outside the Mediterranean region.

There have been few studies of how UV-B exposure of trees interacts with metal availability in soils. Basiouny and Biggs (1975) reported an interaction of UV-B treatment with Zn deficiency (see above) and Dubé and Bornman (1992) reported significant effects on *Picea abies* of cadmium in nutrient solution when combined with a UV-B treatment that were not apparent or were underestimated when either stress factor was applied alone.

Two studies have evaluated how interactions between UV-B and increased atmospheric CO_2 concentration affect trees using controlled environment chambers. Stewart and Hoddinott (1993) exposed seedlings of *Pinus banksiana* in a glasshouse to two levels of UV-B and two levels of CO_2 concentration (Table 2). Total dry weight decreased with the highest UV-B treatment but was unaffected by CO_2 enrichment. Biomass partitioning under high UV-B was shifted in favour of leaf production. Both CO_2 and UV-B treatment decreased dark respiration rate and light compensation point, which implied a potential for greater shade tolerance and a change in competitive ability of the species. However, the authors carefully noted the limitations of the study imposed by the small container size and their fixed level of UV-B irradiation. Similarly, Sullivan and Teramura (1994) exposed *P. taeda* seedlings to three levels of UV-B and two concentrations of CO_2. They observed a clear growth response to elevated CO_2 but no effect of the gas on photosynthetic capacity or quantum efficiency. The highest UV-B treatment reduced total biomass by 12% both with and without increased CO_2. However, biomass allocation was altered by the combined treatment and was allocated preferentially to shoots at ambient CO_2 level but towards the roots in elevated CO_2. The interactive effects of UV-B and elevated CO_2 have recently been reviewed by Sullivan (1997).

Other components of forest ecosystems

Pathogens

Pathogens are major determinants of plant fitness (Burdon, 1987). Therefore, any effects of UV-B radiation on pathogenic organisms have the potential to influence the responses of plants to increased UV-B. The impacts of UV-B on pathogens can take two forms: either directly, by effects on the pathogen itself, or indirectly, through effects on the host which then influence interactions with the pathogen (Manning &

Tiedemann, 1995; Paul, this volume). For example, increased leaf thickness (Fig. 1a) is likely to affect the penetration of pathogens into leaf epidermal layers. Reduced photosynthesis may lead to reductions in rust and mildew diseases, which are normally associated with initially healthy leaves, and increases in the production of light-activated phototoxins (particularly the isoflavonoids and sesquiterpene phytoalexins) in leaves and stems in response to UV radiation may also lead to reductions in a wide range of bacterial and fungal pathogenic infections (Downum, 1992). Changes in phenology (such as altered timing of bud burst) may also influence interactions with pathogens (Manning & Tiedemann, 1995). There are few studies of UV-B effects on pathogens of trees but many examples of relevant effects on other plants.

Many studies have shown direct effects of UV-B radiation on the germination of fungal spores. For example, Semenuik and Stewart (1981) demonstrated that a four-fold increase in UV-B$_{\mathrm{THIM}}$ (biologically effective UV-B weighted with the generalised plant action spectrum of Caldwell (1971) fitted with the mathematical function of Thimijan, Carns and Campbell (1978) and normalised at 300 nm above ambient levels), reduced the infection of a cultivated *Rosa* sp. by black spot (*Diplocarpon rosae* Wolf.) between 6 and 18 h after inoculation with the fungus, which was attributed to the deleterious effects of UV-B radiation on germ tube extension of the fungus. Similarly, Rasanayagam *et al.* (1995) found that the exposure of conidia of *Septoria triciti* and *S. nodorum* on agar media to continuous UV-B radiation at 18 kJ m^{-2} d^{-1} UV-B$_{\mathrm{DNA}}$ reduced the germ tube extension of both species. In addition, Maddison and Manners (1973) demonstrated that the exposure of *Puccinia graminis* to 0.5 W cm^{-2} UV-B reduced urediniospore germination to < 1% of controls. Only one study has apparently been made to differentiate direct and indirect effects of UV-B radiation on plant–pathogen interactions. The irradiation of a disease-susceptible *Cucumis* variety with 11.6 kJ m^{-2} d^{-1} UV-B$_{\mathrm{PAS}}$ for 3 d prior to inoculation with the pathogens *Colletotrichum lagenarum* and *Cladosporium cucumerinum* led to greater incidence of infections on cotyledons of the plant, but post-infection UV-B treatment for 3 d did not affect disease incidence (Orth, Teramura & Sisler, 1990). In this case, the response of the plant to UV-B radiation appears to have mediated subsequent fungal development in host tissues, but the actual mechanism by which this occurred was not determined. Although neither of the latter three studies were carried out on woodland plants, representatives of *Septoria, Puccinia, Colletotrichum and Cladosporium* occur on both tree and understorey species in forests (Ellis & Ellis, 1985) and so it is plausible that UV-B radiation may alter the fitness of forest vegetation in some cases, depending upon the relative sensitivity of the host and its pathogen to short-wave radiation.

Although the study by Orth *et al.* (1990) demonstrated that UV-B radiation could affect fungal response indirectly by effects on the plant host, only one study has shown that the exposure of tree species to UV-B radiation during growth results in increases in the abundance of pathogenic fungal infections. Zeuthen (1995) demonstrated that the exposure of pedunculate oak (*Quercus robur* L.) to 15 and 30% elevations of UV-B radiation above ambient levels increased the infection of leaves by the oak mildew pathogen *Microsphaera alphitoides*. As *M. alphitoides* is a major determinant of fitness in oak saplings, this study may have important implications for seedling recruitment into oak populations in high UV-B environments, particularly at forest margins.

The response of below-ground pathogens to elevated UV-B radiation is possibly also significant in determining plant response to elevated UV-B. Klironomos and Allen (1995) examined the response of sugar maple (*Acer saccharum*) to a photon flux density of 6 μmol m^{-2} s^{-1} UV-B radiation in growth cabinets and found that irradiation did not affect plant biomass but that it significantly affected the fungi associated with the roots of the plant, presumably via altered plant metabolism. Of note was a 69% reduction in the frequency of arbuscules of arbuscular mycorrhizal (AM) fungi in roots, which are mutualistic fungi that improve the nutrition of many plant species by facilitating P inflow into roots, and a 150% increase in the frequency of non-mycorrhizal hyphae in roots. It was broadly concluded that UV-B radiation altered the plant–soil system from a closed, mycorrhizal-dominated system to an opportunist-open, saprobe/pathogen-dominated one. None of the organisms isolated in this study was clearly identified as a pathogen, but the principle applies that by reducing the frequency of AM fungi in roots, pathogenic organisms may then become more abundant (Newsham, Fitter & Watkinson, 1995). Further work needs to be carried out in this area, particularly to determine whether the abundances of ectomycorrhizal fungi, which are commonly found on the roots of temperate tree species, are affected by exposure to UV-B radiation. However, given that mycorrhizal fungi play a pivotal role in the cycling of nutrients in forests (Read, 1991), it is possible that there are wide-ranging implications of the study by Klironomos and Allen (1995) for the functioning of forest ecosystems.

Leaf epiphytes

The impact of UV-B radiation on the abundance of leaf surface microfloras also needs to be taken into account when assessing plant response to UV-B radiation. Leaf surfaces are habitats for a multitude of organisms which exist upon plant exudates and which are typically non-phytopathogenic. Many of these organisms are yeasts belonging to the genera *Sporobolomyces*, *Cryptococcus* and *Tilletiopsis*, as well as filamentous fungi such as *Cladosporium* and *Alternaria* spp. and bacteria, such as pseudomonads and xanthomonads (Last & Deighton, 1965). The response of these organisms to UV-B radiation is of significance since they have been shown to interact with aphids (Dik, Fokkema & van Pelt., 1992), bacteria (McCormack, Wildman & Jeffries, 1994) and fungal pathogens (Zhou, Reeleder & Sparace, 1991; Urquhart, Menzies & Punja, 1994) on leaf surfaces and they may therefore modify plant response to UV-B radiation. A recent study in summer and autumn 1995 in the UK has indicated that the yeasts *Sporobolomyces roseus* Kluy. et van Niel. and *Aureobasidium pullulans* are both significantly reduced, mainly on adaxial leaf surfaces of lammas leaves of *Q. robur* by a 30% increase in erythemally weighted UV-B radiation (K.K. Newsham *et al.*, unpublished observations). This is in general accordance with the data reported in Ayres *et al.* (1996), who have shown that the elimination of the UV-B portion of the spectrum from ambient solar radiation results in significant increases in the abundance of the yeasts *S. roseus* and *Bullera alba* on leaf surfaces of *Q. robur*. Whilst it is difficult to make precise predictions as to the ramifications of these effects of UV-B radiation on leaf surface micro-organisms, it is plausible that plant–pathogen interactions may well be affected, depending on the relative sensitivity of the pathogens and the leaf surface epiphytes to UV-B radiation. A further possibility is that patterns of decomposition may since some leaf epiphytes, for example, *A. pullulans*, are also responsible for the decomposition of abscised plant material (Smith & Wieringa, 1953).

Decomposers

The sustainability of forest ecosystems relies to a large extent upon the mineralisation of nutrients from fallen plant parts back into the soil by decomposer organisms. Any effects of UV-B radiation on litter decomposers may therefore affect primary productivity of forests by reducing the release of nutrients from litters. A recent study by Gehrke *et al.* (1995) has shown that the exposure of leaves of *Vaccinium* spp. to UV-B radiation either prior to, or after, abscission can alter patterns of decomposition. The mass loss rate of decomposing *V. uliginosum* litter in the laboratory under 10 kJ m^{-2} d^{-1} UV-B$_{THIM}$ was not affected, but significant 12 and 8% reductions in α-cellulose and lignin were recorded in irradiated litter compared with that decomposing under ambient conditions. More significantly, the abundances of the decomposer fungi associated with decaying litter were significantly affected by irradiation. *Mucor hiemalis, Truncatella truncata* and *Penicillium brevicompactum* were commonly isolated from litter decomposing under ambient conditions using an agar plating technique, but only *P. brevicompactum* was unaffected in abundance by exposure to UV-B radiation for 62 d; irradiation significantly reduced the abundance of *M. hiemalis* by 65% and *T. truncata* was eliminated completely from irradiated litter. It was suggested by Gehrke *et al.* (1995) that differences in UV-B screening pigments between the three fungi may have been responsible for their relative sensitivities to UV-B radiation. However, it is much more likely that the high reproductive rate of *P. brevicompactum* determined its apparent resistance to UV-B. *Penicillium* spp. are widely recognised as 'ruderal' (*sensu* Grime, 1977) fungi which tend to dominate in stressed environments because of their ability to reproduce rapidly. The exposure of *Vaccinium* litter to 10 kJ m^{-2} d^{-1} UV-B$_{THIM}$ also led to an increase in the number of fragments of leaves which did not yield fungal colonies when plated into agar media. This latter observation has been confirmed in a study into the direct effects of a 30% increase in erythemally weighted radiation UV-B radiation on the decay of *Q. robur* litter, but few changes in fungal communities in decomposing litter have been found (Newsham *et al.*, 1997*b*). Decomposer fungi are often localised within decaying host tissues, making direct effects of UV-B radiation on their abundance unlikely except at unrealistically high fluxes.

Whilst the direct effects of UV-B radiation on decomposition may be more relevant in heathland or open woodlands where *Vaccinium* spp. commonly occur, most decomposing leaves in forests will be exposed to doses of UV-B radiation much lower than those that are incident upon the forest canopy (see leaf transmission and distribution of UV-B in canopies, above). Therefore Gehrke *et al.* (1995) also exposed growing *Vaccinium* plants to UV-B, simulating a 15% ozone depletion scenario, and then decomposed the resulting leaf litter in a zero UV-B environment in the laboratory, to test the hypothesis that changes in leaf chemistry in response to UV-B would subsequently alter decomposition rate. The mass loss rate of decaying leaves previously irradiated with UV-B was reduced by 6%, but microbial respiration rate was unaffected. In accordance with the direct effects of UV-B radiation on the decomposition of *Vaccinium* leaves, α-cellulose was reduced by 14% in previously irradiated leaves and, in addition, the ratio of α-cellulose: lignin was reduced by 19% and tannins were increased by 50%, relative to leaves grown under ambient conditions prior to abscission (Gehrke *et al.*, 1995). This latter response may well account for the lower decay rate of these litters, since decomposition rate is largely determined by the level of polyphenols in decaying leaf litters (Minderman, 1968).

Invertebrates

Few studies have been published on the effects of UV-B radiation on the behaviour and abundance of invertebrates. Direct effects of UV-B on these organisms are much less likely than on leaf surface micro-organisms because of their larger sizes, the fact that many invertebrates (particularly the litter-decomposing organisms) inhabit low UV-B environments and the fact that many herbivorous invertebrates that feed on foliage are not active during the day. However, it is conceivable that indirect effects of UV-B on the behaviour and abundance of these organisms occur, akin to the indirect effects of UV-B radiation on pathogens (see above). For example, increased leaf thickness may alter patterns of insect herbivory, whilst the effects of light-activated photo-toxins may also play a significant role in determining the response of invertebrates to UV-B radiation; Downum (1992) lists ten different classes of phytochemicals which are known to be toxic to a wide range of insect genera. In experiments with caterpillars of the cabbage looper moth (*Trichoplusia ni*), the exposure of rough lemon (*Citrus jambhiri*) plants to elevated UV-B biologically effective radiation at 6.4 kJ m^{-2} d^{-1} led to 137% and 82% increases in the phototoxins psoralen and bergapten in young leaves, which were associated with 10% lower sur-vivorship of larvae after 12 d of growth (McCloud & Berenbaum, 1994).

Studies of *Quercus robur* saplings showed that exposure to lamps filtered with cellulose diacetate (which transmits UV-B and UV-A radiation) and to lamps filtered with polyester (which transmits only UV-A radiation) resulted in the same levels of herbivory by chewing insects, typically moth larvae, under both energised lamp/filter combi-nations, but that this level was 70–76% higher than under unenergised ambient arrays (Fig. 2a and Newsham *et al.*, 1996). However, herbivory leading to the formation of skeletonised areas in which cells between the cuticle are consumed was increased under UV-B (Fig. 2). By con-trast, in a study into the effects of a range of UV-B radiation fluxes (3.3–8.0 kJ m^{-2} d^{-1} UV-B$_{PAS}$ on herbivory of pea (*Pisum sativum*) by the larvae of the moth *Autographa gamma*, Hatcher and Paul (1994) demonstrated that irradiation led to increased N content of plant tissues, which led to a subsequent reduction in the amount of irradiated plant material consumed by larvae of the moth. Although *Pisum* spp. do not

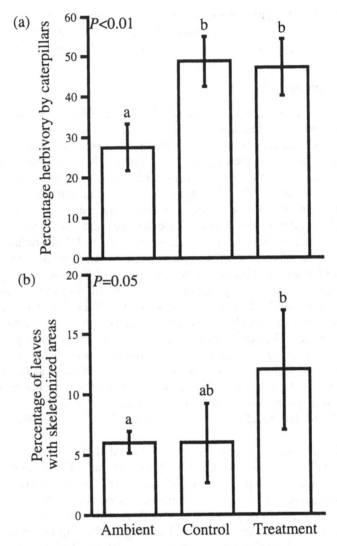

Fig. 2 Effects of ambient, polyester-filtered control (UV-A) and cellulose diacetate-filtered treatment (UV-A and UV-B) levels of radiation
from fluorescent lamps on (a) percentage level of herbivory by chewing insects and (b) percentage of leaves with skeletonized areas on
Quercus robur saplings. See Fig. 1 legend for details of treatments
and notation. (Modified from Newsham *et al.*, 1996 and K.K. Newsham, unpublished observations).

occur in woodlands, closely related species such as vetches do occur in the understorey, and it is therefore plausible that such effects might occur in high UV-B environments such as in rides or at canopy edges. In a recent study, the exposure of Gambel oak (*Quercus gambelii*) to 9.7 kJ m^{-2} d^{-1} UV-B_{PAS} led to an 11% increase in isoprene emissions from saplings (Harley et al., 1996). Although the function of isoprenes has recently been suggested to be a protective measure against high temperatures (Sharkey & Singsaas, 1995), it is conceivable that increases in the release of other volatile organic compounds from foliage could increase the attraction of invertebrates to irradiated plants.

In addition to effects on above-ground organisms, UV-B radiation may also exert effects on invertebrates via changes in plant metabolism. In their study into the effects of UV-B radiation exposure on the below-ground communities associated with *Acer saccharum* roots, Klironomos and Allen (1995) found that exposure of plants to elevated UV-B fluxes led to approximately 110% increases in populations of both collembola and non-oribatid mites, relative to plants grown in visible light. In addition to the nutrient-cycling roles of collembola and mites as comminutors of decomposing plant litter (Swift, Heal & Anderson, 1979), these organisms are also responsible for the severing of AM fungal hyphae attached to roots (McGonigle & Fitter, 1988), which limits plant response, and corroborates the observations made by Klironomos & Allen (1995) that UV-B exposure reduced the abundance of AM fungi in roots (see above). It is clear from this example that the effects of UV-B radiation on food webs in soil are complex and it is not yet possible to reach conclusions about effects on below-ground communities of invertebrates without further studies.

Conclusions

The potential mechanisms by which UV-B impacts forest trees have been demonstrated in many studies (Tables 1, 2). However, the quantification of observed effects both in field experiments and in relation to specific levels of ozone depletion has been limited to only a few studies (Sullivan & Teramura, 1992; Sullivan et al., 1994b; Petropoulou et al., 1995). None of these studies has utilised lamp modulation (McLeod, 1997) to vary the UV-B supplementation in relation to ambient levels, although one recent study of trees has used this approach (Newsham et al., 1996). Extrapolation of experimental data on saplings to predict effects on mature trees in the forest will always be difficult and subject to many assumptions. However, effects on young saplings will directly affect their competitive ability and survival in natural and plantation forests. Sullivan and Teramura (1992) tentatively concluded from the results of their square-wave field supplementation of *Pinus taeda* that pulp production in the USA might be affected if ozone depletion continues. However, there is currently too little quantitative information about UV-B impacts on forest tree species to permit estimation of impacts on global forest productivity.

The functioning of forest ecosystems depends upon a wide variety of components such as trees, soil and leaf microbes and insects (Fig. 3), all of which may be impacted by UV-B. The importance of UV-B effects on ecosystem components other than plants is becoming appar-

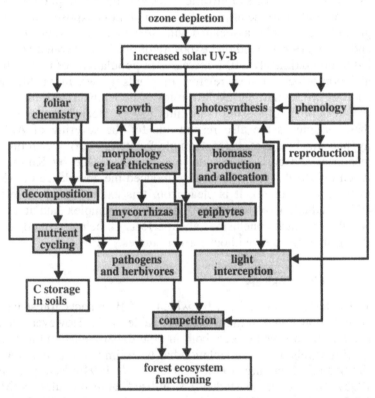

Fig. 3. Potential consequences of increases in solar UV-B radiation on components of forest ecosystems and ecosystem functioning. Shaded boxes indicate where effects of UV-B radiation have been demonstrated on ecosystem components. (Modified from Fig. 3 of Caldwell, Teramura & Tevini, 1989.)

ent in a number of studies: mycorrhizal fungi, insects and leaf surface fungi have all been shown to be affected by UV-B. Given the wide range of interacting factors determining the functioning of forest ecosystems, even small effects of UV-B in one component of the system may have impacts elsewhere (Fig. 3). Trees have long life cycles and may therefore show cumulative effects of higher UV-B over many years. In particular, it is not yet known how increased UV-B may affect carbon storage in forest soils. Currently, predictions of how elevated UV-B will affect forests are further confounded by the lack of knowledge of how drought, temperature and elevated CO_2 may interact with increases in UV-B. Much future research must be accomplished if we are to understand how forests may respond to future changes in UV-B and other components of global environmental change.

References

Allen, L.H., Jr., Gausman, H.W. & Allen, W.A. (1975). Solar ultraviolet radiation in terrestrial plant communities. *Journal of Environmental Quality*, **4**, 285–94.

Ayres, P.G., Gunasekera, T.S., Rasanayagam, M.S., Paul, N.D. (1996). Effects of UV-B radiation (280–320 nm) on foliar saprotrophs and pathogens. In *Fungi and Environmental Change*. (Frankland, J.C., Magan, N. & Gadd, G.M. eds), pp. 32–50. Cambridge University Press, Cambridge.

Basiouny, F.M. & Biggs, R.H. (1975). Photosynthetic and carbonic anhydrase activities in Zn-deficient peaches exposed to UV-B radiation. In *Impacts of Climatic Change on the Biosphere*, CIAP Monograph 5, part 1 (Appendix B), chapter 4, pp. 73–82. Department of Transportation. Washington, DC.

Björn, L.O. & Teramura, A.H. (1993). Simulation of daylight ultraviolet radiation and effects of ozone depletion. In *Environmental UV Photobiology*. (Young, A.R., Björn, L.O., Moan, J. & Nultsch, W., eds), pp. 41–71. Plenum Press, New York.

Blumthaler, M. & Ambach, W. (1990). Indication of increasing solar ultraviolet-B radiation flux in alpine regions. *Science*, **248**, 206–8.

Bogenrieder, A. & Klein, R. (1982). Does solar UV influence the competitive relationship in higher plants? In *The Role of Solar Ultraviolet Radiation in Marine Ecosystems*. (Calkins, J. ed.), pp. 641–9. Plenum, New York.

Bornman, J.F. & Teramura, A.H. (1993). Effects of ultraviolet-B radiation on terrestrial plants. In *Environmental UV Photobiology*. (Young, A.R., Björn, L.O., Moan, J. & Nultsch, W., eds), pp. 427–71. Plenum Press, New York.

276   A.R. McLEOD AND K.K. NEWSHAM

Brown, M.J., Parker, G.G. & Posner, N.E. (1994). A survey of ultra-violet-B radiation in forests. *Journal of Ecology*, **82**, 843–54.

Burdon, J.J. (1987). *Diseases and Plant Population Biology*. Cambridge University Press, Cambridge.

Caldwell, M.M. (1971). Solar ultraviolet radiation and the growth and development of higher plants. In *Photophysiology*. (Giese, A.C., ed.), pp. 131–77. Academic Press, New York.

Caldwell, M.M., Teramura, A.H. & Tevini, M. (1989). The changing solar ultraviolet climate and the ecological consequences for higher plants. *Trends in Ecology and Evolution*, **4**, 363–7.

Caldwell, M.M., Teramura, A.H., Tevini, M., Bornman, J.F., Björn, L.O. & Kulandaivelu, G. (1995). Effects of increased solar ultraviolet radiation on terrestrial plants. *Ambio*, **24**, 166–73.

Cannell, M. (1995). *Forests and the Global Carbon Cycle in the Past, Present and Future*. European Forest Institute, Finland.

Clark, J.B. & Lister, G.R. (1975). Photosynthetic action spectra of trees. II. The relationship of cuticle structure to the visible and ultra-violet spectral properties of needles from four coniferous species. *Plant Physiology*, **55**, 407–13.

Day, T.A. (1993). Relating UV-B radiation screening effectiveness of foliage to absorbing-compound concentration and anatomical characteristics in a diverse group of plants. *Oecologia*, **95**, 542–50.

Day, T.A., Vogelman, T.C. & Delucia, E.H. (1992). Are some plant life forms more effective than others in screening out ultraviolet-B radiation? *Oecologia*, **92**, 513–19.

Day, T.A., Martin, G. & Vogelmann, T.C. (1993). Penetration of UV-B radiation in foliage: evidence that the epidermis behaves as a non-uniform filter. *Plant, Cell and Environment*, **16**, 735–41.

Day, T.A., Howells, B.W. & Rice, W.J. (1994). Ultraviolet absorption and epidermal-transmittance spectra in foliage. *Physiologia Plantarum*, **92**, 207–18.

DeLucia, E.H., Day, T.A. & Vogelmann, T.C. (1991). Ultraviolet-B radiation and the rocky mountain environment: measurement of incident light and penetration into foliage. *Current Topics in Plant Biochemistry and Physiology*, **10**, 32–48.

Dik, A.J., Fokkema, N.J. & van Pelt, J.A. (1992). Influence of climatic and nutritional factors on yeast population dynamics in the phyllosphere of wheat. *Microbial Ecology*, **23**, 41–52.

Dillenburg, L.R., Sullivan, J.H. & Teramura, A.H. (1995). Leaf expansion and development of photosynthetic capacity and pigments in *Liquidambar styraciflua* (Hamamelidaceae) – effects of UV-B radiation. *American Journal of Botany*, **82**, 878–85.

Downum, K.R. (1992). Light-activated plant defence. *New Phytologist*, **122**, 401–20.

Dubé, S.L. & Bornman, J.F. (1992). Response of spruce seedlings to simultaneous exposure to ultraviolet-B radiation and cadmium. *Plant Physiology and Biochemistry*, **30**, 761–7.

Ellis, B.E. & Ellis, J.P. (1985). *Microfungi on Land Plants*. Croom Helm, London.

Fernbach, E. & Mohr, H. (1992). Photoreactivation of the UV light effects on growth of Scots pine (*Pinus sylvestris* L.) seedlings. *Trees*, **6**, 232–35.

Gehrke, C., Johanson, U., Callaghan, T.V., Chadwick, D. & Robinson, C.H. (1995). The impact of ultraviolet-B radiation on litter quality and decomposition processes in *Vaccinium* leaves from the subarctic. *Oikos*, **72**, 213–22.

Gold, W.G. & Caldwell, M.M. (1983). The effects of ultraviolet-B radiation on plant competition in terrestrial ecosystems. *Physiologia Plantarum*, **58**, 435–44.

Grace, J. (1983). *Plant–Atmosphere Relationships*. Chapman & Hall, London.

Grammatikopoulos, G., Karabourniotis, G., Kyparissis, A., Petropoulou, Y. & Manetas, Y. (1994). Leaf hairs of olive (*Olea europaea*) prevent stomatal closure by ultraviolet-B radiation. *Australian Journal of Plant Physiology*, **21**, 293–301.

Green, A.E.S., Sawada, T. & Shettle, E.P. (1974). The middle ultraviolet reaching the ground. *Photochemistry and Photobiology*, **19**, 251–9.

Grime, J.P. (1977). Evidence for the existence of three primary strategies in plants and its relevance to ecological and evolutionary theory. *American Naturalist*, **111**, 1169–94.

Hatcher, P.E. & Paul, N.D. (1994). The effect of elevated UV-B radiation on herbivory of pea by *Autographa gamma*. *Entomologia Experimentalis et Applicata*, **71**, 227–33.

Harley, P., Deem, G., Flint, S. & Caldwell, M.M. (1996). Effects of growth under elevated UV-B on photosynthesis and isoprene emission in *Quercus gambelii* and *Mucana pruriens*. *Global Change Biology*, **2**, 149–54.

Karabourniotis, G., Papadopoulos, K., Papamarkou, M. & Manetas, Y. (1992). Ultraviolet-B radiation absorbing capacity of leaf hairs. *Physiologia Plantarum*, **86**, 414–18.

Karabourniotis, G., Kyparissis, A. & Manetas, Y. (1993). Leaf hairs of *Olea europaea* protect underlying tissues against ultraviolet-B radiation damage. *Environmental and Experimental Botany*, **33**, 341–345.

Kaufmann, M.R. (1978). The effect of ultraviolet (UV-B) radiation on Engelmann spruce and lodgepole pine seedlings. UV-B Biological and Climatic Effects Research (BACER), Final Report EPA-IAG-D-0168, Environmental Protection Agency, Washington, DC.

Kerr, J.B. & McElroy, C.T. (1993). Evidence for large upward trends of ultraviolet-B radiation linked to ozone depletion. *Science*, **262**, 1032–4.

Klironomos, J.N. & Allen, M.F. (1995). UV-B-mediated changes on below-ground communities associated with the roots of *Acer saccharum*. *Functional Ecology*, **9**, 923–30.

Kossuth, S.V. & Biggs, R.H. (1981). Ultraviolet-B radiation effects on early seedling growth of Pinaceae species. *Canadian Journal of Forest Research*, **11**, 243–8.

Krupa, S.V. & Kickert, R.N. (1989). The greenhouse effect: the impacts of carbon dioxide (CO_2), ultraviolet-B (UV-B) radiation and ozone (O_3) on vegetation (crops). *Environmental Pollution*, **61**, 263–393.

Last, F.T. & Deighton, F.C. (1965). The non-parasitic microflora on the surfaces of living leaves. *Transactions of the British Mycological Society*, **48**, 83–99.

McCloud, E.S. & Berenbaum, M.R. (1994). Stratospheric ozone depletion and plant–insect interactions: effects of UV-B radiation on foliage quality of *Citrus jambhiri* for *Trichoplusia ni*. *Journal of Chemical Ecology*, **20**, 525–39.

McCormack, P.J., Wildman, H.G. & Jeffries, P. (1994). Production of antibacterial compounds by phylloplane-inhabiting yeasts and yeastlike fungi. *Applied and Environmental Microbiology*, **60**, 927–31.

McGonigle, T.P. & Fitter, A.H. (1988). Ecological consequences of arthropod grazing on VA mycorrhizal fungi. *Proceedings of the Royal Society of Edinburgh*, **94B**, 25–32.

McKinlay, A.F. & Diffey, B.L. (1987). A reference action spectrum for ultra-violet induced erythema in human skin. In *Human Exposure to Ultra-violet Radiation: Risks and Regulations*. (Passchier, W.F. & Bosnjakovic, B.F.M., eds), pp. 83–7. Elsevier, Amsterdam.

McLeod, A.R. (1997). Outdoor supplementation systems for studies of the effects of increased UV-B radiation. *Plant Ecology*, **128**, 1–16.

Maddison, A.C. & Manners, J.G. (1973). Lethal effects of artificial ultraviolet radiation on cereal rust uredospores. *Transactions of the British Mycological Society*, **60**, 471–94.

Manetas, Y., Petropoulou, Y., Stamatakis, K., Nikolopoulos, D., Levizou, E., Psarus, G. & Karabourniotis, G. (1997). Beneficial effects of enhanced UV-B radiation under field conditions: improvement of needle water relations and surviving capacity of *Pinus pinea* L. seedlings during the dry Mediterranean summer. *Plant Ecology*, in press.

Manning, W.J. & Tiedemann, A.V. (1995). Climate change: potential effects of increased atmospheric carbon dioxide (CO_2), ozone (O_3) and ultraviolet-B (UV-B) radiation on plant diseases. *Environmental Pollution*, **88**, 219–45.

Minderman, G. (1968). Addition, decomposition and accumulation of organic matter in forests. *Journal of Ecology*, **56**, 355–62.

Mirecki, R.M. & Teramura, A.H. (1984). Effects of ultraviolet-B irradiance on soybean. *Plant Physiology*, **74**, 475–80.

Naidu, S.L., Sullivan, J.H., Teramura, A.H. & DeLucia, E.H. (1993). The effects of ultraviolet-B radiation on photosynthesis of different aged needles in field-grown loblolly pine. *Tree Physiology*, **12**, 151–62.

Newsham, K.K., Fitter, A.H. & Watkinson, A.R. (1995). Multifunctionality and biodiversity in arbuscular mycorrhizas. *Trends in Ecology and Evolution*, **10**, 407–11.

Newsham, K.K., McLeod, A.R., Greenslade, P.D. & Emmett, B.A. (1996). Appropriate controls in outdoor UV-B supplementation experiments. *Global Change Biology*, **2**, 319–24.

Newsham, K.K., Low, M.N.R., Greenslade, P.D., McLeod, A.R. & Emmett, B.A. (1977*a*). UV-B radiation influences the abundance and distribution of phylloplane fungi on *Quercus robur*. *New Phytologist*, in press.

Newsham, K.K., McLeod, A.R., Roberts, J.D., Greenslade, P.D. & Emmett, B.A. (1997*b*). Direct effects of elevated UV-B radiation on the decomposition of *Quercus robur* leaf litter. *Oikos*, **79**, in press.

Orth, A.B., Teramura, A.H. & Sisler, H.D. (1990). Effects of ultraviolet-B radiation on fungal disease development in *Cucumis sativus*. *American Journal of Botany*, **77**, 1188–92.

Petropoulou, Y., Kyparissis, A., Nikolopoulos, D. & Manetas, Y. (1995). Enhanced UV-B radiation alleviates the adverse effects of summer drought in two Mediterranean pines under field conditions. *Physiologia Plantarum*, **94**, 37–44.

Rasanayagam, M.S., Paul, N.D., Royle, D.J. & Ayres, P.G. (1995). Variation in responses of *Septoria triciti* and *S. nodorum* to UV-B radiation *in vitro*. *Mycological Research*, **99**, 1371–7.

Read, D.J. (1991). Mycorrhizas in ecosytems. *Experientia*, **47**, 376–90.

Schnitzler, J-P., Jungblut, T.P., Feicht, C., Kofferlein, M., Langebartels, C., Heller, W. & Sandermann Jr., H. (1996*a*). UV-B induction of flavonoid biosynthesis in Scots pine (*Pinus sylvestris* L.) seedlings. *Trees Structure and Function*, in press.

Schnitzler, J-P., Jungblut, T.P., Heller, W., Kofferlein, M., Hutzler, P., Heinzmann, U., Schmelzer, E., Ernst, D., Langebartels, C. & Sandermann Jr, H. (1996*b*). Tissue localization of UV-B-screening pigments and of chalcone synthase mRNA in needles of Scots pine seedlings. *New Phytologist*, **132**, 247–58.

Searles, P.S., Caldwell, M.M. & Winter, K (1995). The response of five tropical dicotyledon species to solar ultraviolet-B radiation. *American Journal of Botany*, **82**, 445–53.

Semenuik, P. (1978). Biological effects of ultraviolet radiation on plant growth and development in florist and nursery crops. In *The Impacts of Ultraviolet-B Radiation on Biological Systems: A Study Related to Stratospheric Ozone Depletion*. Research Report EPA-IAG-DG-0168, pp. 1–18. Washington, DC: Environmental Protection Agency.

Semenuik, P. & Stewart, R.N. (1981). Effects of ultraviolet (UV-B) irradiation on infection of roses by *Diplocarpon rosae*, Wolf. *Environmental and Experimental Botany*, **21**, 45–50.

Setlow, R.B. (1974). The wavelengths in sunlight effective in producing skin cancer: a theoretical analysis. *Proceedings of the National Academy of Sciences, USA*, **71**, 3363–6.

Sharkey, T.D. & Singaas, E.L. (1995). Why plants emit isoprene. *Nature*, **374**, 769.

Skaltsa, H., Verykokidou, E., Harvala, C., Karabourniotis, G. & Manetas, Y. (1994). UV-B protective potential and flavonoid content of leaf hairs of *Quercus ilex*. *Phytochemistry*, **37**, 987–90.

Smith, J. & Wieringa, K.T. (1953). Microbiological decomposition of litter. *Nature*, **171**, 794–5.

Stewart, J.D. & Hoddinott, J. (1993). Photosynthetic acclimation to elevated atmospheric carbon dioxide and UV irradiation in *Pinus banksiana*. *Physiologia Plantarum*, **88**, 493–500.

Sullivan, J.H. (1994). Temporal and fluence responses of tree foliage to UV-B radiation. In *Stratospheric Ozone Depletion/UV-B Radiation in the Biosphere*, (Hilton-Biggs, R. & Joyner, M.E.B., eds), pp. 67–76. Springer-Verlag, Berlin.

Sullivan, J.H. (1997). Effects of increasing UV-B radiation and atmospheric CO_2 on photosynthesis and growth: implications for terrestrial ecosystems. *Plant Ecology*, in press.

Sullivan, J.H. & Teramura, A.H. (1988). Effects of ultraviolet-B radiation on seedling growth in the *Pinaceae*. *American Journal of Botany*, **75**, 225–30.

Sullivan, J.H. & Teramura, A.H. (1989). The effects of ultraviolet-B radiation on loblolly pine. 1. Growth, photosynthesis and pigment production in greenhouse-grown seedlings. *Physiologia Plantarum*, **77**, 202–7.

Sullivan, J.H. & Teramura, A.H. (1992). The effects of ultraviolet-B radiation on loblolly pine. 2. Growth of field-grown seedlings. *Trees Structure and Function*, **6**, 115–20.

Sullivan, J.H. & Teramura, A.H. (1994). The effects of UV-B radiation on loblolly pine. 3. Interation with CO_2 enhancement. *Plant Cell and Environment*, **17**, 311–17.

Sullivan, J.H., Teramura, A.H., Adamse, P., Kramer, G.F., Upadhyaya, A., Britz, S.J., Krizek, D.T. & Mirecki, M. (1994a). Comparison of the response of soybean to supplemental UV-B radiation supplied by either square-wave or modulated irradiation systems. In *Stratospheric Ozone Depletion/UV-B Radiation in the Biosphere.* (Hilton-Biggs, R. & Joyner, M.E.B., eds), pp. 211–20. Springer-Verlag, Berlin.

Sullivan, J.H., Teramura, A.H. & Dillenburg, L.R. (1994*b*). Growth and photosynthetic responses of field-grown sweetgum (*Liquidambar styraciflua*; Hamamelidaceae) seedlings to UV-B radiation. *American Journal of Botany*, **81**, 826–32.

Sullivan, J.H., Howells, B.W., Ruhland, C.T. & Day, T.A. (1996). Changes in leaf expansion and epidermal screening effectiveness in *Liquidambar styraciflua* and *Pinus taeda* in response to UV-B radiation. *Physiologia Plantarum*, **98**, 349–57.

Swift, M.J., Heal, O.W. & Anderson, J.M. (1979). *Decomposition in Terrestrial Ecosystems.* Studies in Ecology, Vol. 5. Blackwell Scientific Publications, Oxford.

Teramura, A.H. & Sullivan, J.H. (1991). Field studies of UV-B radiation effects on plants: case histories of soybean and loblolly pine. In *Impact of Global Climatic Changes on Photosynthesis and Plant Productivity.* (Abrol, Y.P., Wattal, P.N., Gnanam, A., Govindjee, Ort, D.R. & Teramura, A.H., eds), pp. 147–61. Oxford & IBH Publishing Co. Pvt. Ltd, New Delhi.

Thimijan, R.W., Carns, H.R. & Campbell, L.E. (1978). Radiation sources and related environmental control for biological and climatic effects of UV research (BACER). Final report EPA-IAG-D6-0168. Environmental Protection Agency, Washington, DC.

Urquhart, E.J., Menzies, J.G. & Punja, Z.K. (1994). Growth and biological control activity of *Tilletiopsis* species against powdery mildew (*Sphaerotheca fuliginea*) on greenhouse cucumber. *Phytopathology*, **84**, 341–51.

Yang, X., Miller, D.R. & Montgomery, M.E. (1993). Vertical distributions of canopy foliage and biologically active radiation in a defoliated/refoliated hardwood forest. *Agricultural and Forest Meteorology*, **67**, 129–46.

Yang, X., Heisler, G.M., Montgomery, M.E., Sullivan, J.H., Whereat, E.B. & Miller, D.R. (1995). Radiative properties of hardwood leaves to ultraviolet irradiation. *International Journal of Biometeorology*, **38**, 60–6.

Zeuthen, J. (1995). Influence of UV-B radiation on oak mildew. *Skoven*, **12**, 490–3.

Zhou, T., Reeleder R.D. & Sparace, S.A. (1991). Interactions between *Sclerotinia sclerotiorum* and *Epicoccum purpurascens*. *Canadian Journal of Botany*, **69**, 2503–10.

S.A. MOODY, D.J.S. COOP and N.D. PAUL

Effects of elevated UV-B radiation and elevated CO_2 on heathland communities

Introduction

Penetration of harmful ultraviolet-B (UV-B) radiation (280–320 nm) to the earth's surface is limited by stratospheric ozone. However, ozone depletion, caused by the emission of synthetic chlorofluorocarbons and related compounds, is currently estimated at 4–5% per decade at UK latitudes (Stolarski *et al.*, 1992; Herman, McPeters & Larko, 1993). If the Montreal Protocol continues to be implemented, a possibility that remains uncertain (Greene, 1995; Jordan, 1995), concentrations of chlorine and bromine compounds may reach a maximum in the stratosphere around 1998. Thus, ozone depletion may peak within the next decade, followed by a slow recovery over the next 50 years (Madronich *et al.*, 1995). Despite this more optimistic outlook, other factors such as exaggerated springtime ozone loss in the Arctic due to global warming (Austin, Butchart & Shine, 1992) may need to be considered. In addition, yearly mean ozone depletion figures conceal considerable variation in depletion with season, with maximal losses in late winter and spring (Niu *et al.*, 1992; Herman, McPeters & Larko, 1993; Reinsel *et al.*, 1994), although at that time of year the level of UV-B is relatively low. This evidence suggests that the potential for increasing levels of UV-B over Northern latitudes will remain a problem for many years to come. Many studies on the effects of UV-B on plants have been conducted in controlled environment cabinets or greenhouses, where the levels of UV-A (320–400 nm) and PAR (400–700 nm) may be low compared to the field. In addition, many of these experiments have concentrated on crops (Krupa & Kickert, 1989). Few studies have considered longer-lived species over several growing seasons. It is, therefore, increasingly recognised that long-term field experiments are necessary to adequately determine the threat to natural ecosystems posed by increasing UV-B.

Another major anthropomorphic change to the atmosphere is the increase in CO_2, contributing to what is commonly referred to as the 'greenhouse effect'. Studies considering the effect of elevated CO_2 on plant growth and physiology are now well under way, and the effects of CO_2 on ecosystem processes ranging from herbivory to decomposition are being investigated (Cotrufo, Ineson & Rowland, 1994; Cotrufo & Ineson, 1995; Watt *et al.*, 1995; Doherty, Salt & Holopainen, in press). However, little work to date has considered interactions between UV-B and CO_2 (Teramura, Sullivan & Ziska, 1990; Rozema *et al.*, 1990; Sullivan & Teramura, 1994). Recently several field studies have been initiated to consider the impact of elevated UV-B, and in some cases elevated UV-B/CO_2 interactions, on heathlands (Björn *et al.*, 1996). Heaths are areas of land in which the vegetation is usually dominated by dwarf shrubs, and are widespread in parts of Europe, southern South Africa, south-east Australia, some oceanic islands and some mountain ranges (Tubbs, 1985). In many cases they have been formed by human disturbance. Typically, the felling of primary woodland has resulted in a nutrient-poor soil through years of erosion and nutrient leaching down the soil profile, and ensures that tree growth can no longer be supported. In some parts of Europe, once heaths had formed, activities such as burning, turf cutting and grazing have guaranteed their existence. In addition to 'man-made' heathlands, similar vegetation types can be found further north in Arctic regions where the land is predominantly free from human influence. Heathlands can be formed on many different soil types, in Britain these ranging from the drier acid sands and podzols of lowland dry heath to the shallow peats and humic soils of subalpine and montane heaths (Rodwell, 1991). In many heaths in Britain and the rest of western Europe, the plant community can be dominated by heather (*Calluna vulgaris*), other ericaceous species and gorse (*Ulex* spp.). Other sub-shrubs, herbs, lichens and mosses are found to a greater or lesser extent depending on soil types and climatic conditions.

Many plant species found growing on heathlands are adapted to nutrient-poor, often acidic, soils. These species are often long-lived, and in some evergreens (such as *C. vulgaris*) individual leaves are retained for several years. The potential therefore exists for long-term cumulative effects of UV-B or CO_2. It is possible that UV-B/CO_2 induced changes in plant growth and physiology, timing of processes such as flowering, or the rate of decomposition and nutrient cycling could considerably alter heathland communities. Two projects examining the effects of atmospheric change on heathland ecosystems have been running at the Air Pollution and Climate Change Unit (APCCU) at Lancaster University, UK since 1993. The first considered the effects of elevated CO_2, elevated UV-B, and their interaction, on the growth and physiology of *Calluna vulgaris* (L.) Hull (heather, an evergreen dwarf shrub) and

Vaccinium myrtillus L. (bilberry, a deciduous dwarf shrub). The second concentrated on the effects of elevated UV-B on the litter quality and decomposition of *Calluna vulgaris* and *Rubus chamaemorus* L. (cloudberry, a deciduous herb). This chapter reviews the few studies that have been made of the effects of elevated UV-B, and in some cases elevated CO_2, on physiology, decomposition and herbivory of heathland plants, and includes details of studies made at Lancaster. The methodology of giving supplementary UV-B, especially in terms of action spectra, is crucial in defining the dosages delivered, and so our approach is also described in some detail.

UV-B supplementation to heathland systems

The techniques used at Lancaster for enhancing UV-B have also been used by HRI Wellesbourne (see Corlett *et al.*, this volume) and Sheffield University's UCPE Climate Change Impacts Laboratory in Buxton. The technique, which has been described elsewhere (Mepsted *et al.*, 1996), involves giving supplementary UV-B using Philips TL40/12RS UV-B fluorescent tubes wrapped in 100 μm cellulose diacetate sheets. Controls were provided by unpowered bulbs (to allow for shading effects of the structures). UV-B enhancement was provided in a modulated fashion (in which elevated UV-B is constantly altered in relation to ambient) by controlling lamp output via a PC-based data acquisition and control system (LanDACS, Lancaster University, Lancaster, UK). UV-B was measured and controlled using two BW100-UV-B sensors (Vital Technologies, Bolton, Canada) placed under an ambient and an elevated treatment. The sensors were calibrated weekly in terms of Caldwell's plant action spectrum (normalised to 1 at 300 nm: PAS300: Caldwell, 1971) against daylight and lamp spectra using a double monochromator spectroradiometer. Supplemental UV-B was provided throughout the year by wrapping lamps with shade material to reduce their output. Cellulose diacetate filters were only replaced when lamp output could no longer maintain treatments (Mepsted *et al.*, 1996).

Supplementary plant weighted UV-B treatments (PAS300) were based on a 15% ozone depletion. Calculations of UV-B treatments were performed using the model of Björn and Murphy (1985), on software developed by E.L. Fiscus and F.L. Booker (University of N. Carolina). The baseline ozone column data used were from the Nimbus VII Total Ozone Mapping System (TOMS), provided by J. Herman (Goddard Space Flight Laboratory, Greenbelt, Maryland) (Herman & Larko, 1994). Calculations of UV-B fluxes for baseline ozone and 15% ozone depletion were performed for each day of the year, taking the mean column from 1979–1990 as the 'zero depletion' baseline.

Two UV-B supplementation regimes were used. Calculations of ozone depletion at 55° N, based on TOMS measurements for the period 1979–1991 (Herman, McPeters & Larko, 1993), were extrapolated to give an annual mean depletion of 15%. A polynomial was fitted to the TOMS data to give day-to-day values of ozone depletion from which mean weekly PAS300 supplements were calculated. Estimated dose increases were superimposed onto the effects of the seasonal variation in ozone column thickness. This treatment was named the 'constant' treatment because it simulated the effects of a constant year-round 15% ozone depletion. The second treatment was designated the 'variable' treatment because, although the mean simulated ozone loss was again 15%, the actual percentage depletion varied over the course of the year. A seasonal nature of ozone depletion is increasingly recognised (Niu et al., 1992; Herman, McPeters & Larko, 1993; Reinsel et al., 1994). Springtime ozone loss has now been related to episodic increases in ground-level UV-B in Europe (Blumthaler et al., 1994; Seckmeyer et al., 1994; Jokela et al., 1995). The variable treatment produces a large percentage addition in spring (although the addition is small in absolute terms) in comparison to the constant treatment.

Plant growth and development studies

For studies of effects of UV-B and CO_2 on growth and physiology, four chambers glazed with 0.05 mm thick clear teflon film were used. Teflon was chosen (rather than glass) for its almost equal transmission (> 95%) of wavelengths between 300 and 800 nm, and transmission of > 90% between 290 and 300 nm. Two of the chambers had elevated CO_2 at 350 ppm above ambient. Three UV-B treatments (ambient, constant and variable) were provided in each chamber by 6 Philips TL40 UV-B lamps arranged in three groups. The constant and variable supplementary UV-B treatments were run using a common power supply to the lamps, with the UV-B sensor, measuring elevated UV-B levels, placed in the centre of a variable treatment block. The constant treatment was simultaneously provided by the careful altering of lamp height and the use of muslin shading material. Mylar polyester screens, placed between the blocks, prevented the spread of unwanted UV-B between treatments. Each chamber contained a drained, peat-filled pit; plants were grown in 5 litre nylon bags containing a mixture (pH 3.9) of 80% moorland peat and 20% acid sand, such that the soil inside the bags was level with that outside. This allowed a continuum of water and nutrients between the rooting volume and the surrounding medium, avoiding the effects of limited pot volume and preventing pot warming (Townend, 1993). Plants were watered frequently with stored rainwater, ensuring that the soil mass was constantly damp. Plants were planted such that they formed a closed canopy by the end of the first growing season. Morphological measurements were made initially and again after one and two growing seasons. Measurements were made of carbon : nitrogen ratios (C : N), leaf chlorophyll concentrations, UV-absorbing pigments, and gas exchange.

Decomposition

Supplementary UV-B was provided separately to plots containing growing plants, and to plots containing decomposing material. In the first case, established plants of *C. vulgaris* and *R. chamaemorus* were transplanted from upland heathland at Moor House National Nature Reserve, North Yorkshire (NY761327, 580 m altitude). These were potted with their native peat in 25-litre pots placed up to their rims in ten peat-filled wooden frames. Frames were designed to maintain an 8 cm deep water table (17 cm from the peat surface) which was restored to capacity once a week using rainwater. This design ensured the plants were grown in conditions as close as possible to those experienced in the field. Two UV-B treatments were supplied by arrays of 8 Philips TL40 UV-B bulbs suspended 1.3 m above the soil surface on an aluminium frame. UV-B sensors were placed at plant height at the south end of a treatment and control array. There were five replicate arrays of two treatments: control (ambient UV-B under arrays of unpowered lamps) and elevated UV-B (the 'variable' treatment as described above). Nitrogen concentration, C : N ratio (as above) and the level of total water soluble phenolics (Folin-Denis method, Allen, 1989) were measured in *Rubus chamaemorus* litter, which was collected as it was ready to fall from the plants (July–October), and in *Calluna vulgaris* litter, which was collected as recently dead material at the end of the second growing season. Living shoots of *C. vulgaris* were analysed for C : N ratio, total water soluble phenolics and free amino acids.

For the decomposition studies, ten replicate arrays, each with 4 UV-B bulbs suspended 90 cm above the soil surface (4 non-functional bulbs for controls), provided the same level of supplemental UV-B as was used for plant growth. Each array was placed above a wooden frame which was filled with sphagnum peat (as above) and covered with a layer of fresh sphagnum moss, and *R. chamaemorus* and *C. vulgaris* litter collected from Moor House. A 10 cm deep water table (5 cm from the peat surface) was maintained with rainwater to ensure the peat did not dry out. This arrangement allowed litter to be decomposed in conditions as close as possible to those experienced in the field by providing soil fauna and flora expected at the field site. Litter to be decomposed (approx. 0.9 g/sample air dry weight) was placed in polypropylene cylinders (10 cm diameter, 2.5 cm deep) covered on the bottom with a 1 mm nylon mesh, and on the top with a 7 mm mesh. This arrangement allowed wind, rain, light, fungal decomposers and the majority of soil fauna to reach the litter. Samples of *R. chamaemorus* litter were collected from the wild (grown under ambient UV-B), and from material grown at Lancaster under both ambient and elevated UV-B. Samples, examined at 3- or 6-month intervals, were dried at 80 °C and weighed to assess mass loss.

Laboratory studies were designed to complement field decomposition data. *Rubus chamaemorus* litter was decomposed in microcosms (Anderson & Ineson, 1982) fitted with UV-transparent teflon lids. Litter (approximately 0.5 g/sample air dry weight) was inoculated with 1 ml of a coarsely sieved peat and leaf macerate from the field site at Moor House. Samples were decomposed under zero, ambient (2.3 kJ m^{-2} d^{-1} PAS300) or elevated (3.2 kJ m^{-2} d^{-1} PAS300) UV-B in controlled environment (CE) cabinets (white light, 36 W m^{-2}, UV-A, 2.5 W m^{-2}, temp, 15 °C). UV-B supplements were provided for approx. 6 h in the middle of a 12 h photoperiod. The litter was remoistened every 2 days. Respiration was measured as CO_2 release and leachates were collected periodically.

Effects of UV-B on phenology and physiology

Reproduction and flowering time

Johanson *et al.* (1995a) reported that a UV-B treatment simulating a 15% ozone reduction over a subarctic heath did not have any effects on the phenology of leaf-bud break, flowering, ripening of fruit, or onset of senescence in several heathland species. By contrast, earlier flowering in plants grown at elevated CO_2 is well documented (Garbutt & Bazzaz, 1984; Slack, 1986; Krupa & Kickert, 1989) and Woodin *et al.* (1992) found that elevated CO_2 greatly advanced flowering in *C. vulgaris*. The implications of such changes in phenology cannot easily be predicted. Changes in the timing of flowering may alter seed production in insect pollinated plants if pollinators are not active when flowers open. This is also likely to have impacts on the pollinators if food sources are restricted or unavailable. Conversely, earlier flowering may increase seed production in cold climates, where many seeds may fail to reach maturation at the end of the growing season. However, the changes found here were generally in the order of a few days and may be rather small in comparison with other climatic factors, such as temperature or duration of snow cover (Johanson *et al.*, 1995*a*). Similar results were obtained from studies carried out at Lancaster; elevated UV-B had little effect on the timing of flowering in *C. vulgaris* in either growing season or the rate of bud burst in *V. myrtillus*, and there were no significant UV-B/CO_2 interactions. Elevated CO_2 advanced the time for plants of *C. vulgaris* to begin flowering in both seasons (Figs. 1 a,b), but in contrast, bud break was slightly delayed in *V. myrtillus* (Fig. 1c). This delay in bud break due to elevated CO_2 is difficult to explain, but may reflect a carry-over either from the previous season or, perhaps, a CO_2 effect on the green photosynthesising stems.

Hellmers and Hesketh (1973) proposed that increases in C : N ratio due to CO_2 supplementation would result in increased production of flowers, and perhaps earlier flowering. This appeared to be the case in *C. vulgaris* since Woodin *et al.* (1992) found both significantly earlier and greater flowering and increased C : N ratio in this species. We also found that elevated CO_2 resulted in earlier and increased flowering in *C. vulgaris* and increased C : N (see below). However, a very clear UV-B/CO_2 interaction on flowering was observed. High UV-B reduced stimulation of flowering by high CO_2 from a 22% increase under ambient UV-B to a 2% increase under the variable treatment and a 16% decrease under constant.

Elevated CO_2 significantly increased the C : N ratio in *V. myrtillus* and the number of reproductive structures. However, the greater number of flowers produced under elevated CO_2 was not reflected in increased fruit. Indeed plants in high CO_2 bore fewer fruits than plants in ambient CO_2, and fruit had a lower dry weight. The reasons for this apparent reduction in the fruit production of *V. myrtillus* are unclear, but may be related to delayed bud-break under high CO_2 (see above). The constant UV-B treatment caused small increases in the flowering of *V. myrtillus* in both seasons (15% and 22%). By contrast the effects of the variable UV-B treatment were inconsistent, a massive (173%) increase in the first season being completely absent in the second. Reproduction by means of seeds is negligible in *V. myrtillus* where vegetative reproduction through the large underground mass of creeping rhizomes predominates populations (Grime, Hodgson & Hunt, 1988). Therefore, from the point of view of reproduction in the field, any effects of CO_2 or UV-B on flower and fruit production may be less important than effects on below-ground biomass (see below).

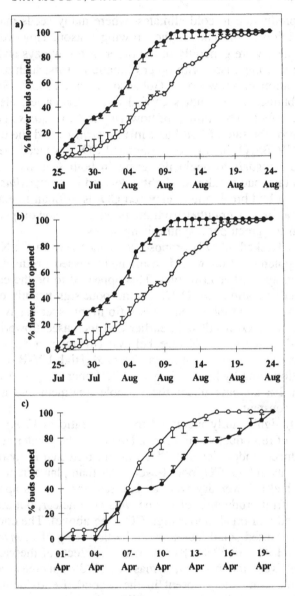

Fig. 1. The progress of flowering in *Calluna vulgaris* in a) 1994 and b) 1995 and c) of vegetative bud-break in *Vaccinium myrtillus* during 1995 in plants grown under either ambient CO_2 (open symbols) or ambient plus 350 ppm (filled symbols). Data are means of 60 plants and bars are one standard error.

Plant growth and morphology

In the field under ambient CO_2, Johanson *et al.* (1995*a*) found that leaf area of *V. myrtillus* was reduced by 14% by a UV-B supplement simulating a 15% ozone depletion, although contrasting responses were found in other *Vaccinium* spp. There was no effect of UV-B on leaf number per shoot in *V. myrtillus* in their study, and at Lancaster too the number of leaves per unit length of stem was unaffected by UV-B treatments. However, elevated CO_2 significantly increased both shoot and root dry weight in season one but only root biomass in season two. Increases in root growth were greater than those in shoot growth, resulting in increased root-to-shoot ratio (R : S), significantly in the second season. By contrast, neither shoot nor root dry weights were significantly affected by elevated UV-B treatment in either season. However, in the second season there was a tendency for both supplementary UV-B treatments to reduce root dry weight, especially at ambient CO_2. As a result, R : S was significantly decreased under increased UV-B. High CO_2 increased total plant leaf area, primarily as a result of an increase in the number of leaves per unit length of stem. The effects of UV-B on leaf area were rather small (< 15%) and varied between ambient and elevated CO_2.

There were no statistically significant responses of *C. vulgaris* to CO_2 and UV-B. Although this may be partly due to marked plant-to-plant variation in some variables, coefficients of variation were not substantially greater than in *V. myrtillus*. Therefore, we conclude that our data show a real difference in sensitivity between the two species studied. This is also apparent when growth responses are considered in percentage terms. In the first growing season elevated CO_2 increased both shoot and root dry weights of *C. vulgaris* (13% and 23%, respectively, cf. 23% and 30% in *V. myrtillus*). In the second season of treatment, root biomass was still increased under elevated CO_2, although by a smaller amount (15%, cf. 25% in *V. myrtillus*) while shoot biomass was decreased (by approx. 10%). This loss of shoot biomass was a function of a change in branching patterns, with increased mortality of older branches under elevated CO_2, perhaps itself a result of increased self-shading. In both seasons, R : S was increased under elevated CO_2.

Photosynthesis and chlorophyll

High CO_2 significantly reduced chlorophyll *a*, chlorophyll *b* and carotenoid concentrations by 10–15% for *V. myrtillus* and *C. vulgaris*. Similar (7–17%), although non-significant, reductions in total chlorophyll were also observed in *Pinus taeda* grown under elevated CO_2

(Sullivan & Teramura, 1994). During the first growing season, UV-B treatment had no effect on the concentrations of chlorophyll a, chlorophyll b, total chlorophyll or carotenoids, or the chlorophyll $a : b$ ratio, calculated on a fresh weight basis, for either V. myrtillus or C. vulgaris. In P. taeda, elevated UV-B had no effect on total chlorophyll, while a significant increase in the chlorophyll $a : b$ ratio was confined to one UV-B treatment (8.8 kJ m^{-2} d^{-1}) under ambient CO_2 (Sullivan & Teramura, 1994).

Gas exchange measurements made on V. myrtillus revealed that elevated CO_2 significantly increased photosynthetic rates, especially under increased UV-B. Thus, supplementary UV-B had little effect on rates of photosynthesis for plants grown in elevated CO_2, but rates were reduced by 20–25% in plants grown in ambient CO_2 (Fig. 2a). Stomatal conductance (Fig. 2b) was significantly reduced by elevated CO_2 while elevated UV-B caused smaller, but still significant reductions. The pattern of the response was similar to that seen for photosynthesis, i.e. the effect of high UV-B was considerably greater for plants grown in ambient CO_2. It is possible that the reductions in photosynthesis resulted from UV-B induced stomatal closure, but examination of the internal CO_2 concentration (Ci) indicated that there was no significant difference between UV-B treatments (Fig. 2c). These data would suggest that stomatal conductance was not limiting the rate of photosynthesis. Rather, the reductions occurred as a direct effect of UV-B on the photosynthetic machinery, and stomatal conductance declined to maintain the ratio of internal to external CO_2 concentrations (C_i/C_a). Therefore the effect of UV-B on stomatal conductance was the result of, and not a cause of, the lowered assimilation rate.

Most recent field studies have found little or no effect of UV-B on photosynthesis (Fiscus & Booker, 1995; Mepsted et al., 1996) and, therefore, the marked response observed in V. myrtillus was rather surprising. However, the effect was consistent and similar results were obtained when the measurements were repeated in the next season. The conclusion that elevated concentrations of CO_2 may protect V. myrtillus against the effects of UV-B on photosynthesis is also consistent with data of Sullivan and Teramura (1994). They found that the effects of UV-B radiation on maximum photosynthetic oxygen evolution were small and inconsistent, and that high CO_2 treatment further reduced the effects of UV-B. By contrast, in an earlier study of rice, wheat and soybean, Teramura et al. (1990) observed small reductions in the rate of photosynthesis for plants exposed to high UV-B, but that high UV-B partly reversed the stimulation of photosynthesis by elevated CO_2.

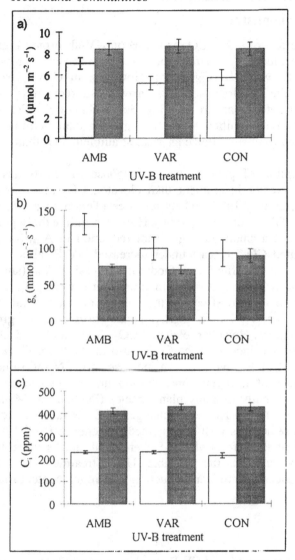

Fig. 2. a) Assimilation rate (μmole CO_2 m^{-2} s^{-1}), b) stomatal conductance (mmol m^{-2} s^{-1}) and c) internal CO_2 concentration of *Vaccinium myrtillus* grown under ambient CO_2 (open columns) or ambient +350 ppm (filled columns) and at ambient UV-B (AMB), or UV-B supplemented assuming either a mean 15% ozone depletion incorporating seasonal variation in loss (VAR) or a constant 15% ozone depletion (CON). See text for more details. All data are means of ten replicate plants and bars are ± one standard error.

294   S.A. MOODY, D.J.S. COOP AND N.D. PAUL

Plant chemistry

There was no effect of either CO_2 or UV-B on UV-absorbing pigments (A_{300}) in the deciduous *V. myrtillus*, or in the current year's growth for the evergreen *C. vulgaris* (Fig. 3a,b). However, in *C. vulgaris* leaf material from the previous season's growth, supplementary UV-B caused a significant increase in A_{300} (Fig. 3c). Although the CO_2/UV-B interaction was not significant (i.e. not greater than additive), this increase was approximately twice as great at ambient CO_2 than at elevated CO_2.

UV-B absorption of plant tissue in response to UV-B has been reported to be greater in ambient than elevated CO_2 in *Pinus taeda* (Sullivan & Teramura, 1994), wheat and rice (Teramura *et al.*, 1990) but not soybean (Teramura *et al.*, 1990). However, unlike these previous studies, we found a small but significant reduction in A_{300} for plants grown in elevated CO_2, at least under increased UV-B, although this may be a function of different methods of expressing A_{300} (since our data are per unit fresh weight rather than per unit dry weight).

Elevated UV-B had no effect on the content of water-soluble phenolics, nitrogen or C : N ratio in *R. chamaemorus* litter or growing shoots of *C. vulgaris*. However, elevated CO_2 significantly reduced the nitrogen content and increased the C : N ratio in shoots of *C. vulgaris* and leaf material of *V. myrtillus*. An increase in C : N ratio due to a relative reduction of nitrogen concentration under elevated CO_2 has been found consistently in many plant species (Peñuelas & Matamala, 1990; Watt *et al.*, 1995). However, changes in C : N ratio due to elevated UV-B are rather less well documented (Hatcher & Paul, 1994). In our investigations the effects of UV-B were small and/or variable. We found that in *V. myrtillus* the 'variable' UV-B treatment, but not the 'constant', produced a significant reduction in nitrogen content irrespec-

---

Fig. 3. Concentration of UV-B absorbing compounds (as A_{300} per unit fresh weight) in a) current year's leaves of *Vaccinium myrtillus*, b) current year's shoots of *Calluna vulgaris*, and c) second season shoots of *Calluna vulgaris*. All plants were grown under ambient CO_2 (shaden columns) or ambient +350 ppm (filled columns) and at ambient UV-B (AMB), or UV-B supplemented assuming either a mean 15% ozone depletion incorporating seasonal variation in loss (VAR) or a constant 15% ozone depletion (CON). See text for more details. All data are means of ten plants and bars are ± one standard error.

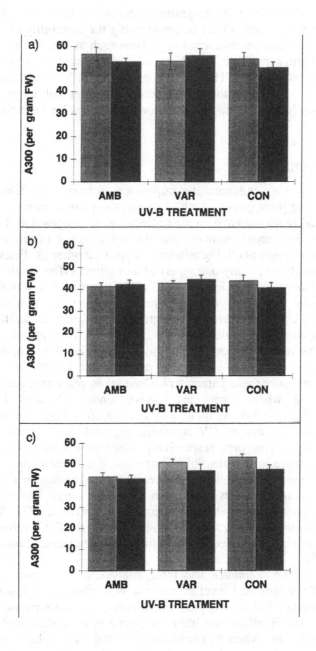

tive of CO_2 treatment, causing a significant increase in C : N ratio. However, C : N ratio is a key factor in determining the suitability of plant material as a resource for consumer organisms such as herbivores (Salt, Major & Whittaker, 1996; Watt *et al.*, 1995) and decomposers (Cotrufo & Ineson, 1995). Therefore, changes in C : N in response to UV-B or CO_2, coupled with those in secondary metabolites such as phenolics, may have wide implications for interactions between trophic levels (Paul *et al.*, 1996).

Decomposition

Elevated UV-B may influence decomposition in two ways. Changes occurring during plant growth, such as changes in plant chemistry, may alter litter quality and so have indirect effects on decomposition. Direct effects during decomposition may occur through effects on decomposer organisms or through photodegradation (Zepp, Callaghan & Erickson, 1995). To date, there is only one report of an indirect effect of elevated UV-B on decomposition. Gehrke *et al.* (1995) recorded a significant reduction in the rate of mass loss of *Vaccinium myrtillus* litter grown under elevated UV-B compared to that grown under ambient conditions when decomposed under ambient UV-B in the field. They attribute this change in the rate of decomposition to changes in plant chemistry caused by exposure to elevated UV-B during plant growth, although they did not measure these parameters. Changes in plant chemistry are often recorded when plants are grown under elevated UV-B (Caldwell & Flint, 1994). As noted above, a significant increase in the C : N ratio and the level of UV absorbing pigments was recorded for *V. myrtillus* and *C. vulgaris*, respectively, when grown under elevated UV-B. An increase in UV absorbing pigments, if it is accompanied by a general increase in phenolics, could result in a reduction in the rate of decomposition. Phenolics are known to reduce fungal growth and are implicated in modifying decomposition rates (Harrison, 1971; Swift, Heal & Anderson, 1979). In addition, an increase in the C : N ratio may have a similar inhibitory effect on plant litter decomposition (Cotrufo & Ineson, 1995).

In studies with *R. chamaemorus* litter, analysis of the nitrogen concentration, C : N ratio and level of water-soluble phenolics showed no effect of elevated UV-B during plant growth on any of these parameters. This lack of UV-B effects on litter chemistry is consistent with our decomposition data. When *R. chamaemorus* litter, from plants grown under ambient and elevated conditions for one growing season, was decomposed under ambient UV-B over a 12-month period in the field, no effect of treatment during growth was recorded for the rate of decomposition measured as mass loss (% weight remaining, elevated 52.0 ± 0.66%, ambient 52.9 ± 0.58%). It seems that, while elevated UV-B may influence decomposition indirectly in some species through changes in plant chemistry during growth, other species, such as *R. chamaemorus*, may not be affected.

The direct effect of elevated UV-B on decomposition is more complex. Inhibitory effects on microbial decomposers may suppress decomposer activity or alter community structure, while direct photo-degradation may accelerate litter breakdown (Zepp *et al.*, 1995). Photodegradation, especially through the photochemical action of UV-B on complex organic molecules such as lignins, may result in the direct breakdown of litter (Moorhead & Callaghan, 1994). Although it has been suggested that photodegradation is responsible for the rapid degradation of surface litter in the Russian steppes (see Moorhead & Callaghan, 1994), few experiments have been conducted to confirm that hypothesis. Gehrke *et al.* (1995) recorded a significant reduction in the amount of lignin in *Vaccinium uliginosum* litter exposed in a microcosm experiment to elevated UV-B when compared to litter decomposed without UV-B. Short-term experiments at Lancaster where dry *R. chamaemorus* leaf discs were exposed to high levels of UV-B for 5 weeks showed a 19% greater loss of mechanical strength in litter exposed to higher UV-B. Although these preliminary results are interesting and would tend to support current theories, they now need to be conducted using more realistic UV-B treatments.

The direct effect of elevated UV-B on decomposition, through UV-B effects on fungal decomposers, is also poorly understood. Most studies of the effects of elevated UV-B on fungi have involved phylloplane or pathogenic species (Ayres *et al.*, 1996), rather than decomposers. However, the capacity for elevated UV-B to inhibit fungal decomposers, or change the species composition is evident in the pioneering study of Gehrke *et al.* (1995). In microcosm experiments using *V. uliginosum* litter, they showed transient but significant reductions in microbial respiration (measured as the rate of CO_2 release) in litter decomposed under elevated compared to zero UV-B. This change in respiration was attributed to changes in the fungal community structure (Gehrke *et al.*, 1995). Microcosm studies are currently being conducted at Lancaster in which UV-B levels are increased from ambient to elevated doses, realistic of expected ozone depletion. Data for changes in the chemical composition of the litter, fungal community structure and nutrient release have not yet been fully analysed. However, we have found no marked or consistent UV-B effects on mass loss and the rate of microbial respiration.

The responses of different fungal species to increasing UV-B have

298   S.A. MOODY, D.J.S. COOP AND N.D. PAUL

been studied *in vitro* using decomposers isolated from *R. chamaemorus* litter. Effects on fungal growth and spore germination were investigated. To quantify the dose responses of fungal decomposers, fungi isolated from *R. chamaemorus* litter were grown on potato dextrose agar in a CE cabinet at 13 °C. Irradiances ranged from 23–27 W m^{-2} for white light and 7.7–14 W m^{-2} for UV-A. The UV-B dose (DNA weighted (UV-B$_{DNA}$) according to Setlow, 1974) was between 0 and 2.06 kJ m^{-2} d^{-1}, and was supplied over a 3-hour period in the middle of a 12-hour day. Effects on fungal growth were assessed by periodically measuring the diameter of growing fungal colonies before colonies reached the edge of the plate. Spore germination was examined by irradiating plates inoculated with a spore suspension and assessing percentage germination. Dose responses were assessed and percentage changes from ambient (1.3 kJ m^{-2} d^{-1} UV-B$_{DNA}$ approximating to mid-summer, clear-sky conditions at Lancaster) to that expected for a 15% ozone depletion (1.7 kJ m^{-2} d^{-1}) were calculated. Increasing UV-B had no stimulatory effect on fungal growth or spore germination in any of the 14 fungal species examined. Mycelial growth was significantly inhibited by increasing UV-B$_{DNA}$ in six out of eight species. The remaining two fungi were unaffected even by the highest dose used (2.06 kJ m^{-2} d^{-1}). Similar results were obtained when a further eight fungi were examined for spore germination, with again six species being significantly inhibited while two were unaffected. Percentage reductions in growth were small, with a maximum of 7% when the UV-B$_{DNA}$ dose was increased from ambient to that expected for a 15% ozone depletion. Spore germination was more severely affected, in one species being reduced by 37%.

These results show that fungal species intimately associated with *R. chamaemorus* decomposition have very different responses to an increase in UV-B from ambient levels to those expected for a 15% ozone depletion. While this work has been done *in vitro*, and it is possible that fungal responses may be different on litter, such reductions in spore germination and subsequent growth may leave some species unable to compete for limiting resources. Between-species variation in response may thus alter the outcome of competitive interactions, perhaps resulting in changes in fungal community structure. Changes in the microbial decomposer community might have knock-on effects not only for litter degradation but for the soil fauna associated with decomposition processes. Fungal interactions with soil fauna may also be influenced by changes in fungal morphology. Our *in vitro* studies have revealed that, in some cases, fungal growth was not only reduced by elevated UV-B, but that the mycelium was very different in structure. Mycelium grown at higher UV-B doses was less diffuse and fluffy in comparison to that growing under Mylar polyester. Such changes in morphology, and their impact on decomposer fauna, merit further investigation.

Herbivory

Natural populations of the heather psyllid (*Strophingia ericae* (Curtis)), were brought to Lancaster from the field site at Moor House on the heather plants. These populations were assessed using Tullgren funnel extraction (described by Salt *et al.*, 1996). Changes in *C. vulgaris* shoot morphology were assessed to determine elevated UV-B effects on the feeding sites of the insects. The distance between leaflets and the angle at which leaflets were subtended to the stem were measured under ×64 magnification for 75 1 cm long shoots from each treatment taken approximately 5 cm from the top of the plant. Before treatment started (season 1), there was no difference in psyllid numbers between plants; in the second, third and fourth seasons the number of psyllids found on plants exposed to elevated UV-B was reduced compared to controls. This reduction (approx. 60%) was significant on the fourth sampling occasion. Elevated UV-B had no effect on shoot morphology, in terms of the distance between leaflets and the angle at which leaflets subtended the stem, or on the nitrogen concentration, C : N ratio or the level of water soluble phenolics. However, increased UV-B tended to reduce shoot amino acids, with greatest differences being recorded for total levels and the concentration of isoleucine (22% and 26%, respectively), with the reduction in isoleucine being significant. The possible effects of elevated UV-B on insect herbivores have been investigated in relation to changes in host chemistry (Yazawa, Shimizu & Hirao, 1992; Hatcher & Paul, 1994; McCloud & Berenbaum, 1994). Thus, it is possible that the reduction seen in psyllid numbers is directly related to the changes recorded for isoleucine, which is an essential amino acid for insects. However, other possibilities such as a direct effect of elevated UV-B on the insects cannot be ruled out (Paul *et al.*, 1996). Further studies on insects in as natural conditions as possible are needed to clarify the effects of elevated UV-B on this aspect of ecosystems.

Conclusions

The effects of UV-B on plant growth are generally small in comparison with the effects of CO_2. Reductions in growth, and increased concentrations of UV-absorbing pigments in response to UV-B occurred after two seasons growth; this highlights the potential for longer-term cumulative UV-B effects in long-lived slow growing plants, as suggested by Johanson *et al.* (1995*b*), and stresses the importance of experiments in which treatments can be maintained for several years.

Although there was no statistically significant (i.e. greater than additive) interactions between CO_2 and UV-B, it was notable that their opposing effects tended to cancel each other. As a result, biomass under high CO_2 plus high UV-B treatments was within 2–3% of that for ambient UV-B plus ambient CO_2. Changes in R : S were one example, both in *C. vulgaris* and *V. myrtillus*, in which opposing effects of CO_2 and UV-B were noted. In several instances the effects of UV-B were reversed by CO_2 treatment and vice versa. It is also possible that interactions between CO_2 and UV-B other than those on plant physiology must be considered. Our results do not yet suggest any major effects of increased UV-B on decomposition in the species we have studied. However, any effect of UV-B or CO_2 on plant litter decomposition could, by altering nutrient availability for plant growth, interact with the direct physiological effects, with major effects on long-term responses of plants and vegetation to climate change. Thus, such changes, even if small in magnitude, could be of considerable importance in heathlands, given that these plant communities are growing on soils with very limited nutrient availability.

Ultimately, possible interactions between elevated CO_2 and increased UV-B highlight the difficulty in predicting long-term changes in natural vegetation from studies that isolate responses to a specific element of climate change. Beyond that, particularly in UK heathlands, such responses must also be considered against management practices. For example, increased partitioning to below-ground organs in *V. myrtillus* under elevated CO_2, or the opposite effect of increased UV-B, may affect regeneration following burning, a frequent management tool in UK heaths. Whilst our studies have not considered competitive interactions, there is undoubtedly potential for altered species balance both between different shrubs and between shrubs and other elements of heathland vegetation, including grasses and bryophytes. Such interactions, in common with many of the processes we have described here, can only be investigated through long-term experiments, which remain essential for any proper analysis of the impacts of climate change in heathland ecosystems.

Acknowledgements

We would like to thank Mark Bacon, Helen Beardall, Linda Hewkin, Jason Queally, Phil Smith and Richard Taylor for technical assistance and Dr Mark Pearson and Dr David Salt for useful comments on the manuscript. Funding for this work was provided by DoE Contract PECD 7/12/21 (NDP), EU Contract EV5V-CT910032 (DC) and NERC GR3 8634 (SAM).

References

Allen, S.E. (1989). *Chemical Analysis of Ecological Materials.* 2nd edn, Blackwell Scientific Publications, Oxford.

Anderson, J.M. & Ineson, P. (1982). A soil microcosm system and its application to measurements of respiration and nutrient leaching. *Soil Biology and Biochemistry,* **14**, 415–16.

Austin, J., Butchart, N. & Shine, K.P. (1992). Possibility of an Arctic ozone hole in a doubled CO_2 climate. *Nature,* **360**, 221–5.

Ayres, P.G., Gunasekera, T.S., Rasanayagam, M.S. & Paul, N.D. (1996). Effects of UV-B radiation (280–320 nm) on foliar saprotrophs and pathogens. In *Fungi and Environment Change* (Gadd, G., Frankland, J. & N. Magan, N., eds), pp. 32–50. Cambridge University Press, Cambridge.

Björn, L.O. & Murphy, T.M. (1985). Computer calculations of solar ultraviolet radiation at ground level. *Physiologia Vegetale,* **23**, 555–61.

Björn, L.O., Callaghan, T.V., Johnsen, I., Lee, J.A., Mantas, Y., Paul, N.D., Sonesson, M., Wellburn, A.R., Coop, D., Heide-Jørgensen, H.S., Gehrke, C., Gwynn-Jones, D., Johanson, U., Kyparissis, A., Levizou, E., Nikolopoulos, D., Petropoulou, Y. & Stephanou, M. (1997). The effects of UV-B radiation on European heathland species. *Plant Ecology,* (in press)

Blumthaler, M.W., Ambach, R., Sibernagl, R. & Staehelin, J. (1994). Erythemal UV-B (Robertson–Berger sunburn meter data) under ozone deficiencies in winter/spring 1993. *Photochemistry and Photobiology,* **59**, 657–9.

Caldwell, M.M. (1971). Solar UV irradiation and the growth and development of higher plants. In *Photophysiology.* (Giese, A.C., ed.), pp. 131–77. Academic Press, New York.

Caldwell, M.M. & Flint, S.D (1994). Stratospheric ozone reduction, solar UV-B radiation and terrestrial ecosystems. *Climate Change,* **28**, 375–94.

Cotrufo, M.F., Ineson, P. & Rowland, A.P. (1994). Decomposition of tree leaf litters grown under elevated CO_2: effect of litter quality. *Plant and Soil,* **163**, 121–30.

Cotrufo, M.F. & Ineson, P. (1995). Effects of enhanced atmospheric CO_2 and nutrient supply on the quality and subsequent decomposition of fine roots of *Betula pendula* Roth. and *Picea sitchensis* (Bong.) Carr. *Plant and Soil,* **170**, 267–77.

Doherty, M., Salt, D.T. & Holopainen, J. (1997). The impacts of climate change and pollution on forest pests. In *Forests and Insects*. (Watt, A.D., Stork, N.E. & Hunter, M.D., eds), 18th Symposium of the Royal Entomological Society. Chapman and Hall, London. In press.

Fiscus, E.L & Booker, F.L. (1995). Is increased UV-B a threat to crop photosynthesis and productivity? *Photosynthesis Research*, **43**, 81–92.

Garbutt, K. & Bazzaz, F.A. (1984). The effects of elevated CO_2 on plants. III. Flower, fruit and seed production and abortion. *New Phytologist*, **98**, 433–46.

Gehrke, C., Johanson, U. Callaghan, T.V., Chadwick, D. & Robinson, C.H. (1995). The impact of enhanced ultraviolet-B radiation on litter quality and decomposition processes in *Vaccinium* leaves from the Subarctic. *Oikos*, **72**, 213–22.

Greene, O. (1995). Emerging challenges for the Montreal Protocol. *The Globe*, **27**, 5–6.

Grime, J.P., Hodgson, J.G. & Hunt, R. (1988). *Comparative Plant Ecology*. Unwin Hyman Ltd, London.

Harrison, F. (1971). The inhibitory effect of oak leaf litter tannins on the growth of fungi in relation to litter decomposition. *Soil Biology and Biochemistry*, **3**, 167–72.

Hatcher, P.E. & Paul, N.D. (1994). The effect of elevated UV-B radiation on herbivory of pea by *Autographa gamma*. *Entomologia Experimentalis et Applicata*, **71**, 227–33.

Hellmers, H. & Hesketh, J.D. (1973). Floral initiation in four plant species growing in CO_2-enriched air. *Environmental Control Biology*, **28**, 35.

Herman, J.R. & Larko, D. (1994). Low ozone amounts during 1992–1993 from Nimbus 7 and Meteor 3 total ozone mapping spectrometers. *Journal of Geophysical Research*, **99**, 3483–96.

Herman, J.R., McPeters, R. & Larko, D. (1993). Ozone depletion at northern and southern latitudes derived from January 1979 to December 1991 total ozone mapping spectrometer data. *Journal of Geophysical Research*, **98**, 12783–93.

Johanson, U., Gehrke, C., Björn, L.O. & Callaghan, T.V. (1995*a*). The effects of enhanced UV-B radiation on the growth of dwarf shrubs in a subarctic heathland. *Functional Ecology*, **9**, 713–19.

Johanson, U., Gehrke, C., Björn, L.O., Callaghan, T.V. & Sonesson, M. (1995*b*). The effects of enhanced UV-B radiation on a subarctic heath ecosystem. *Ambio*, **24**, 106–11.

Jokela, K., Leszczynski, K., Visuir R. & Ylianttlia L. (1995). Increased UV exposure in Finland. *Photochemistry and Photobiology*, **62**, 101–7.

Jordan, A. (1995). Ozone endgames: the implementation of the Montreal Protocol at the sub-national level. *The Globe*, **27**, 6–8.

Krupa, S.V. & Kickert, R.N. (1989). The greenhouse effect: impacts of ultraviolet-B (UV-B) radiation, carbon dioxide (CO_2), and ozone (O_3) on vegetation. *Environmental Pollution*, **61**, 263–93.

McCloud, E.S. & Berenbaum, M.R. (1994). Stratospheric ozone depletion and plant–insect interactions: effects of UVB radiation on foliage quality of *Citrus jambhiri* for *Trichoplusia ni*. *Journal of Chemical Ecology*, **20**, 525–39.

Madronich, S., McKenzie, R.L., Caldwell, M.M. & Björn, L.O. (1995). Changes in ultraviolet radiation reaching the earth's surface. *Ambio*, **24**, 143–52.

Mepsted, R., Paul, N.D., Stephen, J., Corlett, J.E., Nogues, S., Baker, N.R, Jones, H.G. & Ayres, P.G. (1996). Effects of enhanced UV-B radiation on pea (*Pisum sativum* L.) grown under field conditions in the United Kingdom. *Global Change Biology*, **2**, 325–34.

Moorhead, D.L & Callaghan, T. (1994). Effects of increasing ultraviolet B radiation on decomposition and soil organic matter dynamics: a synthesis and modelling study. *Biology and Fertility of Soils*, **18**, 19–26.

Niu, X., Frederick, J.E., Stein, M.L. & Tiao, G.C. (1992). Trends in column ozone based on TOMS data: dependence on month, latitude and longitude. *Journal of Geophysical Research*, **97**, 14661–9.

Paul, N.D., Rasanayagam, M.S., Moody, S.A., Hatcher, P.E. & Ayres, P.G. (1997). The role of interactions between trophic levels in determining the effects of UV-B on terrestrial ecosystems. *Plant Ecology*, **128**, 1–13.

Peñuelas, J. & Matamala, R. (1990). Changes in N and S leaf content, stomatal density and specific leaf area of 14 plant species during the last three centuries of CO_2 increase. *Journal of Experimental Botany*, **14**, 1119–24.

Reinsel, G.C., Tiao, G.C., Wuebbles, D.J., Kerr, J.B., Miller, A.J., Nagatani, R.M., Bishop, L. & Ying, L.H. (1994). Seasonal trend analysis of published ground-based and TOMS total ozone data through 1991. *Journal of Geophysical Research*, **99**, 5449–64.

Rodwell, J.S. (1991). *British Plant Communities. Volume 2, Mires and Heaths*. Cambridge University Press, Cambridge.

Rozema, J., Lenssen, G.M. & Staaij, J.W.M. van de (1990) The effect of increased CO_2 and UV-B radiation on some agricultural and salt marsh species. In *The Greenhouse Effect and Primary Production in European Agro-ecosystems*. (Goudriaan, J., van Keulen, H. & Van Laar, H.H., eds), pp. 68–71. Pudoc, Wageningen.

Salt, D.T., Brooks, G.L. & Whittaker, J.B. (1995). Elevated carbon dioxide affects leaf-miner performance and plant growth in docks (*Rumex* spp.). *Global Change Biology*, **1**, 153–6.

Salt, D.T., Major, E. & Whittaker, J.B. (1996). Population dynamics of root aphids on Sitka spruce at two British sites. *Pedobiologia*, **40**, 1–11.

Seckmeyer, G., Mayer, B., Erb, R. & Bernhard, G. (1994). UV-B in Germany higher in 1993 than in 1992. *Geophysical Research Letters*, **21**, 577–80.

Setlow, R.B. (1974). The wavelengths in sunlight effective in producing skin cancer: a theoretical analysis. *Proceedings of the National Academy of Sciences, USA*, **71**, 3363–6.

Slack, G. (1986). CO_2 enrichment of tomato crops. In *Carbon Dioxide Enrichment of Greenhouse Crops. Vol. 2, Physiology, Yield, and Economics*. (Enoch, H.Z. & Kimball, B.A. eds). CRC Press Inc., Florida, USA.

Stolarski, R., Bojkov, R., Bishop, L., Zerephos, C., Staehelin, J. & Zawodny, J. (1992). Measured trends in stratospheric ozone. *Science*, **256**, 342–9.

Sullivan, J.H. & Teramura, A.H. (1994). The effects of UV-B radiation on loblolly pine. 3. Interaction with CO_2 enhancement. *Plant Cell and Environment*, **17**, 311–17.

Swift, M.J., Heal, O.W. & Anderson, J.M. (1979). *Decomposition in Terrestrial Ecosystems*. Blackwell, Oxford.

Teramura, A.H., Sullivan, J.H. & Ziska, L.H. (1990). Interaction of elevated ultraviolet-B radiation and CO_2 on productivity and photosynthetic characteristics in wheat, rice and soybean. *Plant Physiology*, **94**, 470–5.

Townend, J. (1993). Effects of elevated carbon dioxide and drought on the growth and physiology of clonal Sitka spruce plants (*Picea sitchensis* (Bong.) Carr. *Tree Physiology*, **13**, 389–99.

Tubbs, C. (1985). The decline and present status of the English lowland heaths and their vertebrates. *Focus on Nature Conservation No. 11*. Nature Conservancy Council.

Watt, A.D., Whittaker, J.B., Doherty, M., Brooks, G.L., Lindsay, E. & Salt, D.T. (1995). The impact of elevated atmospheric CO_2 on insect herbivores. In *Insects and Climate Change*. 17th Symposium of the Royal Entomological Society (Harrington, R. & Stork, N.E., eds), pp. 197–217. Academic Press, London.

Woodin, S., Graham, B., Killick, A., Skiba, U. & Cresser, M. (1992). Nutrient limitation of the long term response of heather (*Calluna vulgaris* (L.) Hull) to CO_2 enrichment. *New Phytologist*, **122**, 635–42.

Yazawa, M., Shimizu, T. & Hirao, T. (1992). Feeding response of the silkworm *Bombyx mori* to UV radiation of mulberry leaves. *Journal of Chemical Ecology*, **18**, 561–9.

Zepp, R.G., Callaghan, T.V. & Erickson, D.J. (1995). Effects of increased solar ultraviolet radiation on biogeochemical cycles. *Ambio*, **24**, 181–7.

M.M. CALDWELL

Alterations in competitive balance

Interspecific competition

Ecosystem-level influences of increased solar UV-B (280–320 nm) radiation are only beginning to be explored. Predicting interactions of species in ecosystems based only on studies of plants in isolation is difficult or often impossible. Changes in the competitive balance of higher plants can result when species mixtures are exposed to different levels of UV-B as explained later. How might this come about? Intuitively, one might postulate that if a species is more detrimentally affected by increased solar UV-B than other species, its competitive position in the community would suffer. This might be reasonable if sizeable differences between species in UV-B damage arise. However, a review of the relevant literature suggests that, when tested under field conditions, most higher plants do not suffer appreciable damage when given additional UV-B simulating reasonable changes in the stratospheric ozone layer (Caldwell & Flint, 1993, 1994). While little apparent detrimental effect of elevated UV-B is apparent in monocultures of higher plants, there are several reports that the competitive balance between plant species can be altered when additional UV-B is given to mixtures of species (Fox & Caldwell, 1978; Gold & Caldwell, 1983; Barnes *et al.*, 1988) or even when solar UV is largely excluded (Bogenrieder & Klein, 1982).

Higher plants appear to be reasonably well protected against damage or growth inhibition by UV-absorbing pigments and damage repair systems (Tevini, Braun & Fieser, 1991; Li *et al.*, 1993; Britt, 1995; Britt *et al.*, 1993; Fiscus & Booker, 1995). The argument might still be made that, even if plants are only detrimentally affected to a small degree, this might be sufficient to effect changes in competitive balance. Or alternatively, effects on an early life stage, such as subtle depression of seedling development, might magnify into significant differences in competitive balance later. Such arguments cannot be dismissed. However, apart from detrimental effects, UV-B is thought to elicit morphological changes and alterations of plant secondary chemistry, especially in phenolic constituents (see chapter by Bornman *et al.*, this volume). These changes are much more consistently documented in the literature dealing with plants and UV-B radiation under field conditions than are indications of damage (Caldwell & Flint, 1994). Whether these responses should be categorised as deleterious or simply as changes in chemical and structural allocation is a matter of discussion. However, there are several lines of evidence suggesting these to be allocation changes rather than damage *per se* (Ballaré, Barnes & Kendrick, 1991; Ballaré, Barnes & Flint, 1995; Ballaré *et al.*, 1995; Goto, Yamamoto & Watanabe, 1993; Steinmetz & Wellmann, 1986; Ensminger, 1993). Changes in allocation have the potential to result in competitive balance changes as will be developed below.

Potential mechanisms – mediation by a UV-B receptor

A specific UV-B photoreceptor has been proposed and some evidence presented that it may be a flavin or flavin-like compound (Ballaré *et al.*, 1995*a,b*). (More than a single UV-B photoreceptor is also possible.) Action spectra suggest that appreciable absorption occurs primarily in the UV-B, a very narrow spectral band of solar radiation (Wellmann, 1983; Goto *et al.*, 1993). It is an intriguing question why such a UV-B photoreceptor, or receptors, evolved in plants. Information gleaned by plants through this photoreceptor may well supplement that obtained by the better understood photoreceptors, phytochrome and the blue/UV-A receptor. The UV-B component of sunlight is less tightly coupled with the remainder of the solar spectrum than is, for example, the longer wavelength UV-A (320–400 nm) (Caldwell & Flint, 1994). Thus, information to the plant on UV-B solar flux may not always be garnered from information on flux of other parts of the solar spectrum detected by the better known photoreceptors. This is certainly the case with ozone reduction, the effect of which would only be reflected in the UV-B portion of the spectrum. The increased synthesis of UV-B-absorbing phenolics, thought to be mediated by the UV-B receptor, would have obvious benefit for plants as ozone is reduced. Reduced hypocotyl extension by UV-B has been proposed as a mechanism to reduce exposure of seedlings to sunlight at the soil surface while synthesis of UV-B-absorbing compounds is under way (Ballaré *et al.*, 1995*b*).

Competition for light

While these are plausible advantages of a UV-B photoreceptor system for individual plants, there are several other consequences of the resulting morphological and secondary chemistry changes at the community and ecosystem level. If morphological changes mediated by a UV-B photoreceptor are largely a matter of altered allocation, total growth or productivity of plants may not be changed. This appears to be the case, when monoculture stands are exposed to additional UV-B in the field (Barnes *et al.*, 1988; Barnes, Flint & Caldwell, 1995). Small reductions in internode elongation apparently do not affect the total stand photosynthesis or biomass accumulation (Barnes *et al.*, 1990*a*). Yet, if plants are growing in mixed-species stands, rather than in single-species stands, even rather subtle morphological changes may alter the balance of competition among species (Bogenrieder & Klein, 1982; Gold & Caldwell, 1983). This was tested in a seven-year study of competition between wheat and a common weed of wheat, wild oat (*Avena fatua*) (Barnes *et al.*, 1988; Barnes, Flint & Caldwell, 1995). Over these years, there was a reasonably consistent shift in competitive balance of 50 : 50 mixtures of these two grass species, in favour of the wheat. Testing of potential photosynthesis inhibition under severe conditions did not reveal any detrimental effect of UV-B (Beyschlag *et al.*, 1988). In monocultures of either species, there were no effects on stand yield of the elevated UV-B radiation, nor was there an effect on total mixture yield. Only the proportion of wheat and wild oat was shifted. Interestingly, there were no effects on the timing or rate of seedling emergence in either species due to the supplemental UV-B treatment. It was also apparent that the display of foliage in the canopy was shifted to different heights, largely resulting from somewhat shorter internodes in both species. This internode reduction was more pronounced in the wild oat than in the wheat such that the relative display of foliage with height in the canopy was altered in the mixture (Fig. 1, Barnes *et al.*, 1988; Barnes *et al.*, 1995). From this information, it was postulated that the shift in foliage display by the two species in the mixtures given extra UV-B was sufficient to shift the balance of competition for light to drive photosynthesis.

A quantitative evaluation of the light competition hypothesis

However, are these subtle changes sufficient to quantitatively explain a shift in competition for light? There are two factors that can amplify the importance of even small changes in the relative distribution of foliage in the upper portions of canopies. First, light extinction is exponential in plant canopies and most of the extinction occurs in the uppermost leaf layers. Secondly, the photosynthetic capacity of foliage in the upper leaves is usually much greater than for leaves lower in the

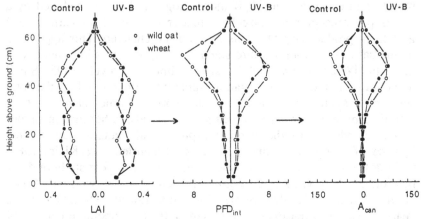

Fig. 1. Vertical canopy profiles of measured leaf area index (LAI), calculated light interception by foliage (PFD$_{int}$, mol photons m^{-2} ground area day^{-1}) and net canopy photosynthesis (A$_{can}$, mmol CO$_2$ m^{-2} ground area day^{-1}) for wheat and wild oat in mixed plantings under control and UV-B supplementation treatments. (From Barnes *et al.*, 1995, 1996.)

canopy. A canopy model was developed to quantitatively evaluate the importance for competition of measured leaf area position changes in the canopies due to the elevated UV-B treatment (Ryel *et al.*, 1990). This model predicts light penetration in plant stands and interception of radiation by leaves and stems of each species in a multi-species mixture for each layer in the canopy. It takes into account the direct beam solar radiation, the diffuse skylight coming from different sectors of the sky hemisphere overhead and diffuse light reflected by and transmitted through the foliage. The radiation predictions are coupled with a single-leaf photosynthesis model. Photosynthesis and transpiration are calculated for each foliage class in as many as 17 canopy layers and summed (weighted by foliage area) to calculate single-layer and total-canopy gas exchange. Foliage class refers to leaves or stems of each species; in this example there are four foliage classes.

Considerable data collection was required to parameterise the model for the specific plants and conditions and to independently validate the model predictions. The canopy architecture of each treatment at several times in the year was evaluated with a fibre-optic point quadrat system (Caldwell, Harris & Dzurec, 1983). This sampling approach could non-destructively provide quantitative information on the distribution of each foliage class in each canopy layer. Single-leaf (or stem) gas

exchange measurements under prescribed temperature and light conditions were collected in the field for intact foliage elements in the
canopy (Beyschlag *et al.*, 1988, 1990; Barnes *et al.*, 1990*a*). This information was used to evaluate the physiological photosynthetic capacity
of leaves and stems at different canopy levels for different times of the
season and to assess actual photosynthesis and transpiration rates for
foliage elements in the canopy under specific conditions.

Validation of the light interception computations was conducted in
mixtures and monocultures of the two species in the field. This involved
measurements with arrays of photodiodes inserted at different positions
and angles in the canopies to assess how well the model was predicting
light on sunlit and shaded foliage located at specific heights and different compass angles in the canopies. It also provided a validation of
model predictions of the fraction of sunlit and shaded foliage in different layers of the canopy. In the model, sunlit and shaded foliage photosynthesis rates are calculated separately and then subsequently summed
to avoid bias in the computations. Photosynthesis and transpiration
measurements of individual leaves and stems in the canopies mentioned
above were used for model validation as well.

Returning to Fig. 1, an example is presented as to how small shifts
in foliage position in the mixed-species canopies resulting from
increased UV-B can translate into differences in intercepted sunlight by
foliage of the two species and finally into differences in photosynthesis
for different canopy layers. As shown in Fig. 1, small changes in foliage
position in the upper canopy layers can be amplified in proportionately
greater changes in light interception and foliage photosynthesis. This is
the result of the exponential nature of light attentuation in canopies.
Thus, high in the canopy, light is rapidly attenuated and uppermost
foliage layers have a distinct advantage. Also, the upper foliage layers
are most competent in photosynthetic capacity. In Fig. 2, the total
canopy photosynthesis is shown over the course of a clear July day to
indicate the effects of the UV-B treatment. The small advantage in daily
carbon gain provided to the wheat plants by their better position in the
canopy resulting from the UV-B treatment would accumulate through
the season and be sufficient to explain the changes in competitive balance found earlier (Barnes *et al.*, 1988).

In this example, the crop plant was favoured by the increased UV-B.
This should not lead to any generalisation concerning the relative
advantages or disadvantages for crop plants in competition with weeds
that might result from ozone reduction. With other species, the situation
could well be reversed. Can any general predictions be made? Shifts in
the competitive balance may often follow rather subtle changes in

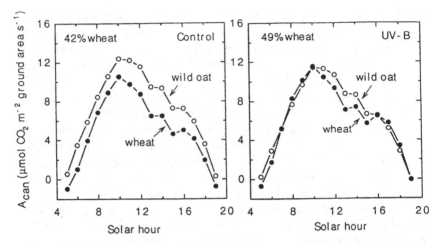

Fig. 2. Mid-season calculated net canopy photosynthesis (A_{can}) for wheat and wild oat in mixed plantings under control and UV-B supplementation treatments. Units as in Figure 1. The percentage of total daily mixture canopy photosynthesis contributed by wheat is also indicated. (From Barnes *et al.*, 1995, 1996).

canopy architecture and, therefore, they would most likely take place in dense canopies where light competition is pronounced and the species are of similar stature. Morphological responses to elevated UV-B vary with species (Barnes *et al.*, 1990b) and some predictions could be made from the relative responsiveness of individual species. Monocots appear to be generally more responsive than dicots (Barnes *et al.*, 1990b).

Other potential mechanisms of interspecific competition change

Apart from differential deleterious effects of UV-B on competing species mentioned earlier, other changes in morphological or chemical allocation, likely mediated by the UV-B receptor, could play a role. For example, there are several reports that root/shoot ratios can be altered by increased UV-B. Since root competition is often more severe than above-ground competition for light (Caldwell, 1987), changes in root/shoot ratio might be predicted to have consequences for competitive balance (Tilman, 1988). When root/shoot ratio changes have been reported as statistically significant, they have been positive and negative with about equal frequency. The magnitude of these changes has been generally less than 15% (Barnes *et al.*, 1993; Sullivan & Teramura,

1989, 1992, 1994; Sullivan, Teramura & Dillenburg, 1994; Ziska & Teramura, 1992,). However, there are a few reports of large root/shoot ratio changes of the order of 30% or more (Ziska & Teramura, 1992; Ziska et al., 1993). Whether these changes in root/shoot ratios would actually eventuate in nature is not known and their consequences for competitive balance would only be speculation at this point. Potentially, UV-B alterations of various products of the phenylpropanoid pathway in plants (Dixon & Paiva, 1995) could lead to changes in plant inter-ference, including those mediated by allelopathy. However, too little is known about the effective role of such chemicals in plant–plant interac-tions to speculate on a specific role that elevated UV-B might play.

A potential role for UV-B in intraspecific competition

Intraspecific competition for light is often termed asymmetric in that larger plants gain advantage disproportionate to their size (Weiner, 1990). This is also sometimes termed one-side competition or snowball competition. The basis of this is that light is basically a unidirectional resource and larger plants shade smaller individuals, but smaller plants have little influence on larger plants. Thus, small height advantages can translate into considerable advantage over time and this also means that size inequality in the monospecific population will develop in dense stands where competition for light is strong. When competition is for other resources, it is considered to be symmetric in nature, that is, larger individuals do not have an advantage that is greater than would be predicted from their size (Weiner, 1990). Ballaré et al. (1994) showed that the phytochrome system tends to counteract the tendency for size inequality to develop in dense populations. Mechanistically, it is thought that, in populations with normal phytochrome function, red light intercepted higher in the canopy would tend to reduce internode length while far-red light resulting from shade deeper in the canopy would tend to increase internode length. Size inequality would develop, but less rapidly than in populations where some aspect of phytochrome function was not effective.

Since the UV-B receptor results in some morphological changes that are rather similar to those effected by phytochrome, Searles et al.. (unpublished observations) recently explored the idea that increased UV-B might also function to reduce size inequality in monospecific stands. Tests of this hypothesis tend to support this contention.

Asymmetric competition has many implications for plant population biology and competition theory (Weiner, 1990). Also, in agricultural practice, size inequality can be important in weed control, canopy fruit and seed development, harvesting schedules and processes, and overall economic yield.

Conclusion

Higher-order effects of elevated solar UV-B in communities and eco-systems are not easily predicted from research on plants in isolation. In the cases presented here, elevated UV-B appears to exert its greatest influence through indirect effects on light competition. This influence only became apparent by research with mixed-species systems, although in retrospect, some of these indirect effects of UV-B on competitive balance might have been predicted from the study of plants in isolation. Similarly, other ecosystem-level effects of elevated UV-B will necessitate work with more complex systems and the accompanying difficulties in sorting out specific influences of UV-B will require new approaches to be most effective.

Acknowledgements

I thank Stephan D. Flint for his review of a draft of the manuscript and his help in compiling some of the information. Some of the research described was supported by grants of the US Environmental Protection Agency, the US Department of Agriculture (CSRS/NRICG grants 92-37100-7630 and 95-37100-1612) and the Utah Agricultural Experiment Station.

References

Ballaré, C.L., Barnes, P.W. &. Kendrick, R.E. (1991). Photomorpho-genic effects of UV-B radiation on hypocotyl elongation in wild type and stable-phytochrome-deficient mutant seedlings of cucumber. *Physiologia Plantarum*, **83**, 652–8.

Ballaré, C.L., Scopel, A.L., Jordan, E.T. & Vierstra, R.D. (1994). Signaling among neighboring plants and the development of size inequalities in plant populations. *Proceedings of the National Academy of Sciences, USA*, **91**, 10094–8.

Ballaré, C.L., Barnes, P.W. & Flint, S.D. (1995a). Inhibition of hypocotyl elongation by ultraviolet-B radiation in de-etiolating tomato seedlings. I. The photoreceptor. *Physiologia Plantarum*, **93**, 584–92.

Ballaré, C.L., Barnes, P.W., Flint, S.D. & Price, S. (1995b). Inhibition of hypocotyl elongation by ultraviolet-B radiation in de-etiolating tomato seedlings. II. Time-course, comparison with flavonoid responses and adaptive significance. *Physiologia Plantarum*, **93**, 593–601.

Barnes, P.W., Ballaré, C.L. & Caldwell, M.M. (1996). Photomorphogenic effects of UV-B radiation in plants: consequences for light competition. *Journal of Plant Physiology*, **148**, 15–20.

Barnes, P.W., Jordan, P.W., Gold, W.G., Flint, S.D. & Caldwell, M.M. (1988). Competition, morphology and canopy structure in wheat (*Triticum aestivum* L.) and wild oat (*Avena fatua* L.) exposed to enhanced ultraviolet-B radiation. *Functional Ecology*, **2**, 319–30.

Barnes, P.W., Beyschlag, W., Ryel, R., Flint, S.D. & Caldwell, M.M. (1990a). Plant competition for light analyzed with a multispecies canopy model. III. Influence of canopy structure in mixtures and monocultures of wheat and wild oat. *Oecologia*, **82**, 560–6.

Barnes, P.W., Flint, S.D. & Caldwell, M.M. (1990b). Morphological responses of crop and weed species of different growth forms to ultraviolet-B radiation. *American Journal of Botany*, **77**, 1354–60.

Barnes, P.W., Maggard, S., Holman, S.R. & Vergara, B.S. (1993). Intraspecific variation in sensitivity to UV-B radiation in rice. *Crop Science*, **33**, 1041–6.

Barnes, P.W., Flint, S.D. & Caldwell, M.M. (1995). Early-season effects of supplemented solar UV-B radiation on seedling emergence, canopy structure, simulated stand photosynthesis and competition for light. *Global Change Biology*, **1**, 43–53.

Beyschlag, W., Barnes, P.W., Flint, S.D. &. Caldwell, M.M. (1988). Enhanced UV-B irradiation has no effect on photosynthetic characteristics of wheat (*Triticum aestivum* L.) and wild oat (*Avena fatua* L.) under greenhouse and field conditions. *Photosynthetica*, **22**, 516–25.

Beyschlag, W., Barnes, P.W., Ryel, R., Caldwell, M.M. & Flint, S.D. (1990). Plant competition for light analyzed with a multispecies canopy model. II. Influence of photosynthetic characteristics on mixtures of wheat and wild oat. *Oecologia*, **82**, 374–80.

Bogenrieder, A. & Klein, R. (1982). Does solar UV influence the competitive relationship in higher plants? In *The Role of Solar Ultraviolet Radiation in Marine Ecosystems*. (Calkins, J., ed.), pp. 641–49. Plenum Press, New York.

Britt, A.B. (1995). Repair of DNA damage induced by ultraviolet radiation. *Plant Physiology*, **108**, 891–6.

Britt, A.B., Chen, J.J., Wykoff, D. & Mitchell, D. (1993). A UV-sensitive mutant of *Arabidopsis* defective in the repair of pyrimidine-pyrimidinone (6–4) dimers. *Science*, **261**, 1571–4.

Caldwell, M.M. (1987). Competition between root systems in natural communities. In *Root Development and Function*. (Gregory, P.J., Lake, J.V. & Rose, D.A., eds) pp. 167–85. Cambridge University Press, Cambridge.

Caldwell, M.M. & Flint, S.D. (1993). Implications of increased solar UV-B for terrestrial vegetation. In *The Role of the Stratosphere in*

314   M.M. CALDWELL

Global Change. (Chanin, M. L., ed.), pp. 495–516. Springer-Verlag, Heidelberg.

Caldwell, M.M. & Flint, S. D. (1994). Stratospheric ozone reduction, solar UV-B radiation and terrestrial ecosystems. *Climatic Change*, **28**, 375–94.

Caldwell, M.M., Harris, G. W. & Dzurec, R.S. (1983). A fiber optic point quadrat system for improved accuracy in vegetation sampling. *Oecologia*, **59**, 417–18.

Dixon, R. & Paiva, N.L. (1995). Stress-induced phenylpropanoid metabolism. *The Plant Cell*, **7**, 1085–97.

Ensminger, P.A. (1993). Control of development in plants and fungi by far-UV radiation. *Physiologia Plantarum*, **88**, 501–8.

Fiscus, E.L. & Booker, F.L. (1995). Is increased UV-B a threat to crop photosynthesis and productivity? *Photosynthesis Research*, **43**, 81–92.

Fox, F.M. & Caldwell, M.M. (1978). Competitive interaction in plant populations exposed to supplementary ultraviolet-B radiation. *Oecologia*, **36**, 173–90.

Gold, W.G. & Caldwell, M.M. (1983). The effects of ultraviolet-B radiation on plant competition in terrestrial ecosystems. *Physiologia Plantarum,* **58**, 435–44.

Goto, N., Yamamoto, K.T. & Watanabe, M. (1993). Action spectra for inhibition of hypocotyl growth of wild-type plants and of the hy2 long-hypocotyl mutant of *Arabidopsis thaliana* L. *Photochemistry and Photobiology*, **57**, 867–71.

Li, J.Y., Oulee, T.M., Raba, R., Amundson, R.G. & Last, R.L. (1993). *Arabidopsis* flavonoid mutants are hypersensitive to UV-B irradiation. *Plant Cell*, **5**, 171–9.

Ryel, R.J., Barnes, P.W., Beyschlag, W., Caldwell, M.M. & Flint, S.D. (1990). Plant competition for light analyzed with a multispecies canopy model. I. Model development and influence of enhanced UV-B conditions on photosynthesis in mixed wheat and wild oat canopies. *Oecologia*, **82**, 304–10.

Steinmetz, V. & Wellmann, E. (1986). The role of solar UV-B in growth regulation of cress (*Lepidium sativum* L.) seedlings. *Photochemistry and Photobiology*, **43**, 189–93.

Sullivan, J.H. & Teramura, A.H. (1989). The effects of ultraviolet-B radiation on loblolly pine. 1. Growth, photosynthesis and pigment production in greenhouse-grown seedlings. *Physiologia Plantarum* **77**, 202–7.

Sullivan, J.H. & Teramura, A.H. (1992). The effects of ultraviolet-B radiation on loblolly pine. 2. Growth of field-grown seedlings. *Trees*, **6**, 115–20.

Sullivan, J.H. & Teramura, A.H. (1994). The effects of UV-B radiation on loblolly pine. 3. Interaction with CO_2 enhancement. *Plant Cell and Environment*, **17**, 311–17.

Sullivan, J.H., Teramura, A.H. & Dillenburg, L.R. (1994). Growth and photosynthetic responses of field-grown sweetgum (*Liquidambar styraciflua*; Hamamelidaceae), seedlings to UV-B radiation. *American Journal of Botany*, **81**, 826–32.

Tevini, M., Braun, J. & Fieser, G. (1991). The protective function of the epidermal layer of rye seedlings against ultraviolet-B radiation. *Photochemistry and Photobiology*, **53**, 329–33.

Tilman, D. (1988). *Plant Strategies and the Dynamics and Structure of Plant Communities*. Princeton University Press, Princeton.

Weiner, J. (1990). Asymmetric competition in plant populations. *Trends in Ecology and Evolution*, **5**, 360–4.

Wellmann, E. (1983). UV radiation in photomorphogenesis. In *Encyclopedia of Plant Physiology. Photomorphogenesis.* Volume 16B (New Series). (Shropshire, W.J. & Mohr, H., eds), pp. 745–56. Springer-Verlag, Berlin.

Ziska, L.H. & Teramura, A.H. (1992). CO_2 enhancement of growth and photosynthesis in rice (*Oryza sativa*). Modification by increased ultraviolet-B radiation. *Plant Physiology*, **99**, 473–81.

Ziska, L.H., Teramura, A.H., Sullivan, J.H. & McCoy, A. (1993). Influence of ultraviolet-B (UV-B) radiation on photosynthetic and growth characteristics in field-grown cassava (*Manihot esculentum* Crantz). *Plant Cell and Environment*, **16**, 73–9.

N.D. PAUL

Interactions between trophic levels

Introduction

The effects of UV-B on interactions between plants and their consumers, herbivores and the micro-organisms that cause disease or bring
about decomposition of dead tissues, have been the subject of speculation for a number of years (Caldwell *et al.*, 1989), and are increasingly
seen to be of potential importance in assessing the consequences of
ozone depletion on agriculture and natural ecosystems. Unfortunately,
the possible importance of host–consumer interactions in determining
the impacts of ozone depletion remain very difficult to assess owing to
the scarcity of relevant experimental investigations. It has been hypoth-
esised that increasing UV-B will alter plant–consumer interactions
owing to changes in secondary plant metabolism (for example, Caldwell
et al., 1995). While this hypothesis points to a rather consistent decrease
in herbivory, disease or decomposition, no such consistency is apparent
in the few studies which have been published. Indeed, current data tends
to highlight the great diversity in responses to UV-B. In this chapter
the diversity of responses which have been observed, and the mechanisms which might underlie these variations, will be considered, and
will concentrate here on the effects of UV-B on fungi which occur as
saprotrophs in the phylloplane and, especially, which cause plant disease, since a) there is slightly more published literature relating to the
UV-B responses of these organisms, and b) they have been the main
focus of studies at Lancaster. The more general issue of how the choice
of action spectra influences the interpretation of experimental investigations of plant–microbe interactions will also be considered.

UV-B effects on plant disease: a study in diversity

The effects of UV-B on the interaction between cultivated wheat
(*Triticum aestivum*) and fungal pathogens of the imperfect genus *Septoria* have been studied in detail at Lancaster. *Septoria tritici* (perfect
stage *Mycosphaerella graminicola*), and *S. nodorum* (perfect stage

Leptosphaeria nodorum) are the causal agents of, respectively, leaf blotch and glume blotch, both diseases causing significant yield loss in commercial wheat crops. The wheat-*S. tritici* system appears to be a relatively straightforward example of how UV-B may affect plant disease. During spring in the field, a simple square wave treatment, which increased ambient DNA-weighted UV-B (UV-B$_{DNA}$: Setlow, 1974) by 40%, significantly reduced the number of *S. tritici* lesions that developed following artificial inoculation. In order to separate the effects of increased UV-B on the host and pathogen, wheat was grown at either 1.4 kJ m^{-2} d^{-1} or 2.0 kJ m^{-2} d^{-1} UV-B$_{DNA}$ (equivalent to modelled daily doses for summer at 55° N under clear-sky conditions and 'ambient' ozone and a 15% ozone depletion, respectively) pre- and/or post-inoculation with a UK isolate of *S. tritici*. Pre-inoculation UV-B had no effect on infection. By contrast, high UV-B post-inoculation resulted in a highly significant inhibition of infection. These responses could be explained by the direct effects of UV-B on the pathogen since the germination of conidia and germ tube growth are strongly inhibited by UV-B both *in vitro* (Rasanayagam *et al.*, 1995) and on leaf surfaces (Paul *et al.*, 1997). Thus, changes in infection are dependent only on the response of the pathogen; changes in the host, if they occurred, appeared to have no overall effect.

Unfortunately, the simplicity of UV-B effects on the wheat-*S. tritici*, interaction appears to be the exception rather than the rule. Indeed, even in this system, it was evident that responses to UV-B could be altered by a number of factors. First, the response to UV-B observed in the field during spring did not occur in summer. Although the reasons for this discrepancy are not clear, one possibility was that higher temperatures during summer, especially at night, permitted infection to be completed during darkness. In effect, the vulnerable stages of pathogen development were not exposed to UV-B. Secondly, there was very marked variation in UV-B response between different *S. tritici* isolates, even those collected from a small area of southern England (Rasanayagam *et al.*, 1995; Paul *et al.*, 1997). Many isolates were not affected by UV-B, and this insensitivity was carried through to a lack of response in disease development.

Variation in exposure to UV-B and genetic variability in the UV-B response of the pathogen are not unique to *S. tritici*. Indeed, these two factors seem likely to be important in determining the effects of increasing UV-B on a wide range of plant diseases. A third factor, the variability of plant responses to UV-B (Caldwell *et al.*, 1995) which applies to several host characteristics which influence disease development, may also be significant. Therefore, these three factors can form a basis on which to consider the great variation evident in the published studies of UV-B effects on plant–pathogen interactions (see Table 1).

Variation in exposure to UV-B

It is clear that many phytopathogenic fungi are UV-B sensitive (Maddison & Manners, 1973; Semeniuk & Stewart, 1981; Bashi & Aylor, 1983; Caesar & Pearson, 1983; Rotem, Wooding & Aylor, 1985; Rotem & Aust, 1991). However, cases where the sensitivity of a pathogenic micro-organism to UV-B damage is reflected in reduced infection of the host are in the minority, and in many cases elevated UV-B has been found to increase disease even where the pathogen is known to be sensitive (Table 1). The apparent contradiction between the inherent UV-B sensitivity of pathogens, often determined *in vitro*, and the effects of elevated UV-B on disease may result from the contrasting exposure of organisms *in vitro* and *in planta*. With respect to exposure, powdery mildews (Erysiphales), which grow epiphytically on the leaf surface, are inevitably exposed to incident UV-B in the field. We have found that increased UV-B inhibited infection of barley by powdery mildew (*Erysiphe graminis*) but that this response was a product only of UV-B after inoculation, perhaps reflecting the direct effects of UV-B on the exposed pathogen (Fig. 1). Fungal pathogens that grow beneath the cuticle (for example, *Diplocarpon roseum*) may also be outside most of the protection conferred by strongly absorbing host plant tissues. By contrast, many pathogens may be protected through much of their development by the rapid attenuation of UV-B within the host plant (Day, Vogelmann & DeLucia, 1992). Even so, any pathogen attacking the above-ground parts of plants is potentially exposed to UV-B during infection or dispersal. However, potential exposure is not necessarily reflected in actual responses in the field, as is evident in the case of *Exobasidium vexans*, the cause of blister blight in tea. This pathogen is known to be sensitive to damage by UV-B in sunlight. Although infection in the field may be limited by UV-B during periods of sunshine, the pathogen often escapes such effects, for example, during periods of cloud or when the host is growing in a shaded canopy. Thus, despite the intense solar UV-B in the tropics, blister blight remains a disease of major economic importance in commercial tea crops (Gunasekera *et al.*, 1997*a*). As noted with *S. tritici*, the responses of UV-B-sensitive pathogens to solar radiation may be a function of many factors which affect the rate and timing of pathogen development.

Table 1. *The effects of UV-B radiation on host plants, pathogenic fungi and disease*

Host	Pathogen	Host[1]	Responses to increased UV-B Pathogen	Disease	Method
Cucumber Tomato	*Botrytis cinerea*	Sensitive Sensitive	Sensitive (spore survival)[2] Photosporogenesis[3]	+[4]	Sunlight filtered using UV-transparent and UV-opaque films.
Cucumber Eggplant	*Sclerotinia sclerotiorum*	Sensitive Sensitive	Sensitive (spore survival)[5] Photosporogenesis[6]	+[6]	Sunlight filtered using UV-transparent and UV-opaque films.
Tomato and others	*Alternaria solani* and others	Sensitive	Sensitive (spore survival) [7]	+[8]	Sunlight using UV-transparent and UV-opaque films.
Cucumber	*Colletotrichum lagenarium*	Sensitive	Not known	+ Pre-inoc[10] 0 Post-inoc[10]	CA and polyester filtered UV-B lamps in the glasshouse.
Cucumber	*Cladosporium cucumerinum*	Sensitive	Sensitive[9]	+ Pre-inoc[10] 0 Post-inoc[10]	CA and polyester filtered UV-B lamps in the glasshouse.
Sugar beet	*Cercospora beticola*	Sensitive	Not known	+†[11]	Filtered UV-B lamps in controlled environment chambers
Rice	*Pyricularia grisea*	Sensitive cvs[12]	Not known	0/+† Pre-inoc[12]	Filtered UV-B lamps in a glasshouse.
Wheat	*Septoria tritici*	Tolerant[13]	Sensitive (spore development)[13] Photosporogenesis	0 Pre-inoc[14] − Post-inoc[14]	Filtered UV-B lamps in controlled environment chambers and in the field.

Wheat	*Puccinia recondita*	Tolerant	Sensitive (spore survival)[15]	0/+[16]	Filtered UV-B lamps in the field.
Wheat	*Septoria nodorum*	Tolerant[13,14]	Sensitive (spore development)[13]	+ Pre-inoc[17] 0 Post-inoc[17] 0[16,17]	Filtered UV-B lamps in controlled environment chambers and in the field.
Tea	*Exobasidium vexans*	Not known	Sensitive[18]	–[18]	Sunlight filtered using UV-transparent and UV-opaque films.
Faba bean	*Uromyces viciae-fabae*	Sensitive	Tolerant	0[19]	Sunlight filtered using UV-transparent and UV-opaque films.
				0[19]	Filtered UV-B lamps in the field.

Key + 0 and − imply that increased UV-B increases, has no effect or decreases the severity of infection, respectively. Where known, effects of pre- and post-inoculation UV-B treatments are separated. †shows that increased UV-B increased host damage. The individual responses of host and pathogen are shown where known. Assessment of host sensitivity is based on Krupa & Kickert (1989) unless otherwise stated: sensitive = marked reductions in growth under elevated UV-B, tolerant = no marked responses. Pathogen responses are similarly divided into sensitive and tolerant, and photomorphogenic responses are also noted. References: [1] Krupa & Kickert (1989): [2] Rotem & Aust (1991): [3] Honda & Yunoki (1978): [4] Honda, Toki & Yunoki (1977): [5] Caesar & Pearson (1983): [6] Honda & Yunoki (1977): [7] Rotem *et al.* (1985): [8] Sasaki & Honda (1985): [9] Owens & Krizek (1980): [10] Orth *et al.* (1990): [11] Panagopolous, Bornman & Björm (1992): [12] Finckh, Chavez & Teng (1993): [13] Rasanayagam *et al.* (1995): [14] Rasanayagam *et al.* (1997): [15] Maddison & Manners (1973): [16] Biggs & Webb (1986): [17] see Fig. 2: [18] Gunasekera *et al.* (1997a): [19] Gunasakera (1996).

322   N.D. PAUL

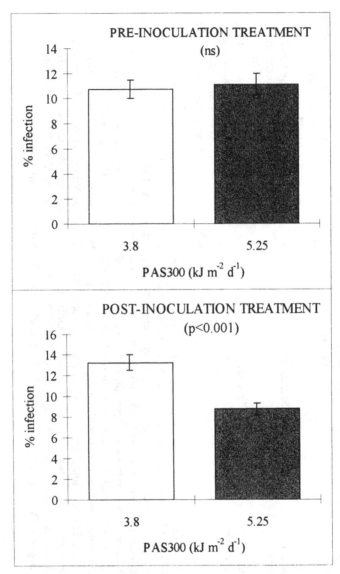

Fig. 1 The effects of pre- and post-inoculation UV-B treatments on infection of barley by powdery mildew (*Erysiphe graminis*). Plants were grown in CE chambers at approximately 800 mmol m^{-2} s^{-1} PAR and PAS300 doses of 3.8 kJ m^{-2} d^{-1} (open columns) or 5.35 kJ m^{-2} d^{-1} (shaded columns) either before or after inoculation. Data are means of 12 replicate plants ± SE.

Genetic variability in the UV-B response of pathogens

There is substantial variation in the magnitude of UV-B response between isolates within species (Rasanayagam *et al.*, 1995; Paul *et al.*, 1997), as well as between different species (Maddison & Manners, 1973; Rotem *et al.*, 1985). Increased UV-B may also have contrasting responses on different developmental stages of the same pathogen since UV-B stimulates both sexual and asexual reproduction in a range of fungi, even some where early developmental stages have been shown to be damaged by this waveband (Table 1). In their recent review, Manning and Teidemann (1995) postulated that stimulation of sporulation might explain why elevated UV-B increased disease in ten, and decreased disease in only five, out of 17 diseases caused by pathogenic fungi. Recent studies of necrotrophs have also shown stimulation of disease by elevated UV-B (Finckh, Chavez & Teng, 1993). However, in these investigations, responses were to UV-B treatment *before* inoculation, excluding the possibility that UV-B acted through photomorphogenetic stimulation of sporulation, and suggesting that changes in the host were of primary importance.

Variation in host responses to UV-B

UV-B-absorbing flavonoids and many phenolic compounds believed to play a key role in resistance to consumers share a common biosynthetic origin in the phenylpropanoid pathway (Hahlbrock & Scheel, 1989). As noted above, the widely observed increase in the concentration of flavonoids in response to increased UV-B has led to the hypothesis that increasing UV-B will result in increases in phenolic-based resistance (for example, Caldwell *et al.*, 1995). This is certainly the case in some systems (see Paul *et al.*, 1997), but not all. Increased UV-B reduced alkaloids in *Aquilegia caerulea* (Larson, Garrison & Carlson, 1990) and furanocoumarins in *Citrus jambhiri* (Asthana *et al.*, 1993) and had no effect on concentrations of tannins and lignin in leaf litter of *Vaccinium* spp. (Gehrke *et al.*, 1995) or on the concentration of tannins in green leaf tissue and fresh litter of *Calluna vulgaris* and *Rubus chamaemorus* (Moody *et al.*, this volume). Clearly, there is no simple rule that increased UV-B results in increases in phenolic defence compounds that are directly related to plant resistance to consumer organisms (Paul *et al.*, 1997). Indeed, it is possible to hypothesise that UV-B protection and disease resistance might be in competition for substrates of the phenylpropanoid metabolism, opening the possibility that increased UV-B protection might leave plants more vulnerable to infection by pathogens. This hypothesis is consistent with the several studies of the

324   N.D. PAUL

effects of UV-B before inoculation with a pathogen which have shown an increase in infection (Orth, Terramura & Sisler, 1990; Finckh, Chavez & Teng, 1993). However, host resistance is not wholly a function of phenolic metabolism, and the effects of UV-B on other host characteristics may be highly significant for plant–pathogen interactions (Paul et al., 1996). For example, UV-B is known to alter the quantity and chemical composition of leaf surface wax deposits (Tevini & Steinmuller, 1987; Barnes et al., 1994, 1996) and Orth, Teramura and Sisler (1990) suggested that changes in leaf surface properties in cucumber caused by increased UV-B before inoculation might account for increased disease development.

Studies at Lancaster of the effects of UV-B on the interaction between wheat and *Septoria nodorum* exemplifies a system in which changes in disease are strongly influenced by changes in the host. We have shown that spore germination and germ tube growth of this fungus are sensitive to UV-B, not only *in vitro* (Rasanayagam et al., 1995) but *in planta* on the leaf surface. None the less, in terms of disease development, the effects of post-inoculation UV-B treatments on the pathogen appeared to be small compared with changes occurring in the host before inoculation, which resulted in *increased* disease development (Fig. 2). Perhaps because of the opposite effects of UV-B on host and pathogen, increasing UV-B above ambient in the field had no effect on *S. nodorum* infection, either in spring or summer (Fig. 2).

In summary, there is variation in pathogen sensitivity to UV-B, in pathogen exposure to UV-B, especially at key developmental stages, and in the UV-B responses of the host which affect disease development. In nature, all three elements may interact, for example, UV-B-induced changes in leaf surface properties might increase or decrease the rate of pathogen development and so modify its exposure to damaging radiation. It is, therefore, not surprising that effects of UV-B on plant disease are so varied. Clearly, such variation prevents any simple prediction of changes in plant disease as a result of stratospheric ozone depletion. Any attempt at such predictions is further complicated by the great uncertainty over the spectral responses relevant to plant–microbe interactions.

Action spectra and dose–responses

The magnitude of biological responses to ozone depletion can be considered as the product of two elements: 1) the relationship between ozone loss and ground-level radiation, weighted according to the appropriate action spectrum (the radiation amplification factor: RAF) and 2)

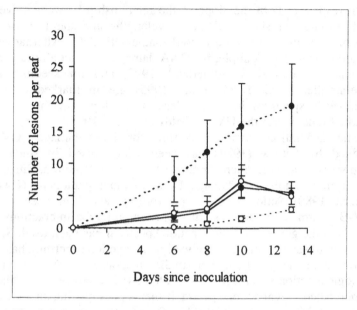

Fig. 2 Infection of winter wheat (cv. Riband) to inoculation with *Septoria nodorum* to low (1.4 kJ m^{-2} d^{-1}) and high (2.0 kJ m^{-2} d^{-1}) UV-B$_{DNA}$ provided before and/or after inoculation. Data are means of 12 replicate plants ± SE. Key: open and closed symbols; high and low pre-inoculation UV-B, respectively, and solid and dotted lines high and low post-inoculation UV-B, respectively.

the relationship between weighted dose and biological response (the biological amplification factor: BAF). Unfortunately, neither of these elements is well defined with respect to the responses of ozone loss on plants and microbes. Most investigations, confronted with the massive diversity of response at the biochemical, physiological and ecological scales, have concentrated on the nature rather than the magnitude of responses. This issue is particularly relevant when considering the possible impacts of ozone loss on interactions between plants and microbes, which may well differ markedly both in terms of action spectrum (RAF) and dose–response (BAF).

A number of action spectra pertinent to plant–consumer interactions have been published. For investigations of plant responses to ozone depletion, the issue of the appropriate choice of plant action spectrum has been widely debated and is considered in detail elsewhere in this volume (see chapter by Holmes). Briefly, the very widely used generalised plant action spectrum of Caldwell (1971) is subject to question with respect to both the slope of the relationship between wavelength and response at shorter UV-B wavelengths and the lack of UV-A response in the most widely used mathematical fit. Alternative plant action spectra, for example, for DNA damage in intact alfalfa seedlings (Quaite, Sutherland & Sutherland, 1992) and for seedling growth (Steinmüller in Caldwell et al., 1995) are in marked contrast to Caldwell's spectrum in having less steep slopes in the UV-B and marked tails within the UV-A (Caldwell et al., 1995).

The generalised action spectrum for DNA-damage (UV-B$_{DNA}$) defined by Setlow (1974) has been widely used in studying the responses of aquatic micro-organisms to ozone depletion. Since UV-B radiation is known to damage DNA in plant pathogens (Owens & Krizek, 1980), Setlow's action spectrum has also been used to weight UV-B treatments in some studies of plant–microbe interactions, while in others the generalised plant action spectrum has been used. Setlow's DNA action spectrum, like Caldwell's plant action spectrum, has a steep response in the UV-B, and a smaller response to UV-A than many biological action spectra (Caldwell et al., 1995). However, on the basis of published fungal action spectra, it seems possible that an action spectrum with less response in the UV-B and more in the UV-A, perhaps something approximating to Quaite et al.'s (1992) spectrum, may be appropriate for fungi (Paul et al., 1997).

Choice of action spectrum has profound implications for the design and interpretation of experiments. RAF is a function of, amongst several factors solar elevation (and so of latitude, season and time of day) and of action spectrum (Madronich et al., 1995). Thus, different action spectra result in contrasting relationships between ozone loss and increases in weighted radiation. For example, in summer at UK latitudes, a 15% ozone loss would result in increases of 30–40% in UV-B$_{DNA}$ or radiation weighted using Caldwell's generalised plant action spectrum (PAS300), but no more than a 20% increase in radiation weighted according to Quaite et al's (1992: UV-B$_{QUAITE}$), and only a 7% increase if Steinmüller's seedling growth spectrum (in Caldwell et al., 1995: UV-B$_{ST}$) is used. (This is because UV-B$_{QUAITE}$ and UV-B$_{ST}$ have a lower biological weighting for the shorter wavelengths of UV-B, which increase most with ozone depletion, than PAS300 action.) Intuitively, experimental treatments seeking to reproduce the effects of a given ozone loss will have far greater percentage increases in UV-B if PAS300 or UV-B$_{DNA}$ are used than if UV-B$_{QUAITE}$ or UV-B$_{ST}$ are used (since the weighting of the shorter wavelengths is greater in the former than in the latter). Caldwell et al. (1986) discussed this issue with specific reference to PAS300 and UV-B$_{DNA}$ and termed it 'RAF error'. However, Caldwell *et al.* (1986) concluded that when comparing PAS300 and UV-B$_{DNA}$, 'RAF error' was a less severe problem than 'enhancement errors' which resulted from the contrasting spectral quality of sunlight and the artificial UV-B source used experimentally, as discussed below.

Sunlight is relatively rich in UV-A compared with UV-B, especially in winter, and UV-A does have some biological activity. As a result, *absolute* weighted doses are actually far greater for action spectra with significant UV-A tails than for PAS300 or UV-B$_{DNA}$ (Table 2). In the most extreme comparison each milliwatt of biologically effective radiation calculated by UV-B$_{DNA}$ is equivalent to 102 mW if calculated with UV-B$_{ST}$, in which UV-A wavelengths have a relatively high weighting.

Table 2. *The relationships between irradiances of biologically effective radiation from summer sunlight calculated using different action spectra*

	Summer sunlight Irradiance equal to 1 W m^{-2} of X					
X	PAS300	UV-B$_{THIM}$	UV-B$_{ERY}$	UV-B$_{DNA}$	UV-B$_{QUAITE}$	UV-B$_{ST}$
PAS300	–	1.47	0.87	0.37	4.93	37.54
UV-B$_{THIM}$	0.68	–	.59	.25	3.34	25.47
UV-B$_{ERY}$	1.15	1.70	–	0.42	6.69	43.26
UV-B$_{DNA}$	2.73	4.02	2.36	–	13.44	102.30
UV-B$_{QUAITE}$	0.20	0.30	0.18	0.07	–	7.61
UV-B$_{ST}$	0.03	0.04	0.02	0.01	0.13	–

Data are based on a sunlight spectrum for Lancaster (approx. 55° N, 2° W) measured spectroradiometrically at solar noon on August 1st under sunny but not perfect 'clear-sky' conditions.
Key:
PAS300: Caldwell's generalised plant action spectrum, normalised to 300 nm.
UV-B$_{THIM}$: a modified mathematical fit to Caldwell's generalised action spectrum, incorporating a UV-A tail.
UV-B$_{ERY}$: the standard CIE erythemal (sunburn) action spectrum (McKinlay & Diffey, 1987).
UV-B$_{DNA}$: Setlow's (1974) action spectrum for DNA damage.
UV-B$_{QUAITE}$: Quaite, Sutherland and Sutherland's (1992) action spectrum for DNA damage in intact alfalfa seedlings.
UV-B$_{ST}$: the whole plant action spectrum of Steinmüller (in Caldwell *et al.*, 1995).

Since the ratio of UV-A:UV-B is at its lowest in summer, the comparison actually represents a minimum; in winter one milliwatt of UV-B$_{DNA}$ is equivalent to 478 mW UV-B$_{ST}$.

In comparison with daylight, the radiation emitted from artificial UV-B lamps widely used experimentally is rich in shorter wavelength UV-B, even when filtered with cellulose acetate (CA). Therefore the differences in weighted dose calculated using the different action spectra (Table 3) are far less pronounced for the Phillips TL40 UV-B tube than for daylight. For example, one milliwatt of UV-B$_{DNA}$ is equivalent to only 3.1 mW UV-B$_{ST}$. The spectral composition of radiation from lamps is also relatively stable. Photodegradation of CA results in a slight increase in the ratio of UVA:UV-B (since ageing causes a greater decline in transmission at longer wavelengths, Mepsted *et al.*, 1996) but this has little effect on the inter-relationships of different action spectra. For example, the ratio of UV-B$_{DNA}$: UV-B$_{ST}$ increases from 3.1 for new CA to 3.6 after 80 h exposure with the filter wrapped directly around the lamp, while changes resulting from lamp ageing are even smaller.

Therefore, the contrasting relationships between action spectra in daylight and under artificial illumination combine to produce 'enhancement error'. For example, assume an experimental treatment, using a UV lamp source, designed to provide 100% of ambient UV-B$_{DNA}$. The

Table 3. *The relationships between irradiances of biologically effective radiation from an artificial UV-B source calculated using different action spectra*

X	PAS300	UV-B$_{THIM}$	UV-B$_{ERY}$	UV-B$_{DNA}$	UV-B$_{QUAITE}$	UV-B$_{ST}$
		Cellulose acetate filtered TL40 Irradiance equal to 1 W m^{-2} of X				
PAS300	–	1.01	0.58	0.82	1.33	2.53
UV-B$_{THIM}$	0.99	–	0.58	0.81	1.32	2.50
UV-B$_{ERY}$	1.74	1.76	–	1.42	2.32	4.41
UV-B$_{DNA}$	1.22	1.24	0.70	–	1.63	3.10
UV-B$_{QUAITE}$	0.75	0.76	0.43	0.61	–	1.90
UV-B$_{ST}$	0.40	0.40	0.23	0.32	0.53	–

Data are based on spectroradiometric measurements of Philips TL40-12 lamps, Starna Ltd, Romford, UK wrapped in un-aged cellulose diacetate (100 μm thick Clarifoil™, Courtaulds Ltd, Derby, UK).
Key: see Table 2.

major contributor to the biologically effective dose would be the shorter UV-B wavelengths of the lamp, and the high weighting of the action spectrum in that region. This treatment, however, would provide only 12% of UV-B$_{QUAITE}$ and only 3% of ambient UV-B$_{ST}$, since these action spectra have less steep slopes in the UV-B, and there is little UV-A emitted from the lamp. Enhancement errors for comparisons between a range of action spectra (Table 4) highlight the great discrepancies that appear when treatments based on one action spectrum are re-expressed on the basis of another.

In summary, the choice of action spectrum affects both the percentage changes in biologically effective radiation calculated for a given ozone depletion, and also substantially alters the relationship between experimental doses and those occurring in the field. This, in turn, has significant implications for the interpretation of UV-B experiments using artificial UV-B sources, which can be illustrated with reference to investigations conducted at Lancaster.

Dose–responses of pigment accumulation in cultivated pea

We investigated the UV-B dose response of cultivated pea using four doses based on PAS300 (Gonzalez, 1996). A variety of physiological and morphological variables were quantified, but here the absorbance at 300 nm (A_{300}) in mature leaves will be dealt with since, as noted

Table 4. *The effect of using different action spectra on the dose from an artifical UV-B source (enhancement error sensuo Caldwell* et al., *1986) expressed as the percentage of maximum summer dose (the value using X is 100%)*

	% of summer maximum dose produced by a					
X	PAS300	UV-B$_{THIM}$	UV-B$_{ERY}$	UV-B$_{DNA}$	UV-B$_{QUAITE}$	UV-B$_{ST}$
PAS300	100	69	66	223	27	7
UV-B$_{THIM}$	145	100	96	324	39	10
UV-B$_{ERY}$	151	104	100	336	41	10
UV-B$_{DNA}$	45	31	30	100	12	3
UV-B$_{QUAITE}$	370	254	245	824	100	25
UV-B$_{ST}$	1481	1018	981	3301	400	100

Data are derived from figures presented in Tables 2 and 3.
Key: see Table 2.

above, it is possible that changes in UV-B absorbing compounds may be pertinent to plant–consumer interactions. The treatments aimed to cover a wide range of doses from 2.3 kJ m^{-2} d^{-1} PAS300, (approximating to mean ambient cloud-corrected dose for summer in the southern UK) to 4.6 kJ m^{-2} d^{-1} (approximately the mid-summer peak dose under clear sky in Southern UK), 6.9 kJ m^{-2} d^{-1} (approximating to these idealised clear-sky conditions but simulating approximately 22% stratospheric ozone depletion), and the highest dose, 9.2 kJ m^{-2} d^{-1} was chosen as a deliberately extreme treatment for temperate latitudes, equivalent to an ozone depletion of approximately 45%. Although some UV-A was provided in the controlled environment chambers used for this study, the ratio of UV-B dose between the four treatments $(1 : 2 : 3 : 4)$ is effectively constant regardless of what action spectrum is used to weight the radiation, and biological responses can be simply interpreted on this basis. Therefore, A$_{300}$ was increased by 23% by a four-fold increase in the UV-B dose, regardless of action spectrum (Fig. 3a). Where interpretation is critically dependent on the choice of action spectrum is when experimental doses are considered relative to those in sunlight (Fig. 3b). The increase in A$_{300}$ would be interpreted as resulting from an increase from approximately 50% to 200% of the maximum mid-summer dose in terms of PAS300; in terms of UV-B$_{QUAITE}$, which gives a lower weighting at the shorter UV-B wavelengths, the same increase in A$_{300}$ results from an increase from approximately 14% to 54% of midsummer maximum and for UV-B$_{ST}$, from only 3.5 to 14% (Fig. 3b).

Interpreted on the basis of PAS300, these data indicate that UV-B absorbing pigments accumulate progressively over the environmental range, including the higher doses expected given ozone depletion, and statistically the data can be fitted with both a linear regression ($r^2 = 0.49$) and a non-linear regression ($r^2 = 0.51$) in which pigment accumulation is saturated at doses 2–2.5 times that of ambient. The use of UV-B$_{QUAITE}$ or UV-B$_{ST}$ requires a very different interpretation: very marked increases in UV absorbance would seem to occur in response to relatively small increases in weighted radiation well within the environmental range. Assessment of changes which might occur in response to increases above the current ambient are constrained by the low doses provided experimentally. Simple extrapolation of the linear trend is flawed, and not only on statistical grounds. Such extrapolation of the growth data obtained in the same experiment would lead to the nonsense prediction that pea plants would be killed at UV-B$_{QUAITE}$ or UV-B$_{ST}$ doses well within the range currently experienced in the field. Alternatively, extrapolation of the non-linear regression, although equ-

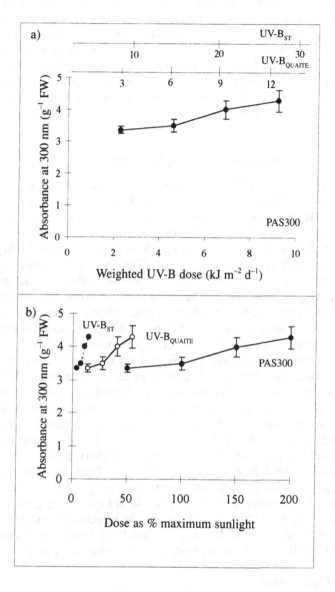

Fig. 3 Response of UV-B absorbances in mature leaves of cultivated pea grown at increasing UV-B dose expressed in terms of a) absolute doses of either PAS300, UV-B$_{QUAITE}$ or UV-B$_{ST}$ or b) doses relative to the maximum midsummer dose for the three action spectra. Data are means of twelve replicate plants ± SE. (Modified from Gonzalez, 1996.)

ally flawed statistically, provides a more intuitively correct prediction, that pigments accumulate primarily in response to increases in dose within a range expected under ambient conditions with season and weather conditions. In this case, increases in UV-B resulting from ozone depletion would not be expected to have significant effects on pigment accumulation. Although only a single investigation, we found no accumulation of UV-B absorbing pigments in pea in the field when ambient PAS300 was increased by approximately 25% (Mepsted et al., 1996).

Dose responses of phylloplane micro-organisms

We investigated the UV-B dose response of a number of phylloplane yeasts using UV-B$_{DNA}$ doses ranging from well below the maximum expected in the UK to approximately 2.5 times this maximum (Gunasekera et al., 1997b). I will consider the responses of UK isolates of *Sporidiobolus* sp. and *Bullera alba*. Dose responses were highly non-linear, especially for *Sporidiobolus* sp. There was little or no response in cells per colony to doses less than approximately 1.5 kJ m^{-2} d^{-1} UV-B$_{DNA}$ (Fig. 4a), which is near the maximum expected summer dose for the UK. This result was interpreted as reflecting the adaptation of the micro-organisms to local conditions, that is, selection had favoured genotypes with UV-B protection or repair mechanisms sufficient to confer tolerance to local conditions, but no more (Gunasekera et al., 1997b). Thus, the effects of seasonal variation in ambient UV-B$_{DNA}$ would be rather small, only *Bullera alba* showing a marked response (Fig. 5). *B. alba* would also be vulnerable to increases in UV-B$_{DNA}$ resulting from ozone loss. No such interpretation is tenable if doses are expressed in terms of UV-B$_{QUAITE}$ rather than UV-B$_{DNA}$. In this case the threshold in the dose response occurs at approximately 10% of the maximum expected in summer (Fig. 4b). This interpretation based on UV-B$_{QUAITE}$ implies (a) that both yeasts are strongly influenced by seasonal changes in ambient UV-B and would not survive in exposed sites for much of the year (Fig. 5) and (b) that the effects on these organisms of any small increase in UV-B$_{QUAITE}$ resulting from ozone depletion would be small in comparison with existing variation (Fig. 5).

Conclusions

It is still too soon to state with confidence that action spectra such as UV-B$_{QUAITE}$ or UV-B$_{ST}$ are more appropriate than PAS300 or UV-B$_{DNA}$ in studies of either plants or their consumers. Indeed, given the great diversity of organisms involved in plant–consumer interactions it is

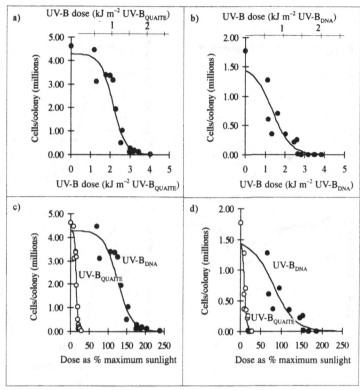

Fig. 4 Response of colony size (cells colony⁻¹) in *Sporidiobolus* sp. (a,c) and *Bullera alba* (b,d) to increasing UV-B dose expressed in terms of absolute doses of either UV-B $_{DNA}$ or UV-B$_{QUAITE}$ (a,b) or doses relative to the maximum midsummer dose for the two action spectra (c,d). Each point represents data for a single Petri dish or, in the case of colony size, the mean of 5–10 colonies per plate. Modified from Gunasekera *et al.* (1997*b*).

unlikely that any one action spectrum, whether PAS300, UV-B$_{DNA}$, UV-B$_{QUAITE}$, UV-B$_{ST}$ or any other, can be used as a standard in all studies of such interactions. Even within the studies I have discussed here, observations on the wheat–*S. tritici* interaction can be comfortably re-interpreted on the basis of UV-B$_{QUAITE}$ rather than UV-B$_{DNA}$ but the same re-interpretation applied to studies of the phylloplane yeasts leads to paradoxes. For example, interpretation using UV-B$_{QUAITE}$ suggests that both *Sporidiobolus* sp. and *Bullera alba* would be unable to survive on exposed leaf surfaces for much of the year (Fig. 5). Such paradoxes, indicating that UV-B$_{QUAITE}$ is inappropriate to these organisms and/or

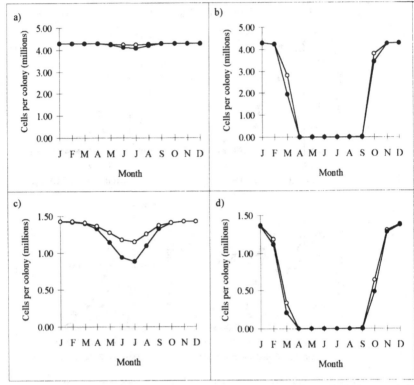

Fig. 5. Predicted changes in the colony size of *Sporidiobolus* sp. (a,b) and *Bullera alba* (c,d) to variation in UV-B due to season and ozone depletion when UV-B doses are weighted using either UV-B $_{DNA}$ (a,c) or UV-B$_{QUAITE}$. (b,d). Data are derived from the dose responses shown in Fig. 4. UV-B doses are monthly means of doses calculated for each day of the year for Lancaster UK (approx. 55° N, 2° W) using the model of Björn and Murphy (1985) in the form of software kindly provided by E.L. Fiscus & F.L. Booker (University of North Carolina). Ambient ozone for each day of the year was taken as the mean between January 1979 and April 1993 in the Nimbus 7 Total Ozone Mapping Spectrophotometer (TOMS) data for 55° N (data provided by Dr J. Herman, Goddard Space Flight Laboratory, Greenbelt, Maryland: Herman and Larko, 1994). Calculations were repeated with a constant ozone depletion of 15%. Correction for cloud was estimated on the basis of sunshine hours recorded at Lancaster using the formula:

UVcloud = (UVclear*(SUNo/SUNp))
          +(UVclear*((SUNo/SUNp)−1)*0.43)

that laboratory dose responses cannot be extrapolated to the field, may be valuable in focusing future research. However, if action spectra such as Quaite *et al.*'s (1992) or Steinmüller's (in Caldwell *et al.*, 1995) prove to be correct, much of the existing data on the responses of plants and microbes to UV-B will need to be reassessed. Comparing PAS300 and UV-B$_{DNA}$, Caldwell *et al.* (1986) concluded that the need for such reassessment was largely a product of enhancement error rather than RAF error, and this seems to apply to the wider range of action spectra compared here. Indeed, in many cases enhancement error may be so large that experiments have used doses well below those occurring in the field, even under ambient ozone, and so providing little information pertinent to ozone depletion. Such potential limitations in the existing data inevitably constrain any objective assessment of the possible effects on plant–pathogen interactions of increases in UV-B radiation resulting from stratospheric ozone depletion. However, it would suggest that significant changes in plant disease, and perhaps in other plant–consumer interactions, would be the exception rather than the rule, especially when variation in exposure and between different genotypes is taken into account. On the other hand, it seems increasingly possible that variation in UV-B, whether resulting from season, weather conditions or microsite, may be a significant factor influencing plant–microbe interactions in the field.

Acknowledgements

I am grateful to Andy McLeod, Sandra Moody and Raquel Gonzalez for many useful discussions and their comments on the manuscript. I am also grateful for the financial support provided by the Biology and Biotechnology Research Council, Natural Environment Research Council, the UK Department of Environment (Contracts PECD 7/12/21 and EPG 1/1/19) and the EU (Contract EV4V-CT96-0208).

Fig. 5 caption contd.

where UVcloud and UVclear are cloud-corrected and clear-sky UV-B, respectively, SUNo is recorded sunshine hours, SUNp the potential hours of daylight and 0.43 is taken as the average % of UV-B reaching the ground under 'non-sunshine' conditions. Other environmental variables were held constant.
Key: open symbols, ambient ozone; closed symbols, 15% ozone depletion.

References

Asthana, A., McCloud, E.S., Berenbaum, M.R. & Tuveson, R.W. (1993). Phototoxicity of *Citrus jambhiri* to fungi under enhanced UV-B radiation: role of furanocoumarins. *Journal of Chemical Ecology*, **19**, 2813–30.

Barnes, J.D., Paul, N.D., Percy, K., Broadbent, P., McLaughlin, C., Mullineaux, P., Criessen, G. & Wellburn, A.R. (1994). Effects of UV-B radiation on wax biosynthesis. In *Air Pollutants and the Leaf Cuticle*. (Percy, K, Cape, J.N. & Jagels R., eds), NATO-ASI Series, pp. 195–204. Springer-Verlag, Berlin.

Barnes, J.D., Percy, K.E., Paul, N.D., Broadbent, P., McLaughlin, C.K., Mullineaux, P.M., Creissen, G. & Wellburn, A.R. (1996). The influence of UV-B radiation on the physico-chemical nature of tobacco (*Nicotinia tabacum* L.) leaf surfaces. *Journal of Experimental Botany*, **47**, 99–109.

Bashi, E. & Aylor, D.E. (1983). Survival of detached sporangia of *Peronospora destructor* and *Peronospora tabacina*. *Phytopathology*, **83**, 1135–39.

Biggs, R.H. & Webb, P.G. (1986). Effects of enhanced ultraviolet-B radiation of yield and disease incidence for wheat under field conditions. In *Stratospheric Ozone Reduction, Solar Ultraviolet Radiation and Plant Life; Workshop on the Impact of Solar Ultraviolet Radiation upon Terrestrial Ecosystems: 1. Agricultural Crops*. (Worrest, R.C. & Caldwell, M.M., eds), pp. 303–11. Springer-Verlag, New York.

Björn L.O. & Murphy T.M. (1985). Computer calculation of solar ultraviolet radiation at ground level. *Physiologia Vegetale*, **23**, 555–61.

Caesar, A.J. & Pearson, R.C. (1983). Environmental factors affecting survival of ascospores of *Sclerotinia sclerotiorum*. *Phytopathology*, **73**, 1024–30.

Caldwell, M.M. (1971). Solar UV irradiation and the growth and development in higher plants. In *Photophysiology*. (Giese, A.C., ed.), pp. 131–77. Academic Press, New York.

Caldwell, M.M., Camp, L.B., Warner, C.W. & Flint, S.D. (1986). Action spectra and their key role in assessing biological consequences of solar UV-B radiation change. In *Stratospheric Ozone Reduction, Solar Ultraviolet Radiation and Plant Life; Workshop on the Impact of Solar Ultraviolet Radiation upon Terrestrial Ecosystems: 1. Agricultural Crops*. (Worrest, R.C. & Caldwell, M.M., eds), pp. 87–111. Springer-Verlag, New York.

Caldwell, M.M., Teramura, A.H. & Tevini, M. (1989). The changing solar ultraviolet climate and the ecological consequences for higher plants. *Trends in Evolution and Ecology*, **4**, 363–67.

Caldwell, M.M., Teramura, A.H., Tevini, M., Bornman, J.F., Björn,

L.O. & Kulandaivelu, G. (1995). Effects of increased solar ultraviolet radiation of terrestrial plants. *Ambio*, **24**, 166–73.

Day, T.A., Vogelmann, T.C. & DeLucia, E.H. (1992). Are some life forms more effective than others in screening out ultraviolet-B radiation? *Oecologia*, **92**, 513–19.

Finckh, M.R., Chavez, A.Q & Teng, P.S. (1993). Effects of enhanced UV-B radiation on the susceptibility of rice to rice blast. *Agriculture, Ecosystems and Environment*, **52**, 223–33.

Gehrke, C., Johanson, U., Callaghan, T.V., Chadwick, D., Robinson, C.H. (1995). The impact of enhanced ultraviolet-B radiation on litter quality and decomposition processes in *Vaccinium* leaves from the subarctic. *Oikos*, **71**, 213–22.

Gonzalez, R. (1996). PhD thesis, University of Lancaster.

Gunasekera, T.S. (1996). Effects of UV-B (290–320 nm) radiation on micro-organisms on the leaf surface. PhD thesis, University of Lancaster.

Gunasekera, T.S., Paul, N.D. & Ayres, P.G. (1997*a*). The effects of ultraviolet-B (UV-B: 290–320 nm) radiation on blister blight disease of tea (*Camellia sinensis* L.). *Plant Pathology*, (in press).

Gunasekera, T.S., Paul, N.D. & Ayres, P.G. (1997*b*). Responses of phylloplane yeasts to UV-B (290–320 nm) radiation: inter species differences in sensitivity. *Mycological Research*, in press.

Hahlbrock, K. & Scheel, D. (1989). Physiology and molecular biology of phenylpropanoid metabolism. *Annual Review of Plant Physiology*, **40**, 347–69.

Herman, J.R. & Larko, D. (1994). Low ozone amounts during 1992–1993 from Nimbus 7 and Meteor 3 total ozone mapping spectrometers. *Journal of Geophysical Research*, **99**, 3483–96.

Honda, Y., Toki, T. & Yunoki, T. (1977). Control of grey mould of greenhouse cucumber and tomato by inhibiting sporulation. *Plant Disease Reporter*, **61**, 1041–4.

Honda, Y. & Yunoki, T. (1977). Control of *Sclerotinia* disease of greenhouse eggplant and cucumber by inhibition of development of apothecia. *Plant Disease Reporter*, **61**, 1036–40.

Honda, Y. & Yunoki, T. (1978). Action spectrum for photomorphogenesis in *Botrytis cinerea* Pers ex. Fr. *Plant Physiology*, **61**, 711–13.

Krupa, S.V. & Kickert, R.N. (1989). The greenhouse effect – impacts of ultraviolet-B (UV-B) radiation, carbon dioxide (CO_2) and ozone (O_3) on vegetation. *Environmental Pollution*, **61**, 263–393.

Larson, R.A., Garrison, W.J. & Carlson, R.W. (1990). Differential responses of alpine and non-alpine *Aquilegia* species to increased UV-B radiation. *Plant Cell and Environment*, **13**, 983–7.

McKinlay, A.F. & Diffey, B.L. (1987). A reference spectrum for ultraviolet induced erythema in human skin. *CIE Journal*, **6**, 17–22

Maddison, A.C. & Manners, J.G. (1973). Lethal effects of artificial ultraviolet radiation on cereal rust uredospores. *Transactions of the British Mycological Society*, **60**, 471–94.

Madronich, S., McKenzie, R.L., Caldwell, M.M. & Björn, L.O. (1995). Changes in ultraviolet radiation reaching the earth's surface. *Ambio*, **24**, 143–52.

Manning, W.J. & Teidemann, A.V. (1995). Climate change: potential effects of increased atmospheric carbon dioxide (CO_2), ozone (O_3) and ultraviolet-B radiation (UV-B) on plant diseases. *Environmental Pollution*, **88**, 219–45.

Mepsted, R., Paul, N.D., Stephen, J., Corlett, J.E., Nogués, S., Baker, N.R., Jones, H.G. & Ayres, P.G. (1996). Effects of enhanced UV-B radiation on pea *(Pisum sativum* L.) grown under field conditions in the United Kingdom. *Global Change Biology*, **2**, 325–34.

Orth, A.B., Teramura, A.H. & Sisler, H.D. (1990). Effects of ultraviolet-B radiation on fungal disease development in *Cucumis sativus*. *American Journal of Botany*, **77**, 1188–92.

Owens, V.O. & Krizek, D.T. (1980). Multiple effects of UV radiation (265–330 nm) on fungal spore emergence. *Photochemistry and Photobiology*, **32**, 41–9.

Panagopoulos, I., Bornman, J.F. & Björn, L.O. (1992). Response of sugar beet plants to ultraviolet-B (280–320 nm) radiation and *Cercospora* leaf spot disease. *Physiologia Plantarum*, **84**, 140–5.

Paul, N.D, Rasanayagam, M.S., Moody, S.A., Hatcher, P.E. & Ayres, P.G. (1997). The role of interactions between trophic levels in determining the effects of UV-B on terrestrial ecosystems. *Plant Ecology*, **128**, 1–13.

Quaite, F.E., Sutherland, B.M. & Sutherland, J.C. (1992). Action spectrum for DNA damage in alfalfa lowers predicted impact of ozone depletion. *Nature*, **358**, 576–8.

Rasanayagam, M.S. Paul, N.D., Royle, D.J. & Ayres, P.G. (1995). Variation in responses of spores of *Septoria tritici* and *S. nodorum* to UV-B irradiation *in vitro*. *Mycological Research*, **99**, 1371–7.

Rotem, J. & Aust, H.J. (1991). The effect of ultraviolet and solar radiation and temperature on survival of fungal propagules. *Journal of Phytopathology*, **133**, 76–84.

Rotem, J., Wooding, B. & Aylor, D.E. (1985). The role of solar radiation, especially ultraviolet, in the mortality of fungal spores. *Phytopathology*, **75**, 510–14.

Sasaki, T. & Honda, Y. (1985). Control of certain diseases of greenhouse vegetables with ultraviolet absorbing films. *Plant Disease*, **69**, 530–3.

Semeniuk, P. & Stewart, R.N. (1981). Effects of ultraviolet (UV-B) irradiation on infection of roses by *Diplocarpon rosae*. *Environmental and Experimental Botany*, **21**, 45–50.

Setlow, R.B. (1974). The wavelengths in sunlight effective in producing skin cancer: a theoretical analysis. *Proceedings of the National Academy of Sciences, USA*, **71**, 3363–6.

Steinmüller, D. & Tevini, M. (1985). Action of ultraviolet radiation (UV-B) upon cuticular waxes in some crop plants. *Planta*, **164**, 557–64.

Tevini, M. & Steinmüller, D. (1987). Influence of light, UV-B radiation, and herbicides on wax biosynthesis of cucumber seedlings. *Journal of Plant Physiology*, **131**, 111–21.

Index